MECHANICS AND PHYSICS OF POROUS SOLIDS

MECHANICS AND PHYSICS OF POROUS SOLIDS

Olivier Coussy
Laboratoire Central des Ponts et Chaussées
Ecole des Ponts ParisTech
Université Paris-Est
http://navier.enpc.fr/~coussy/

A John Wiley and Sons, Ltd., Publication

This edition first published 2010
© 2010 John Wiley & Sons, Ltd

Registered Office
John Wiley & Sons Ltd, The Atrium, Southern Gate, Chichester, West Sussex, PO19 8SQ, United Kingdom

For details of our global editorial offices, for customer services and for information about how to apply for permission to reuse the copyright material in this book please see our website at www.wiley.com.

The right of the author to be identified as the author of this work has been asserted in accordance with the Copyright, Designs and Patents Act 1988.

All rights reserved. No part of this publication may be reproduced, stored in a retrieval system, or transmitted, in any form or by any means, electronic, mechanical, photocopying, recording or otherwise, except as permitted by the UK Copyright, Designs and Patents Act 1988, without the prior permission of the publisher.

Wiley also publishes its books in a variety of electronic formats. Some content that appears in print may not be available in electronic books.

Designations used by companies to distinguish their products are often claimed as trademarks. All brand names and product names used in this book are trade names, service marks, trademarks or registered trademarks of their respective owners. The publisher is not associated with any product or vendor mentioned in this book. This publication is designed to provide accurate and authoritative information in regard to the subject matter covered. It is sold on the understanding that the publisher is not engaged in rendering professional services. If professional advice or other expert assistance is required, the services of a competent professional should be sought.

Library of Congress Cataloging-in-Publication Data

Coussy, Olivier.
  Mechanics and physics of porous solids / Olivier Coussy.
      p. cm.
  Includes bibliographical references and index.
  ISBN 978-0-470-72135-3 (cloth)
  1. Porous materials.   I. Title.
  TA418.9.P6C683 2010
  620.1′16–dc22   2009049316

A catalogue record for this book is available from the British Library.

ISBN: 978-0-47072135-3 (Hbk)

Typeset in 10/12pt Times by Aptara Inc., New Delhi, India

To my wife Sandra
for putting up with it

Contents

Preface			xiii
1	**The Strange World of Porous Solids**		**1**
2	**Fluid Mixtures**		**11**
2.1	Chemical Potential		11
	2.1.1	Free energy and chemical potential	11
	2.1.2	Equilibrium of mixture composition	13
2.2	Gibbs–Duhem Equation		13
	2.2.1	Derivation of the Gibbs–Duhem equation	13
	2.2.2	Molar Gibbs–Duhem equation	14
	2.2.3	Ideal gases	15
	2.2.4	Real gases	16
	2.2.5	Partial molar property	16
2.3	Ideal Mixtures		17
	2.3.1	Ideal gas mixture	17
	2.3.2	Ideal mixture definition	19
	2.3.3	Entropy of mixing	19
	2.3.4	Ideal solution – Raoult's law	21
	2.3.5	Dilute ideal solution – Henry's law	21
	2.3.6	Osmotic pressure	23
	2.3.7	Electrostatics and excess of osmotic pressure	25
	2.3.8	Reactive ideal mixture	32
	2.3.9	Unconditional stability of ideal solutions	34
2.4	Regular Solutions		36
	2.4.1	Regular solutions and the interaction between molecules	36
	2.4.2	Conditional stability of regular solutions	37
	Further Reading		40
3	**The Deformable Porous Solid**		**41**
3.1	Strain		42
	3.1.1	Deformation gradient and displacement	42

		3.1.2 Strain tensor	43

3.2 Stress — 45
- 3.2.1 The hypothesis of local contact forces — 45
- 3.2.2 The action–reaction law — 46
- 3.2.3 The Cauchy stress tensor — 47
- 3.2.4 Local mechanical equilibrium — 49
- 3.2.5 Symmetry of stress tensor — 50

3.3 Strain Work — 51
- 3.3.1 Mechanical energy balance — 51
- 3.3.2 Strain work in infinitesimal transformations — 51

3.4 From Solids to Porous Solids — 52
- 3.4.1 Porosity and deformation — 52
- 3.4.2 Pore pressure and stress partition — 54
- 3.4.3 Free energy balance for the fluid–solid mixture — 55
- 3.4.4 Free energy balance for the porous solid — 56
- 3.4.5 Strain work and the effective stress 'principle' — 57

Further Reading — 59

4 The Saturated Poroelastic Solid — 61

4.1 The Poroelastic Solid — 61
- 4.1.1 The linear poroelastic solid — 61
- 4.1.2 Microporoelasticity — 63
- 4.1.3 The nonlinear poroelastic solid — 68

4.2 Filling the Porous Solid — 71
- 4.2.1 Filling by a compressible fluid — 72
- 4.2.2 Filling by a mixture containing gas bubbles – bubble pressure — 72
- 4.2.3 Undrained poroelasticity — 75

4.3 The Thermoporoelastic Solid — 78
- 4.3.1 The linear thermoporoelastic solid — 78
- 4.3.2 The linear thermoporoelastic fluid–solid mixture — 79

4.4 The Poroviscoelastic Solid — 80
- 4.4.1 The linear viscoelastic solid matrix — 80
- 4.4.2 The linear poroviscoelastic solid — 82

Further Reading — 83

5 Fluid Transport and Deformation — 85

5.1 Transport Laws — 86
- 5.1.1 Mole and mass conservation — 86
- 5.1.2 Dissipation associated with transport — 86
- 5.1.3 Fick's law — 89
- 5.1.4 Darcy's law — 90

5.2 Coupling the Deformation and the Flow — 96
- 5.2.1 The Navier equation — 97
- 5.2.2 The diffusion equation — 98

5.3 Consolidation of a Soil Layer — 100
- 5.3.1 Consolidation equation — 100

	5.3.2	Early time solution	103
	5.3.3	Any time solution	103
	5.3.4	Layer apparent creep	104
	Further Reading	106
6	**Surface Energy and Capillarity**	**107**
6.1	Physics and Mechanics of Interfaces	108
	6.1.1	Origin of surface energy	108
	6.1.2	Basic approach to van der Waals forces	109
	6.1.3	Surface and interface energy	111
	6.1.4	Surface energy and cohesion	113
	6.1.5	Surface energy and surface stress	114
	6.1.6	Wettability, angle of contact and the Young–Dupré equation	118
	6.1.7	The Laplace equation	121
	6.1.8	Pore invasion and interface energy change	122
	6.1.9	Interface energy and adsorption	125
	6.1.10	The disjoining pressure	129
6.2	Capillarity in Porous Solids	132
	6.2.1	Capillary pressure curve and interface energy	132
	6.2.2	Capillary rise	133
	6.2.3	Porosimetry	135
	6.2.4	Capillary hysteresis	136
6.3	Transport in Unsaturated Porous Solids	140
	6.3.1	Relative permeability	140
	6.3.2	Injection	141
	Further Reading	146
7	**The Unsaturated Poroelastic Solid**	**149**
7.1	Interface Stress as a Prestress	150
	7.1.1	Interface energy and saturated poroelasticity	150
	7.1.2	Adsorption-induced deformation	153
7.2	Energy Balance for the Unsaturated Porous Solid	154
	7.2.1	Lagrangian and Eulerian saturations	154
	7.2.2	Lagrangian saturation and free energy balance	156
7.3	The Linear Unsaturated Poroelastic Solid	159
	7.3.1	Constitutive equations of unsaturated poroelasticity	159
	7.3.2	Unsaturated microporoelasticity	159
	7.3.3	Double porosity approach to the brittle fracture of liquid-infiltrated materials	162
7.4	Extending Linear Unsaturated Poroelasticity	164
	7.4.1	The nonlinear poroelastic solid in unsaturated conditions	164
	7.4.2	Accounting for interface stress effects upon deformation	165
	7.4.3	The linear thermoporoelastic solid in unsaturated conditions	166
	7.4.4	The linear unsaturated poroviscoelastic solid	166
	Further Reading	166

8	**Unconfined Phase Transition**		**167**
8.1	Chemical Potential and Phase Transition		168
	8.1.1	Phase equilibrium law	168
	8.1.2	Chemical potential of a pure substance in any form	168
	8.1.3	Supersaturation	171
	8.1.4	Phase transition as an instability	172
8.2	Liquid–Vapor Transition		175
	8.2.1	The Kelvin equation	175
	8.2.2	Effect of a solute	178
8.3	Liquid–Solid Transition		179
	8.3.1	The Thomson equation	179
	8.3.2	Salt crystallization – the Correns equation	182
8.4	Gas Bubble Formation		185
8.5	Surface Energy and Phase Transition		185
	8.5.1	Nucleation	186
	8.5.2	The precondensed liquid film, the disjoining pressure and the Gibbs adsorption isotherm	190
	8.5.3	The premelted liquid film and crystallization	194
	Further Reading		196
9	**Phase Transition in Porous Solids**		**197**
9.1	In-pore Phase Transition		198
	9.1.1	Liquid saturation, pore-entry radius distribution and phase transition	198
	9.1.2	Spherical stress state and stability of in-pore crystallization	201
	9.1.3	Intermolecular forces and in-pore phase transition	204
9.2	Kinetics and Mechanics of Drying		211
	9.2.1	Continuum approach to drying kinetics	211
	9.2.2	Drying asymptotics	215
	9.2.3	Drying mechanics	221
9.3	Mechanics of Confined Crystallization		223
	9.3.1	Cryogenic swelling	223
	9.3.2	The hydraulic pressure	225
	9.3.3	Freezing and air voids	229
	9.3.4	Weathering and the crystallization of sea-salts	235
	Further Reading		238
10	**The Poroplastic Solid**		**239**
10.1	Basic Concepts of Plasticity		240
	10.1.1	Plastic loading function and flow rule	240
	10.1.2	The principle of maximum plastic work	242
	10.1.3	Hardening plasticity	246
	10.1.4	Dilatancy	250
	10.1.5	Three-dimensional plasticity	253
	10.1.6	Limit analysis and stability of dry sandpiles	254
10.2	From Plasticity to Poroplasticity		256
	10.2.1	The poroplastic solid	256

	10.2.2	*Critical state and the Cam-clay model*	260
	10.2.3	*Capillary hardening and capillary collapse*	263
	10.2.4	*Stability of wet sandpiles*	266
10.3	From Material to Structure		267
	Further Reading		269
11	**By Way of Conclusion**		**271**
	Further Reading		275
Index			**277**

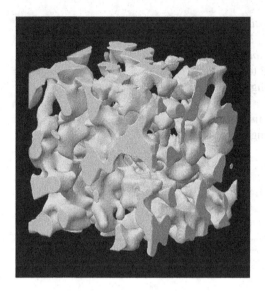

Figure 1 Three dimensional reconstruction of Vycor®glass (courtesy of Pierre Levitz)

Preface

Nowadays it is a nearly impossible task to master all the knowledge required to be innovative in a specialized field, without missing the infinite variety and beauty offered by the many fields of science. In his acceptance talk *Science and the Engineer* for the Timoshenko Award Medal in 1962, Maurice Antony Biot (1905–1985) said that he believed 'that engineers and engineering schools will play an important part in restoring unity and a central viewpoint in the natural sciences. This is because modern engineering, by its very nature, must be synthetic. Specialization, carried to an extreme, is a form of death and decay'.

More than 45 years later the observation of M. A. Biot remains a topical issue. If we do not want to build a Babel Tower of sciences, that will be deserted by the coming generations of students, it becomes more urgent every day to increase the number of interdisciplinary educational programs, which do not focus prematurely on specific disciplines, as well as giving some insights into the history of sciences. As the French philosopher Auguste Comte (1798–1857) said, 'on ne connaît pas complètement une science tant qu'on n'en sait pas l'histoire' (you don't completely know a science if you don't know its history). With this spirit in mind, as well as the modest field that the mechanics and physics of porous solids covers, this book has tried to cross the boundary, from the continuum mechanics of solids, towards the physical chemistry of porous solids. It originates from a course I have been giving for several years as a Professor at Ecole des Ponts ParisTech in the interdisciplinary master of science 'Materials Science for Sustainable Construction'.

When writing this book I have experienced the strength and the accuracy of the celebrated words of the poet John Donne (1572–1631), 'no man is an island'. Indeed, this book has taken advantage of collaborations and discussions with my colleagues and my PhD students covering nearly three decades. Even though some of the results given in this book are due to them, they are too numerous to be named individually. However, I am specially indebted to Sébastien Brisard for his enthusiastic help. He read this book in its early version, commented on it meticulously and suggested many improvements, though any remaining errors are, of course, of my own. I am also pleased to express my gratitude to Matthieu Vandamme, Laurent Brochard and Jean-Michel Pereira for their efficient feedback and perceptive remarks with regard to the topics addressed in this book. The existence of this book would not have been possible without the constant support through the years of my institution, the Laboratoire Central des Ponts et Chaussées.

1

The Strange World of Porous Solids

Buy a bottle of good wine and enjoy it with good friends. Then fill it up with water, seal it, put it in your freezer and ... of course you will forget it overnight. The next morning anxiously open the freezer, and there it is ... you knew it! The bottle has broken into pieces (see Figure 1.1). What is strange about that? 'Nothing' you might say. Everybody knows that water expands when freezing and the bottle broke because it could not withstand the buildup of pressure that this expansion induced. No big deal, but water is a very strange substance because of its hydrogen bonds. Benzene, for instance, unlike water contracts when solidifying and so do many other liquids. In fact, most of them do. Only bismuth, germanium, silica and gallium expand when solidifying like water. So if you fill up a bottle with liquid benzene and freeze it the bottle will not break into pieces. Once again: no big deal.

However, now consider a sample of cement paste. Cement paste is a porous solid that you can 'fill up', or saturate, with a liquid. So let us saturate this sample with benzene and seal it in order to prevent any benzene from escaping. Now bring it below the melting point of benzene ($T = 5.5\,°C = 41.9\,°F$). Since benzene contracts when solidifying, you may legitimately expect that the sample of cement paste will also contract. Wrong! Instead, it will slightly expand, as can be observed in Figure 1.1. Welcome to the strange world of porous solids!

What is going on there? Why does the sample of cement paste expand when benzene freezes, whereas the bottle does not? Because the porous space of the cement paste and that of the bottle are very different. The bottle is a porous solid whose porous space consists of one big pore that is almost as large as the bottle. In contrast, pores in a cement paste have sizes that range from the nanometer scale to the millimeter scale (see Figure 1.2). A pore is a confined environment, and the energy balance allowing the solid to invade a pore involves a surface energy cost because of the solid walls delimiting the pore. The smaller the pore, the more significant this surface energy cost with regard to the volume to be frozen, according to a ratio which is inverse to the pore radius. Because of the Laplace equation, the smaller the pore radius, the higher the pressure in the solid invading the pore and, therefore, the lower the temperature producing the in-pore freezing of the substance. As a result liquid-saturated pores with different sizes freeze at different temperatures, the largest pores freezing first.

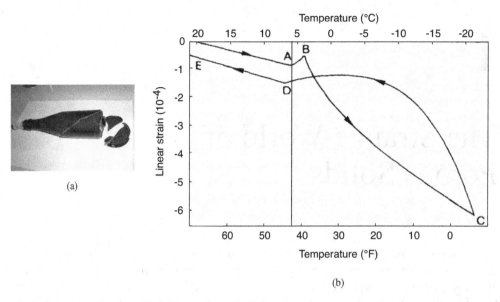

Figure 1.1 (a) A bottle, which is completely filled up with water and sealed, breaks into pieces when freezing. (b) Unlike water, benzene contracts when solidifying; however, from point A to B a benzene-saturated cement paste sample is paradoxically observed to expand slightly when subjected to temperatures below the melting point of benzene, that corresponds to the dashed line $T = 5.5\,°\text{C} = 41.9\,°\text{F}$. (The original data are extracted from Beaudoin, J.J. and MacInnis, C. (1974). The mechanism of frost damage in hardened cement paste, *Cement and Concrete Research*, **4**, 139–147.) See Section 9.3.1 for a detailed quantitative analysis

However, the fact that the liquid in the largest pores freezes first does not explain the swelling of the sample when the liquid contracts as it solidifies. Basically, once formed the solid crystals have to remain in thermodynamic equilibrium with the supercooled liquid which still remains unfrozen in the smallest pores. This equilibrium is achieved when the free energy of both phases is the same. Because the molecules are more ordered in a solid than the jumble of molecules that constitute a liquid, the entropy that measures the disorder is less for the solid than for the liquid. As a result, under the same pressure the free energy of the solid increases less rapidly than that of the liquid when the temperature decreases. As the temperature decreases further below the melting point related to the pore size, in order to offset its free energy difference with the liquid due to its lower entropy, the solid becomes more pressurized than the liquid does. To produce such a pressure increase, some still unfrozen liquid is sucked into the frozen pore and freezes in its turn. The additional freezing made possible by this so-called cryosuction finally produces the observed swelling (see Figure 1.1(b)), that occurs irrespective of the change in density over the liquid—solid phase transition. In short, when pores of various sizes are put in a solid (see Figure 1.2), its behavior can become very intriguing, complex and even counter-intuitive!

We have to realize that porous solids are all around us. They can be natural (like plants, meat, rocks, stones, soils ...), as well as man-made (like gravels, cement, concrete, plaster, filters, gels ...). The human body itself, which is an assembly of bones, muscles, skin and

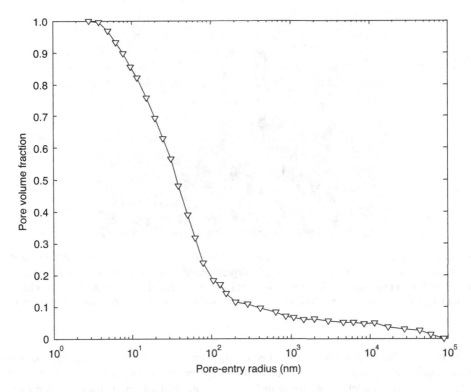

Figure 1.2 Because of their pore size distribution the mechanical behavior of actual porous solids is much more complex than that of the bottle of Figure 1.1. As shown here the size of the pore-entry radius to the volume fraction related to a cement paste can range from the nanometer to the millimiter (courtesy of Véronique Baroghel-Bouny for the data)

so on, is also a complex structure made of porous solids. Since porous solids are everywhere, they affect our day-to-day life.

If you go to the beach with your bucket and spade try building a sandcastle. Maybe something like the beautiful reproduction of Gaudi's Sagrada Familia in Barcelona displayed in Figure 1.3. If the sand is dry, there is no way — everything falls apart. If the sand is too wet, the same. However, if you add only a little bit of water to the sand in your bucket and turn it out your castle does not fall apart. Why? Sand is a porous material because you cannot completely fill the space with rounded grains. Capillary liquid bridges trap the little water you add within the slits between the sand grains. This water becomes strongly depressurized and sticks the grains together, giving cohesion to damp sand. If the sand is too wet the capillary bridges and the cohesion they give to the sand disappear.

Before leaving the beach, have a walk on the wet sand beside the sea. With each step you take you might expect liquid water to be squirted around your footprint in the sand. Surprise! As shown in Figure 1.4, instead of observing this squirt flow, it is exactly the opposite — each step dries out the sand around your footprint. What is this new trick of granular porous solids? Under your foot the sand does not experience only a compression but also a significant shear, particularly on the border of your footprint. Under shear the solid grains become less tangled

Figure 1.3 Capillary hardening — a little water added to initially dry sand gets trapped between the sand grains by capillary bridges. Being strongly depressurized, this water sticks the grains together and gives cohesion to damp sand, making possible this beautiful reproduction on the beach of Gaudi's Sagrada Familia (copyright Thierry Joffroy, CRATerre-ENSAG). For a mechanical analysis see Section 10.2.4

up and the sand becomes more porous. The enlargement of the pore space under your foot causes a liquid depression there which sucks the water from the wet sand around your foot. At the next step the water sucked towards your foot flows into the footprint you leave behind you.

The weather report you consulted before going to the beach predicted hot weather interrupted by rain, so you decided to take your leather jacket with you. If the weather is hot you know

Figure 1.4 Because of the dilatancy of granular materials under shear stress, each step you take in wet sand causes a liquid depression that dries out the wet sand around your foot (courtesy of Yann Goiran). For a modelling of dilatancy see Section 10.1.4

The Strange World of Porous Solids

Figure 1.5 Day after day this statue has endured the endless deposit of salt-spray (courtesy of Leo Pel, Eindhoven University of Technology). The weathering results from the in-pore pressure buildup due to the crystallization of sea-salts induced by the drying following each deposit. For a quantitative analysis see Section 9.3.4

that you will not be bothered by sweating, but if it rains your leather jacket will be the perfect raincoat. Why is this? Because your leather jacket is a porous solid that has been treated with oil: its wettability properties are such that liquid water is nonwetting and can thus not easily go through the layer of leather while water vapor is wetting and can thus easily escape.

Being near the sea, you may also see a statue that looks like the one displayed in Figure 1.5–a statue that has been badly damaged by years and years of weathering. Why does weathering happen? Because stone is a porous solid. Saline solutions are sprayed by the wind onto the surface of the statue and penetrate into the stone by capillarity. Subsequent drying increases the concentration of salt in the residual liquid until sea-salt crystallizes within the stone. Weathering is the result of the endless repetition of imbibition–drying cycles and the subsequent buildup of pressure that the crystals of sea-salt induce.

When you return home you might clean the dishes. Your sponge is hard when you pick it up, but immerse it and it becomes soft. Why? Because the sponge also is a porous solid. This time, rather than its strength, capillary bridges which drying induces within the sponge were increasing its stiffness. Saturating the sponge by immersing it causes those bridges to disappear and thus the increase in stiffness too. Letting the sponge dry again makes the capillary bridges and the increase in stiffness reappear.

The weekend is now over so you go to your laboratory and test a few samples. A material cannot withstand stresses that are arbitrarily high. When the stresses reach a critical value, the internal cohesion of the material is destroyed and the material fails. The maximum stress that the material can withstand characterizes its strength. This strength is a property that is intrinsic to the material and should therefore not depend on the rate of loading. However, if the material that you are testing is a porous solid that contains some fluid, you find out that its strength does depend on the rate of loading. Isn't it strange? If you now extract a gassy sediment (again a porous solid) from the deep seabed and unload it gradually you will observe that failure occurs at some time during the unloading process. Again, isn't it strange?

We are surrounded by porous solids. Since porous solids through which fluids can seep or flow are ubiquitous, they are of interest to a wide range of fields: food engineering, geosciences, civil engineering, building physics, petroleum geophysics, chemical industry, biomechanics and so on. Even though materials and fields are very diverse, all porous solids for all applications have one thing in common: they are subject to the same coupled processes such as freezing and swelling, drying and shrinkage, diffusion of liquids and creep, osmosis and expansion. Such coupled processes occur at the interface between physical chemistry and mechanics.

Since environmental engineering, petroleum geophysics, civil engineering, geotechnical engineering, biomechanics, the food industry and so on, involve processes that pertain to both physical chemistry and solid mechanics, experts in each of those two fields interact regularly. However, the interface between those two fields is not often explored, be it in textbooks or in more advanced books. This may partly explain why the dialog between experts in applied mechanics and in physical chemistry remains difficult. Experts in applied mechanics often think that experts in physical chemistry focus almost exclusively on the physical explanation of the phenomena so that the results are unsuited for quantitative engineering applications. In a similar way physical chemists think that applied mechanics experts focus almost exclusively on the mathematical modeling and thus provide no actual physical explanation for the observed phenomena. In short, an explanation before any equation can provide great help in carrying out a sound modeling. Indeed, would you be blindly confident in your numerical results if they did mimic the experimental results of Figure 1.1, but without having identified the cryosuction process that sucks liquid water towards the already frozen sites? The computer may understand but not you! Albeit the opposite is also true. When two equally attractive explanations of the same phenomenon compete, assembling equations to model the phenomenon can settle the dilemma. For instance, does a wet porous solid dry like a water pond, through vapor molecular diffusion with no significant motion of the liquid water as illustrated in Figure 1.6(a), or does the liquid water move towards the surface of the porous solid where it finally evaporates as illustrated in Figure 1.6(b)?

This book is entitled *Mechanics and Physics of Porous Solids* because the author believes that there is room for more physics in engineering questions that are raised by the mechanics of porous solids. The ambition here is to provide a unique and consistent framework in order to address the large variety of physical phenomena that produce a mechanical effect on porous solids. This book aims to bridge the gap between physical chemistry, which governs what happens at the level of the pore, and solid mechanics, which is the natural frame in which deformations, stresses and fluid transport are addressed and quantified at the macroscopic level of the porous material.

Intended as a first introduction, this book focuses on both the mechanics and the physics of porous solids. It also presents updated developments both in the field of unsaturated deformable porous media and in the field of the mechanics of porous media subject to phase transitions.

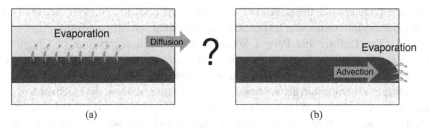

Figure 1.6 Two possible drying mechanisms: (a) the drying is achieved through the inner evaporation of the inner liquid water and the subsequent diffusive transport of the water vapor so produced to the material surface; (b) the drying is achieved through the advective transport of liquid water to the material surface and its subsequent evaporation. See Section 9.2.2 to discover which is the winning drying mechanism

In order to facilitate the aforementioned interdisciplinary dialog, between experts in physical chemistry and experts in the mechanics of solids, the energy approach is chosen, which provides the common language of thermodynamics. Energy considerations will therefore be favored throughout the book.

Master students, Ph.D. students and scientists in various fields should easily be able to tackle the parts of the topics that pertain to their educational background. The expectation is that this book will generate an increasing interest in the parts that do not pertain to this background and enable the readers to expand their knowledge. As a result, various readerships are possible, and people with different backgrounds will hopefully find an interest in consulting this book.

Mechanics and Physics of Porous Solids addresses the mechanics and the physics of deformable porous materials, whose porous space is filled up with one or several fluid mixtures which *interact* with the solid matrix. Including neither this introductory chapter nor the concluding chapter, the book is made up of nine chapters some of which are introduced by historical comments which relate directly to the focus of the chapter. The book progressively combines basic physical and mechanical concepts that apply to fluid mixtures and to solids, in order to provide a comprehensive energy approach of their complex physical interactions. The basic concepts are reintroduced in order for the book to be self-contained. In addition, rather than writing 'as can be easily derived', intermediate calculations are often given, resulting in a substantial number of equations. Depending on their background readers may therefore browse or even skip some basic parts, calculations or chapters. Each chapter ends with indications for further reading. The list of proposed references is in no way exhaustive and priority has been given to books.

Chapter 2, 'Fluid Mixtures', is an invitation to revisit the basic concepts of thermodynamics. This is a basic chapter, and part of it may be skipped by readers having an undergraduate background in thermodynamics and fluid mixtures. After defining the chemical potential, it introduces the fundamental equality of physical chemistry, namely, the Gibbs–Duhem equation, and browses the principal results associated with ideal mixtures. The chapter revisits the electric double-layer theory. It looks into how the swelling of a porous material, filled with an electrolytic solution, originates from the excess of osmotic pressure caused by the presence of electric charges on its internal solid walls. The chapter ends with the analysis of the stability of regular fluid mixtures. These two situations are looked at in detail because they offer archetypal approaches developed throughout the book to address a variety of effects caused by nonlocal intermolecular forces and the associated stability analyses.

Chapter 3, 'The Deformable Porous Solid', is the natural companion of Chapter 2. It provides the relevant thermodynamic framework to elaborate the constitutive equations of a porous solid whose porous space is filled with a fluid mixture. The chapter starts by revisiting the basic concepts of continuum mechanics of solids. The two pillars of the mechanics of solids, strain and stress, are defined and related mechanical energy balances are stated. This part can be skipped by readers having the appropriate background. The second part of the chapter progressively extends these concepts to porous solids. The link between porosity variations and deformation is given first. The chapter then goes on by looking at the stress partition theorem, which splits the stress into two parts, namely, the stress sustained by the solid matrix and the part devoted to the pore pressure. The thermodynamics of a saturated porous solid is then investigated by extending the thermodynamics of fluid mixtures to solid−fluid mixtures. The porous solid itself is the material loaded by the external stress and by the pore pressure applying to its internal solid walls, irrespective of the actual cause producing the pressure. The porous solid defined in such a way is a closed system, whose thermodynamics can be carried out by extracting its energy balance from that related to the previous solid−fluid mixture. This extraction is achieved by fictitiously removing the bulk fluid mixture saturating the porous space, and whose own energy balance is independently captured by its related Gibbs−Duhem equation.

Chapter 4, 'The Saturated Poroelastic Solid', explores in detail the constitutive equations of both linear and nonlinear poroelastic solids. Particular attention is given to the presence of the same substance in the form of both gas bubbles and solute. The chapter gives some insights into microporoelasticity, which provides a mean field assessment of the macroscopic poroelastic properties from the porosity and the properties of the solid and fluid components. The chapter ends up by accounting for thermal effects and by considering delayed behaviors that the viscoelasticity of the solid matrix induces.

Chapter 5, 'Fluid transport and Deformation', analyzes how the transport of a fluid within a porous solid, and the deformation of the latter upon both the external stress and the pore pressure, are coupled phenomena. The chapter first replaces the derivation of transport laws in the context of continuum thermodynamics, and it shows how they can be derived by combining thermodynamic restrictions and an up-scaling dimensional analysis. Both molecular diffusion and advective transport laws in a porous solid are examined. When considering the gas transport, the chapter includes the possible sliding of the molecules on the internal walls of the porous network. The coupling of the deformation of the porous solid with the viscous flow of the fluid saturating the porous space is looked at later on in detail. The approach reveals that the diffusion equation governing the flow of a poorly compressible fluid in a poroelastic solid is universal, with a diffusion coefficient independent of the particular problem under consideration. The coupling of the flow and the deformation is finally illustrated through the one-dimensional theory of the consolidation of a porous solid layer.

Chapter 6, 'Surface Energy and Capillarity', investigates how the interface energy between two saturating fluids and the internal solid walls delimiting the porous space govern the imbibition and the drainage of a porous solid. After recalling the intermolecular origin of van der Waals forces, the chapter shows how intermolecular forces are the source of the cohesion of a substance and the interface energy between two substances. Special attention is given to making the distinction between surface energy and surface stress. This distinction becomes crucial when examining the equilibrium of a solid–fluid interface. The chapter continues by successively deriving the Young–Dupré and Laplace equations, which govern the equilibrium of the triple line between three substances and the equilibrium of the interface between two substances, respectively. A further analysis looks at how interfacial energy can be modified by adsorption and how intermolecular forces can induce a disjoining pressure in thin liquid films lying on a solid substrate. The Laplace equation and the surface energy balance combine to give the microscopic interpretation of the capillary pressure curve which provides the saturation of the wetting fluid in a porous solid. In relation with the pore size distribution, this capillary curve turns out to be the key state function to capture the macroscopic effects due to capillarity — possibly hysteretic — on porous solids. After addressing transport laws in unsaturated conditions, the chapter ends by analyzing how advection and diffusion compete in the injection of a wetting phase in a porous medium.

Chapter 7, 'The Unsaturated Poroelastic Solid', combines the concepts and the results derived in Chapters 4 and 6, in order to explore the deformation of poroelastic solids whose porous space is invaded by a nonwetting phase at the detriment of a wetting phase. The chapter first reconsiders poroelasticity in saturated conditions with regard to surface energy, which induces a prestress depending on the nature of the saturating fluid. The concept of Lagrangian saturation is then introduced. This concept is used to perform separate energy balances regarding the interfaces and the deformation, respectively. The energy balance related to the deformation provides the appropriate framework to establish the constitutive equations of unsaturated poroelasticity, both linear and nonlinear. The derivation of unsaturated poroelastic properties can then be performed from microporoelasticity. An illustration of unsaturated poroelastic constitutive equations is given by detailing how the strength of water-infiltrated porous solids depends upon the loading rate. The chapter ends by extending constitutive equations of saturated thermoporoelasticity and poroviscoelasticity to unsaturated conditions.

Chapter 8, 'Phase Transition in Free Space', proposes a survey of the basic laws governing the thermodynamics of unconfined phase transitions. The chapter starts by recalling the role of the chemical potential regarding phase transition. Special attention is given to determining the chemical potential of a solid elastic phase. The concept of supersaturation, as the driving force of the phase transition, is defined, and its general expression is derived as a function of the pressures of the mother and daughter phases. The chapter continues by recalling how a phase transition results from an instability through the opposing effects of thermal agitation and molecular attraction. Standard equilibrium laws related to various phase transitions are revisited. The derivation of the Kelvin equation governing evaporation and condensation is first derived, including the effect of a solute. The Thomson equation governing melting and crystallization is then investigated, with a focus on the role of the elastic energy stored in the solid phase. Salt crystallization is addressed through the determination of Correns equation. Gas bubble formation is examined in the same way. The chapter ends by studying the role of the surface energy in the nucleation process. It also examines this role in the formation of a precondensed or premelted liquid film on solid substrates, in relation with both the adsorption process and the work produced by the disjoining pressure.

Chapter 9, 'Phase Transition in Porous Solids', looks at how the in-pore confinement and associated surface effects do affect a phase transition and, in turn, what can be the mechanical effects of in-pore phase transition upon the porous solids. It starts by investigating the effect of confinement upon phase transition at the pore scale, and how phase transition in a porous solid is governed by the pore size distribution. A further stability analysis shows that, despite the shear stress that an elastic solid crystal can withstand, a crystal that can form in a pore must be subjected to a spherical stress state. Extending the analysis which ended Chapter 8, interfacial energy effects are also shown to result in the existence of precondensed and premelted films at the surface of the internal walls delimiting the porous volume. The chapter then addresses the drying of materials, regarding both the moisture transport processes and the associated mechanical effects such as drying shrinkage and drying stiffening. The chapter ends with the mechanics of confined crystallization. Using the constitutive equations of unsaturated poroelasticity, the mechanical behavior of porous solids subjected to freezing is studied. Special attention is given to cryogenic swelling and to the role of air voids which act both as expansion reservoirs and cryopumps for the liquid water expelled from the freezing sites. The chapter is completed by analyzing the mechanics of stones subjected to salt crystallization in the same way.

Chapter 10, 'Poroplasticity', surveys how standard plasticity can be extended to porous solids. First of all the basic concepts of plasticity are revisited. The principle of maximal plastic work, that turns out to be a principle of maximal production of entropy, is introduced by analyzing the mechanical behavior of a simple friction element. It is shown how this principle guarantees both the stability of plastic systems and the uniqueness of the stress prevailing in their various parts, resulting from their external loading. Hardening plasticity is then addressed and capillary hysteresis is shown to be typical of such a behavior. Further, the exploration of two-dimensional systems allows us to analyze the dilatancy of granular materials. The standard cohesion−frictional model is then formulated in the framework of three-dimensional plasticity. The first part of the chapter ends with the analysis of the stability of dry sandpiles. The second part of the chapter is devoted to poroplasticity, and investigates how plasticity can be extended to porous solids exhibiting an irreversible behavior. The effective stress is first shown driving the plastic strains of porous solids whose solid matrix does not undergo plastic volumetric changes. In saturated conditions the plastic behavior of porous solids can then be captured by replacing the stress by the effective stress in usual models of plastic solids. Accounting for the hardening plastic behavior of a large variety of saturated porous solids, the Cam-clay plastic model is then presented in detail. The chapter goes on by investigating how capillary effects can alternatively cause the hardening or the collapse of a granular porous material under unsaturated conditions. The chapter ends by analyzing how capillary hardening quantitatively strengthens wet sandpiles with regard to the dry situation.

Finally, a concluding chapter, 'By Way of Conclusion', emphasizes how the various scales have to be linked to each other in most analyses of porous materials and structures.

Because of the interdisciplinary nature of topics it encompasses *Mechanics and Physics of Porous Solids* was not originally intended to be a textbook relative to a specific discipline, even though some parts of it may be used for this purpose. Because of the introductory level of exposition adopted, but also because of the various new results presented in this book, *Mechanics and Physics of Porous Solids* is neither a research book about some exhaustive state of the art, nor a review of updated literature on sophisticated topics. It rather aims at exploring the frontier existing between the physical chemistry and the mechanics of porous solids, and at providing new insights into their fruitful combination. May the readers of this book, whatever their background, enjoy the view of the strange world of porous solids from the bridge that this combination offers. As Frederik Paulsen said, 'the true journey is not to look for new landscapes, but for a fresh look'.

2

Fluid Mixtures

If various names are associated with the discovery of the laws of thermodynamics, their extension to heterogeneous systems subjected to any transformation is the achievement of a single person, Josiah Williard Gibbs (1839–1903). In 1876 and 1878 he published *On the Equilibrium of Heterogeneous Substances*, the historical paper which marked the birth of chemical thermodynamics. However, its austere mathematical form made the reading of this 300 page paper difficult, and several decades were needed before its major significance was finally recognized. In his *Étude sur les travaux thermodynamiques de J Willard Gibbs* published in 1887, Pierre Duhem (1861–1916) contributed significantly to the extension of Gibbs' work, and his name is attached to one of the fundamental equations of physical chemistry, namely, the Gibbs–Duhem equation.

Throughout the chapters of this book we will explore the mechanics and the physics of porous solids, whose porous space is saturated by one or several fluid mixtures. The first stage of this exploration, before addressing the coupling between these fluid mixtures and the porous solid, will be to revisit the standard thermodynamics of fluid mixtures. This part of thermodynamics is mainly governed by the Gibbs–Duhem equation, which is one of the key elements for studying the thermodynamics of fluid–solid mixtures in the next chapter. Having introduced chemical potential, this chapter explores the Gibbs–Duhem equation and its consequences for mixtures. The chapter revisits the electric double-layer theory. It looks into how the swelling of a porous material, filled with an electrolytic solution, originates from the excess of osmotic pressure caused by the presence of electric charges on its internal solid walls. The chapter ends with the analysis of the stability of regular fluid mixtures. These two situations are looked at in detail because they offer typical approaches developed throughout this book to address a variety of effects caused by nonlocal intermolecular forces and the associated stability analyses.

2.1 Chemical Potential

2.1.1 Free energy and chemical potential

The various thermodynamic potentials form a sort of club, whose members are linked to each other by energy and entropy balances. In the physical chemistry of fluid mixtures one of them plays a major role, namely, the chemical potential. Its admission to the club gives us the opportunity to revisit the basic laws of thermodynamics briefly and to introduce a few notations.

When a certain amount of fluid undergoes a volume change dV, at temperature T and pressure p, the first and second laws of thermodynamics combine to provide the change dE of its internal energy E in the form

$$dE = -pdV + TdS. \tag{2.1}$$

The energy balance (2.1) assumes no dissipation so that, according to the second law, the infinitesimal heat supply is equal to TdS, where S is the fluid entropy. According to the first law, the energy balance (2.1) states that the change of internal energy dE is the sum of the amount of heat TdS and the mechanical work $-pdV$ supplied to the fluid. If we take the entropy as the reference potential, from the energy balance (2.1) we derive the fluid state equations in the form

$$\frac{p}{T} = \frac{\partial S}{\partial V}; \quad \frac{1}{T} = \frac{\partial S}{\partial E}. \tag{2.2}$$

Alternatively, we can introduce the free energy $F = E - TS$ so that Equation (2.1) can be rewritten in the equivalent form

$$dF = -p\,dV - S\,dT. \tag{2.3}$$

The free energy F is the energy which is available to produce work. Since Equation (2.1) has assumed no dissipation, the energy balance (2.3) states that the infinitesimal mechanical work $-pdV$ will be fully stored in the form of free energy during isothermal evolutions. The quantity $-SdT$ accounts for the deterioration of current free energy F with regard to its further transformation into work. The loss $-SdT$ of free energy is due to the increase in the disorder of fluid molecular motion induced by the increase in thermal agitation associated with the temperature change dT. This thermal agitation affects their coordinated motion, which is needed to produce work.

The balance of free energy (2.3) applies to a closed fluid system, that is, to an amount of fluid consisting of the same molecules at any time. Consider now a fluid mixture formed of various components individually referred to by the subscript J. Furthermore, let N_J be the current number of moles of component J contained in the amount of fluid occupying the volume V. If dN_J moles are added to this fluid amount during the volume change dV, the related free energy supply is $\mu_J dN_J$, where μ_J stands for the molar chemical potential related to component J. According to this very definition of chemical potential μ_J, the balance of free energy (2.3) related to closed fluid systems extends to fluid mixtures in the form

$$dF = -pdV - SdT + \sum_J \mu_J dN_J. \tag{2.4}$$

Let us now introduce the Gibbs free energy G (also called free enthalpy) of the fluid mixture defined by

$$G = F + pV. \tag{2.5}$$

With the help of Equation (2.5) we transform the energy balance (2.4) into the following equivalent form

$$dG = V\,dp - S\,dT + \sum_J \mu_J\,dN_J. \tag{2.6}$$

The energy balance (2.6) holds whatever reversible transformation occurs. As a result, dG is an exact differential form and G depends only on the intensive state variables p and T and the extensive state variables N_J. We write

$$G = G(p, T, N_1, N_2, \ldots). \tag{2.7}$$

The volume V, the entropy S and the chemical potential μ_J are then energy conjugate variables of the pressure p, the temperature T and the mole number N_J, respectively, according to

$$V = \frac{\partial G}{\partial p}; \quad S = -\frac{\partial G}{\partial T}; \quad \mu_J = \frac{\partial G}{\partial N_J}. \tag{2.8}$$

2.1.2 Equilibrium of mixture composition

Now let two fluid mixtures be in such a contact that one of their components, referred to by the subscript J, is allowed to be freely exchanged between them. In a spontaneous transformation dN_J moles of this component can leave one mixture, and join the other one, whereas the pressure and the temperature, as well as the number of moles of the other components, remain constant. Because of the conservation of the number of moles we have $dN_J = -dN'_J$ where the symbol $'$ distinguishes one of the mixtures from the other. The second law of thermodynamics states that the overall change $d(G + G')$ in the Gibbs free energy undergone during this spontaneous transformation cannot be positive. Applying Equation (2.6) separately to each mixture and adding the resulting equations, we obtain

$$d(G + G') = (\mu_J - \mu'_J)\,dN_J \leq 0. \tag{2.9}$$

With regard to component J the equilibrium is reached when the transformation may occur in whatever direction. From Equation (2.9) we therefore derive the equilibrium condition

$$\mu_J = \mu'_J. \tag{2.10}$$

At equilibrium the values of the chemical potential of the same component in both mixtures are therefore equal.

2.2 Gibbs–Duhem Equation

2.2.1 Derivation of the Gibbs–Duhem equation

As expressed by Equation (2.7), the Gibbs free energy G of a mixture depends on both intensive state variables, namely p and T, and extensive state variables, namely, N_J. As any energy, the Gibbs free energy G is an extensive quantity: the scaling of the amount of all components by a given positive factor λ will result in the increase of G by the same factor providing the remaining intensive state variables are held constant. In other words, G is a homogeneous function of the first order with respect to its arguments N_J. This may be expressed in the following mathematical form

$$\lambda > 0: \quad G(p, T, \lambda N_1, \lambda N_2, \ldots) = \lambda G(p, T, N_1, N_2, \ldots). \tag{2.11}$$

Deriving Equation (2.11) with respect to λ, and later on adopting the value $\lambda = 1$, G is shown to satisfy Euler's identity relative to homogeneous functions of the first order, namely,

$$G = \sum_J N_J \frac{\partial G}{\partial N_J}. \tag{2.12}$$

Substitution of the last of the state Equations (2.8) in Euler's identity (2.12) provides

$$G = \sum_J N_J \mu_J. \tag{2.13}$$

By substituting Equation (2.13) in Equation (2.6), we finally obtain the celebrated Gibbs–Duhem equation:

$$V dp - S dT - \sum_J N_J d\mu_J = 0. \tag{2.14}$$

The Gibbs–Duhem equation is a cornerstone of physical chemistry because it applies to any fluid mixture and to any reversible transformation, with no further assumption. Indeed, in the next chapter, it will enable us to extract the free energy balance restricted to the porous solid from the free energy balance relative to the fluid–solid mixture considered as a whole.

2.2.2 Molar Gibbs–Duhem equation

It is often useful to relate the behavior of the mixture to the behavior of one of the moles forming it. A mole is conventionally made up of $\mathcal{N}_A \simeq 6.022 \times 10^{23}$ mol^{-1} molecules, $\mathcal{N}_A$ being the Avogadro number. Nowadays $\mathcal{N}_A$ is defined by saying 0.012 kg of ^{12}carbon contains $\mathcal{N}_A$ atoms. Letting N be the total number of moles, the molar volume $\overline{V}$ and the molar entropy $\overline{S}$ of the fluid mixture are then defined by

$$N = \sum_J N_J; \quad \overline{V} = V/N; \quad \overline{S} = S/N. \tag{2.15}$$

Furthermore, let $\overline{N}_J$ be the molar fraction relative to component J, which is defined by

$$\overline{N}_J = N_J/N; \quad \sum_J \overline{N}_J = 1. \tag{2.16}$$

With the help of definitions (2.15) and (2.16), when dividing the Gibbs–Duhem Equation (2.14) by N, we obtain the Gibbs–Duhem equation relative to one mole, namely,

$$\overline{V} dp - \overline{S} dT - \sum_J \overline{N}_J d\mu_J = 0. \tag{2.17}$$

A pure fluid substance corresponds to a fluid mixture with a single component. Restricting the subscript J to 1 in Equation (2.17) and removing the subscript, we obtain the standard energy balance related to a pure substance in the form

$$\overline{V}dp - \overline{S}dT - d\mu = 0, \qquad (2.18)$$

and the associated state equations

$$\overline{V} = \frac{\partial \mu}{\partial p}; \qquad \overline{S} = -\frac{\partial \mu}{\partial T}. \qquad (2.19)$$

The state Equations (2.19) can be inverted with regard to p by considering the molar Helmholtz free energy a related to a pure substance. For a pure substance, according to Equation (2.13), we have $\mu = G/N$. From Equation (2.5) and $a = F/N$ we obtain

$$\mu = a + p\overline{V}. \qquad (2.20)$$

Substitution of Equation (2.20) in Equation (2.18) provides us with the Clausius–Duhem equation:

$$-pd\overline{V} - \overline{S}dT - da = 0. \qquad (2.21)$$

The Clausius–Duhem Equation (2.21) allows us to invert the state Equations (2.19) with regard to pressure in the form

$$p = -\frac{\partial a}{\partial \overline{V}}; \qquad \overline{S} = -\frac{\partial a}{\partial T}. \qquad (2.22)$$

The determination of $\mu(p, T)$ or $a(\overline{V}, T)$ and consequently the explicit formulation of the state equations, can only be achieved by introducing further pieces of information. This is done in the following sections for ideal and real gases.

2.2.3 Ideal gases

When the fluid is a gas at sufficiently low pressure it is dilute, and the sizes of the molecules forming the fluid are negligible with respect to the mean distance between the molecules. The macroscopic behavior of the gas is then governed by the shocks between molecules having no dimensions. This is the limit of ideal gases, whose behavior is governed by the universal state equation

$$p = \frac{RT}{\overline{V}}, \qquad (2.23)$$

where $R = 8.3144 \, \text{J K}^{-1} \, \text{mol}^{-1}$ is the ideal gas constant. If the gas is also thermally ideal, its molar enthalpy h, which is defined by

$$h = \mu + T\overline{S} \qquad (2.24)$$

and thereby satisfies

$$dh = \overline{V}dp + Td\overline{S}, \qquad (2.25)$$

depends linearly on temperature T so that

$$h = \overline{c}_p T. \qquad (2.26)$$

Substitution of Equations (2.23) and (2.26) in the enthalpy balance (2.25) and a subsequent integration provide

$$\overline{S} = \overline{S}_0 + \overline{c}_p \ln \frac{T}{T_0} - R \ln \frac{p}{p_0}. \quad (2.27)$$

In turn Equations (2.24), (2.26) and (2.27) combine to provide the chemical potential μ of an ideal gas in the form

$$\mu(p, T) = f(T) + RT \ln p, \quad (2.28)$$

where $f(T)$ is a function depending on temperature only.

2.2.4 Real gases

In addition to the collisions, the molecules of a real gas are subjected to intermolecular forces. To account for these forces and for the finite dimensions of real molecules, the ideal gas Equation (2.23) can be modified in the form of the van der Waals equation of real gases:

$$p = -\frac{A}{\overline{V}^2} + \frac{RT}{\overline{V} - B}, \quad (2.29)$$

where B is the so-called molar excluded volume due to the impenetrability of the molecules of finite dimensions; the first term, where A is proportional to the interaction potential between two molecules, accounts at the macroscopic level for the intermolecular forces operating at the microscopic level. These forces are addressed in more detail in Chapter 6 devoted to capillarity. Their effects on the phase transitions are examined in Chapters 8 and 9.

2.2.5 Partial molar property

Equation (2.12) extends to any extensive quantity $\Pi(p, T, N_1, N_2,...)$ according to

$$\Pi = \sum_J N_J \frac{\partial \Pi}{\partial N_J}. \quad (2.30)$$

Introducing the molar property $\overline{\Pi} = \Pi/N$, we rewrite Equation (2.30) in the form

$$\overline{\Pi} = \sum_J \overline{N}_J \overline{\Pi}_J, \quad (2.31)$$

where the partial molar property $\overline{\Pi}_J$ associated with Π is given by

$$\overline{\Pi}_J = \frac{\partial (N\overline{\Pi})}{\partial N_J}. \quad (2.32)$$

Owing to this very definition of $\overline{\Pi}_J$ the increase dN_J of the number of moles of component J induces the increase $\overline{\Pi}_J dN_J$ of the extensive property Π according to

$$\Pi(p, T, N_1, ..., N_{J-1}, N_J + dN_J, N_{J+1},...) \quad (2.33)$$
$$= \Pi(p, T, N_1, ..., N_{J-1}, N_J, N_{J+1},...) + \overline{\Pi}_J dN_J.$$

This property will prove to be convenient when expressing the balance equation of any extensive property Π in the context of continuum thermodynamics. If $\underline{u}_J$ is the molar flux vector related to component J of a mixture, the number of moles dN_J of component J flowing during time dt, through the surface having dA as area[1] and oriented by the normal unit $\underline{n}$, is $\underline{u}_J \cdot \underline{n}\,dA\,dt$. These moles take away the quantity $\overline{\Pi}_J \underline{u}_J \cdot \underline{n}\,dA\,dt$ of Π. For any closed volume Ω, formed from infinitesimal volumes $d\Omega$, and whose border is the surface $\partial\Omega$, the divergence theorem provides

$$\int_{\partial\Omega} \overline{\Pi}_J \underline{u}_J \cdot \underline{n}\, dA = \int_{\Omega} \nabla \cdot \left(\overline{\Pi}_J \underline{u}_J\right) d\Omega, \tag{2.34}$$

where $\nabla \cdot$ stands for the divergence operator. As a result, $\nabla \cdot \left(\overline{\Pi}_J \underline{u}_J\right)$ represents the quantity of property Π taken away by the moles of the component J, that flow out from the infinitesimal volume $d\Omega$.

With respect to the previous general definition of partial molar property $\overline{\Pi}_J$, and according to the last state Equation (2.8) and to Equation (2.13) for G, the chemical potential μ_J is now identified as the partial molar Gibbs potential $\overline{G}_J$. Another example is provided by the partial molar entropy $\overline{S}_J$ satisfying

$$\overline{S}_J = \frac{\partial (N\overline{S})}{\partial N_J}. \tag{2.35}$$

Using the state equations (2.8) and definition $\overline{S} = S/N$, from Equation (2.35) we derive

$$\overline{S}_J = -\frac{\partial \mu_J}{\partial T}, \tag{2.36}$$

in a consistent way with Equation (2.19) previously found for a pure fluid.

2.3 Ideal Mixtures

When putting together several pure components, the key question is to determine how the chemical potential of a given component is affected by the mixing. The ideal mixture is the simplest mixture, where the only interactions occuring between the molecules that mix are those resulting from the shocks they undergo when they meet. For gaseous ideal mixtures, as we shall see in the next section, it is then easy to derive how the chemical potential of each component is affected by the mixing. The result will be used further as a general definition of any ideal mixture, either gaseous or liquid.

2.3.1 Ideal gas mixture

The macroscopic behavior of an ideal gas is governed by the collisions between molecules of negligible dimensions. Assuming that the mixing of ideal gases does not induce any other kind of interaction than the collisions between the molecules, both the mixture as a whole and

---

[1] In order to make the distinction between the differential operator with respect to time d, as for instance in dN_J, the differential operator with respect to space is denoted by the symbol d, as for instance in dA.

its gaseous components will be ideal gases. According to the ideal gas Equation (2.23) we can therefore write

$$p = N\frac{RT}{V}; \quad p_J = N_J\frac{RT}{V}, \quad (2.37)$$

where V is the volume occupied by both the mixture and its gaseous components, whereas p_J is the pressure that can be associated with the gaseous component J within the mixture. Eliminating RT/V between both equations we obtain

$$p_J = \overline{N}_J\, p. \quad (2.38)$$

As a result, p_J is identified with the partial pressure related to the gaseous component J. Indeed, owing to the second part of Equation (2.16), the sum of partial pressures is equal to the mixture pressure p. This is Dalton's law, namely,

$$\sum_J p_J = p. \quad (2.39)$$

According to Equation (2.28) the chemical potential $\mu_J^*(p_J, T)$ of an ideal gas at pressure p_J satisfies the relation

$$\mu_J^*(p_J, T) = f_J(T) + RT \ln p_J. \quad (2.40)$$

If the mixing of ideal gases does not induce any kind of interaction other than the collision between the molecules, then the Gibbs potential G relative to the mixture will be the sum of the chemical potentials related to the gaseous components weighted by their number of moles. Thus

$$G = \sum_J N_J \mu_J^*(p_J, T). \quad (2.41)$$

Using Equations (2.38) and (2.40) allows us to rewrite Equation (2.41) in the form

$$G = \sum_J N_J \left[\mu_J^*(p, T) + RT \ln \overline{N}_J\right]. \quad (2.42)$$

A comparison between Equations (2.13) and (2.42) allows us to express the partial molar Gibbs potential μ_J in the form

$$\mu_J = \mu_J^*(p, T) + RT \ln \overline{N}_J. \quad (2.43)$$

Interestingly, with the help of the relation

$$\sum_K N_K \frac{\partial \ln \overline{N}_K}{\partial N_J} = 0, \quad (2.44)$$

we can check that Equation (2.43) for μ_J can also be derived by substituting Equation (2.42) for G in the last of the state Equations (2.8).

2.3.2 Ideal mixture definition

Regardless of its gaseous or liquid nature, a fluid mixture is ideal if the chemical potential μ_J of any component J is given by Equation (2.43). In the limit $\overline{N}_J = 1$, Equation (2.43) reduces to $\mu_J = \mu_J^*(p, T)$, so that $\mu_J^*(p, T)$ is identified with the chemical potential of the pure component J at the pressure p of the mixture. Applying the Gibbs–Duhem Equation (2.18) to the evolutions of the pure component J, we obtain

$$d\mu_J^*(p, T) = \overline{V}_J^* dp - \overline{S}_J^* dT, \tag{2.45}$$

where $\overline{V}_J^*$ and $\overline{S}_J^*$ are the molar volume and the molar entropy, respectively, of the pure component at the pressure p of the mixture. Substituting Equation (2.43) in the Gibbs–Duhem Equation (2.17) related to the mixture, and making use of the second part of Equation (2.16) and of Equation (2.45), we derive

$$\overline{V} = \sum_J \overline{N}_J \overline{V}_J^*; \quad \overline{S} = \sum_J \overline{N}_J \overline{S}_J^* - R \sum_J \overline{N}_J \ln \overline{N}_J. \tag{2.46}$$

Owing to the general Equation (2.31) applied to $\overline{\Pi} = \overline{V}$, according to Equation (2.46) the partial molar volume $\overline{V}_J$ is equal to the molar volume $\overline{V}_J^*$ of the pure component J, when the latter is subjected to the mixture pressure p and the temperature T. This means, if each component is at pressure p prior to mixing, and therefore occupies the volume $N_J \overline{V}_J^*$, that no change of the overall volume will be recorded after ideal mixing. In short, a mixture can therefore be considered ideal if the interactions to which the molecules are subjected within a pure component prior to mixing remain unaffected by the mixing. As a result, if a gaseous mixture can generally be considered an ideal mixture, this is often questionable for a liquid mixture with no further assumption.

2.3.3 Entropy of mixing

Before mixing, the Gibbs free energy G^* of a set of pure components J at pressure p and temperature T is given by

$$G^* = \sum_J N_J \mu_J^*(p, T). \tag{2.47}$$

With the help of the second of the state Equations (2.8) applied to G^* and S^*, and the second of the state Equations (2.19) applied to μ_J^* and S_J^*, taking the derivative of Equation (2.47) with respect to T we derive the related molar entropy $\overline{S}^*$ before mixing in the form

$$\overline{S}^* = \sum_J \overline{N}_J \overline{S}_J^*. \tag{2.48}$$

In contrast to what we concluded for the partial molar volume, a comparison between the general Equation (2.31) applied to entropy, that is $\overline{\Pi} = \overline{S}$, and the second part of Equation (2.46) shows that the partial molar entropy $\overline{S}_J$ is not equal to the molar entropy $\overline{S}_J^*$ of the pure component J when the latter is subjected to the mixture pressure p and the temperature T.

Indeed, a comparison of the second part of Equation (2.46) with Equation (2.48) shows that there is an entropy change due to mixing.

The Gibbs free energy G of an ideal mixture is obtained by substituting Equation (2.43) in Equation (2.13), so that we can use Equation (2.42) regardless of the gaseous or liquid nature of the ideal mixture. When the volumes occupied by the pure substances come into contact, the spontaneous mixing is achieved at constant pressure p and temperature T. Using Equations (2.42) and (2.47), we can express the variation $-\Delta G = -(G - G^*)$ of Gibbs free energy due to mixing in the form

$$-\Delta G = -(G - G^*) = -\sum_J RT N_J \ln \overline{N}_J > 0. \tag{2.49}$$

From Equations (2.46), (2.48) and (2.49) we finally obtain the energy balance $-\Delta G = T \Delta S$ where ΔS is the entropy of mixing whose expression is

$$\Delta S = N \left(\overline{S} - \overline{S}^* \right) = -R \sum_J N_J \ln \overline{N}_J > 0. \tag{2.50}$$

Consistent with the second law, the spontaneous production of entropy $\Delta S > 0$ and the related decrease $\Delta G < 0$ in Gibbs free energy available for producing work are associated with the irreversibility of the mixing process – the impossibility of coming back to the initial state where the substances were pure and separated.

The actual origin of the entropy of mixing is statistical and can be broadly outlined as follows. The entropy of an isolated system is given by the Boltzmann entropy equation

$$S = k \ln \mathcal{W}, \tag{2.51}$$

where $\mathcal{W}$ is the number of accessible states and $k = R/\mathcal{N}_A$ is the Boltzmann constant ($k = 1.381 \times 10^{-23}$ J K^{-1}). The number of accessible states related to a single molecule is proportional to the volume V where the fluid is confined. The number of states which are accessible to the $\mathcal{N}$ molecules making up the system is therefore proportional to $V^{\mathcal{N}}$. Since $\mathcal{N} = N \mathcal{N}_A$ we can write

$$S = R \ln V^N, \tag{2.52}$$

where an additive constant depending only on the internal energy of the system was omitted. Substitution of Equation (2.52) in the first part of fluid state Equation (2.2) provides

$$p = RTc, \tag{2.53}$$

where c is the molar concentration defined by

$$c = N/V = 1/\overline{V}. \tag{2.54}$$

State Equation (2.53) eventually duplicates state Equation (2.23) of ideal gases. However, Equation (2.53) has been derived from statistical considerations only. For this reason it applies to both gases and liquids providing the behavior of the liquid is mainly governed by collisions between molecules.

Applying Equation (2.52) to each component, because of the additive character of entropy we have

$$\Delta S = R \sum_J N_J \ln V - R \sum_J N_J \ln V_J > 0, \tag{2.55}$$

where the first term on the right-hand side accounts for the entropy after mixing, while the second term accounts for the entropy before mixing. If each mole occupies the same volume within the pure component J and within the mixture we have $V_J/V = N_J/N = \overline{N}_J$, then we end up with Equation (2.50).

2.3.4 Ideal solution – Raoult's law

An ideal solution is a liquid mixture whose components satisfy Equation (2.43). In a similar way to the equilibrium Equation (2.10), the thermodynamic equilibrium between a component J within the solution and its vapor above the solution is achieved when their respective chemical potentials become equal. When assuming the vapor to be an ideal gas, an expression of its chemical potential $\mu_J^*(p_J, T)$ is provided by Equation (2.40). Then equating expressions (2.40) and (2.43) we obtain

$$\mu_J^*(p, T) + RT \ln \overline{N}_J = f_J(T) + RT \ln p_J, \tag{2.56}$$

where p and p_J are the pressure of the solution and the pressure of component J in vapor form above the solution, respectively. When the solution is formed from the pure component J, the value of $\overline{N}_J$ reduces to 1, and Equation (2.56) provides

$$\mu_J^*(p, T) = f_J(T) + RT \ln p_J^*, \tag{2.57}$$

where p_J^* is the pressure of the vapor in equilibrium with the solution consisting of the pure component J. Subtracting the two previous equations we derive Raoult's law, reading

$$p_J = p_J^* \overline{N}_J. \tag{2.58}$$

The derivation of Raoult's law requires the validity of Equation (2.43) for μ_J on the whole range of $\overline{N}_J$ values. As a result Raoult's law can only be approximate. Indeed, for values of $\overline{N}_J$ departing significantly from 1, the validity of Equation (2.43) becomes questionable: a molecule of the J component within the mixture is mainly surrounded by molecules of the other components and is generally subjected to quite distinct interactions from those prevailing before the mixing. The validity of Raoult's law on the whole range of $\overline{N}_J$ is therefore restricted to mixtures whose components are chemically similar, so that the interactions before and after mixing remain comparable. For instance a mixture of benzene and toluene, which are close hydrocarbons, behaves almost ideally.

2.3.5 Dilute ideal solution – Henry's law

A dilute solution is a solution in which one component, called the solvent, has a molar fraction close to one, whereas the other components, namely, the solutes, therefore have a molar fraction much lower than one. The dilute solutions whose solutes are not subjected to strong electrostatic forces can accurately be considered as ideal solutions.

The solvent molecules are so numerous that the interactions they undergo within the dilute solution remain quite similar to the interactions within the pure solvent. The mixing effect affecting the solvent during mixing reduces to the entropy of mixing. Raoult's law, that is Equation (2.58), applies to the solvent since Equation (2.57) remains accurate for the solvent whose molar fraction is close to 1.

This is no longer true for solutes. They have small molar fractions $\overline{N}_J \ll 1$, and the vapor pressure of the solute can no longer be related to the pressure of the pure solute's vapor in assigning the value 1 to $\overline{N}_J$ in Equation (2.56), so that Equation (2.57), and thus Raoult's law no longer apply. The solute molecules pertaining to a dilute solution are far apart so that they cannot significantly interact. The environment of the solute molecules within the solution is thus quite different from that prevailing for the same molecules in the pure solute. However, each solute molecule can be associated with a composite sphere formed of both the solute molecule and the closest solvent molecules with which the solute molecule interacts. Except for solvent molecules belonging to these composite spheres, the solvent molecules are not affected by the presence of the solute molecules. A dilute solution can therefore be considered as resulting from the mixing of the solute–solvent composite spheres and the other unperturbed solvent molecules, the interactions therefore remaining the same before and after mixing. In the range $\overline{N}_J \ll 1$ Equation (2.43) thus applies to the fictitious component formed from the composite spheres incorporating the solute and having a molar fraction close to $\overline{N}_J$ because of the dilute approximation. Therefore, in the range $\overline{N}_J \ll 1$, Equation (2.56) still holds for such a fictitious component J. In practice, the variations of pressure of the solution p are small enough so that the variations of $\mu_J^*(p, T)$ they induce are negligible. So taking $\mu_J^*(p, T) \simeq f_J^*(T)$ in Equation (2.56), we derive Henry's law:

$$\overline{N}_J \ll 1: \quad p_J = K_H(T) \overline{N}_J, \qquad (2.59)$$

where

$$K_H(T) = \exp\left[f_J^*(T) - f_J(T)\right]/RT \qquad (2.60)$$

is Henry's constant, which is an equilibrium property depending only on temperature and which needs to be determined experimentally. The dilute solutions obeying Henry's law are called ideal dilute solutions. Raoult's and Henry's laws are both illustrated in Figure 2.1.

Henry's law accounts for the various phenomena induced by the sudden variation of gas solubility with the values of the partial pressure prevailing above the solution. In Section 4.2.3 of Chapter 4, which is devoted to saturated poroelasticity, Henry's law will allow us to explain the detrimental effects of successive unloadings that a gassy sediment undergoes in petroleum engineering during its extraction from the deep seabed. It explains, in everyday life, why opening a bottle or a can of soda may cause effervescence. Opening the bottle induces a sudden drop in the partial pressure of carbon dioxide and, therefore, of its molar fraction $\overline{N}_{CO_2}$ in the gaseous drink with respect to its initial value, resulting in the formation of gas bubbles. This will be examined more closely in Sections 8.4 and 8.5.1. Henry's law also explains the embolism resulting from the obstruction of blood vessels by gas bubbles, nitrogen bubbles in particular, accompanying an excessively sudden decompression of the air breathed by divers. At a depth of 40 m, the pressure is five times higher than the atmospheric pressure. Since air is

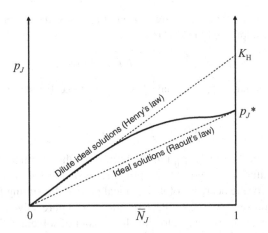

Figure 2.1 Partial pressure p_J of the solute vapor prevailing above the solution plotted against the molar fraction $\overline{N}_J$ of the solute in the solution; a component J playing the role of an almost pure solvent ($\overline{N}_J \simeq 1$) follows Raoult's law, whereas the same component acting as a quite dilute solute ($\overline{N}_J \ll 1$) follows Henry's law

mainly made out of nitrogen, according to Henry's law (Equation (2.59)) the nitrogen molar fraction $\overline{N}_{N_2}$ in the blood is five times higher than its surface value. If the diver comes back to the surface too quickly, his body pressure drops abruptly below the pressure that maintained the nitrogen in solute form in his blood when he was deeper down. As a result, in a similar way to what happens with carbon dioxide when a soda bottle is opened, the nitrogen contained in the blood starts forming bubbles that ultimately produce the embolism. Because of its lower solubility in the blood[2] the use of helium is preferred to that of nitrogen when diluting oxygen in the respiratory tubes of divers.

Conversely, if the partial pressure of a gas above a solution increases, so will its solubility in the solution. The increase of the partial pressure of the carbon dioxide in the atmosphere and, therefore, of its solubility in the oceans has another consequence greater than the much-mediatized global warming effect: the hydrolysis of the extra carbon dioxide absorbed by the oceans causes the waters to acidify progressively. The hydrolysis of carbon dioxide corresponds to the chemical reaction $H_2O + CO_2 \rightarrow H^+ + HCO_3^- \rightarrow 2H^+, CO_3^{2-}$. According to this reaction, calcium carbonate $CaCO_3$ dissolves in seawater whose pH is lower than the present averaged value 8.1 of oceans so that the living organisms partly made of $CaCO_3$ might soon disappear and this would in turn have consequences on the alimentary chain.

2.3.6 Osmotic pressure

Let us now restrict ourselves to the case where a single solute is present, and let $x = \overline{N}_{\text{solute}}$ be its molar fraction so that the molar fraction relative to the solvent is $1 - x$. We assume the

---

[2] At 25 °C the value of constant $K_H(T)$ relative to nitrogen is a little higher than half the value of the constant relative to helium (0.87×10^4 MPa compared with 1.49×10^4 MPa).

solution to be ideal, meaning that according to Equation (2.43) the chemical potential of the solvent noted μ_0 can be expressed in the form

$$\mu_0 = \mu_0^*(p, T) + RT \ln(1 - x). \tag{2.61}$$

If the solution is dilute so that $x \ll 1$, we have $\ln(1 - x) \simeq -x$. From the previous expression we then obtain

$$\mu_0 = \mu_0^*(p, T) - RTx. \tag{2.62}$$

On the right-hand side the first term, $\mu_0^*(p, T)$, represents the chemical potential related to the solvent as if the latter were pure, at the pressure p of the solution, while the second term, $-RTx$, accounts for the decrease of its chemical potential resulting from the entropy of mixing induced by the presence of the solute. Consider now a pure solvent separated from the solution by a semipermeable membrane allowing the transfer of solvent molecules, but not that of the solute molecules. The composition equilibrium Equation (2.10) applies to the solvent in the solution since it is in contact with the pure solvent through the membrane. The chemical potential of the solvent in both compartments must therefore be equal, allowing us to write

$$\mu_0^*(p - \Delta p, T) = \mu_0^*(p, T) - RTx, \tag{2.63}$$

where Δp is the difference in pressure that the membrane has to sustain. The dilute assumption $x \ll 1$ entails $\Delta p / RT \ll 1$ so that Equation (2.63) can be rewritten in the form

$$\frac{\partial \mu_0^*}{\partial p} \Delta p \simeq RTx. \tag{2.64}$$

The first part of the state Equation (2.19) applied to the pure solvent and combined with Equation (2.64) provides

$$\Delta p \simeq RTx / \overline{V}_0. \tag{2.65}$$

The solution being dilute, we have $N_{\text{solute}} \ll N_{\text{solvent}}$ so that $x / \overline{V}_0 \simeq N_{\text{solute}} / V = c$, where c is the solute concentration, and we can rewrite the previous equation in the form of van't Hoff's law

$$\Delta p \simeq RTc. \tag{2.66}$$

The pressure excess Δp undergone by the solution with respect to the pure solvent is termed the osmotic pressure. The osmotic pressure is the extra pressure that is required to establish the thermodynamic equilibrium between the solvent in the solution and the pure solvent. This extra pressure offsets the loss in free energy of the solvent induced by the mixing with the solute. A comparison between Equations (2.53) and (2.66) shows that the dilute solution behaves exactly the same as if the overall pressure p had resulted from the sum of the partial pressures $p - \Delta p$ and $\Delta p = RTc$, associated with the solvent and the solute respectively, the latter behaving as an ideal gas. As illustrated in Figure 2.2, the osmotic pressure can be measured by the difference in heights observed when the thermodynamic equilibrium is achieved between the two compartments separated by the semipermeable membrane.

Osmosis explains the age-old use of salts for the preservation of fish and meat, or that of sugar for the preservation of jams from bacterial proliferation. Bacteria are generally

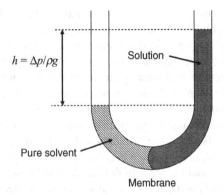

Figure 2.2 Thermodynamic equilibrium between a pure solvent and a solvent in a solution through a semipermeable membrane; the difference Δp between the liquid pressure of the two compartments is the osmotic pressure and is sustained by the membrane

unicellular organisms containing an aqueous solution of ions, aminoacids and proteins. They are protected by a water-permeable membrane allowing nutritive exchanges with the external environment so that they can proliferate in aqueous solutions. In contrast, if the bacterium is immersed in salted solutions or sugar, the intracellular solution is purer and the water leaves the bacterium in order to reestablish the thermodynamic equilibrium between the concentrations of both sides of the membrane. As a result, the bacterium dehydrates and finally dies.

Since van 't Hoff's law (Equation (2.66)) identifies with the law of ideal gases, the osmotic pressure of a mole of solute in a volume of 24.5 l at 25 °C is 1 atm. The salt concentration in our body (blood, tears) is 9 g l^{-1}, or equivalently 0.154 mol l^{-1} of NaCl, and therefore 0.308 mol l^{-1} of Na$^+$ and Cl$^-$ ions. This concentration is equivalent to an osmotic pressure of $24.5 \times 0.308 = 7.35$ atm. This explains why the liquids coming into contact with our internal fluids through tissues much more permeable than our skin, for instance the liquid of lenses in contact with our eyes, must be perfectly balanced with regard to salt concentration.

2.3.7 Electrostatics and excess of osmotic pressure

When a chunk of clay previously saturated with salty water is immersed in a large reservoir of fresh water, it swells. Such a swelling is intriguing because it happens without any change of the external (atmospheric) pressure. In the same vein of puzzling observations, a chunk of dried clay swells less when immersed in salty water than in fresh water. Since the sample is not wrapped in a membrane, in both situations the ions forming the salt or the aqueous solvent can freely escape from the clay or invade it. Therefore the excess of internal pressure originating the swelling cannot be attributed to the osmotic pressure analyzed in the previous section.

In fact, the excess of internal pressure making the sample swell must be associated with the electric charges borne by the platelets forming the solid part of the clay. Because these electric charges have the same sign, the observed swelling might be wrongly attributed to the electrostatic repulsion occurring between the platelets, according to the following scenario. Before the immersion in the reservoir of pure water, this repulsion is partly screened by the elecrolyte. The immersion in the reservoir causes an imbalance in the electrolyte concentration. In order to restore the equilibrium some amount of the electrolyte has to leave the sample, reducing the screening previously provided: swelling ensues. Unfortunately, this interpretation is irrelevant since the electroneutrality has to be ensured on a large scale, so that the required minimum number of counterions will always remain within the sample, in order to offset the electric charges borne by the platelets. In the absence of any thermal agitation, that is, ideally at $T = 0\,\mathrm{K}$, the counterions would even condense on the solid walls carrying the electric charges and delimiting the pore space, causing the electrostatic repulsion of the platelets to vanish completely. This last remark is a first clue to understanding that the excess of osmotic pressure has an entropic origin. Indeed, it arises from the extra thermal agitation opposing the electric potential created by the permanent charges of the platelets and those of the electrolyte.

Basically, since the electric potential has the same sign as the electric charges from which it originates, its contribution to the chemical potential of the counterions is always negative. When the equilibrium is achieved, the chemical potential within the sample has to be equal to the chemical potential of the bulk solution of the large reservoir. A greater positive thermal contribution to the chemical potential then has to offset the negative contribution of the electric potential. This extra contribution is achieved through a greater concentration of counterions of the electrolyte within the sample than the concentration of the electrolyte in the bulk solution of the reservoir. The higher the intensity of the electric potential, the greater the concentration of the counterions, so that the latter form a diffuse layer whose concentration decreases away from the charged platelets where the intensity of the electric potential is the highest. Since the electrolyte in the large reservoir is not subjected to any electrostatic interaction, this results in an overall excess of osmotic pressure for the electrolyte within the sample, which eventually provokes its swelling. Owing to this excess of osmotic pressure, clay platelets remain well separated in fresh river water and, because of their individual lightness, they form a suspension. At the junction of the river with the sea the water becomes salty. As a result, the excess of osmotic pressure vanishes, and the clay platelets flocculate because of the attractive van der Waals forces (see Section 6.1.2) acting between the platelets. This results in darker zones of water forming at the river mouth.

In order to assess the excess of osmotic pressure that can build up in porous solids whose internal walls carry electric charges, let us consider the simplified situation of the slit-pore represented in Figure 2.3, at the origin of the electric double-layer theory associated with this geometry. The solid walls are formed of two infinite parallel planes, which are separated by a distance $2d$ and bear an electric charge of algebraic intensity q per surface unit. If t is the thickness of each solid wall, the separation distance $2d$ and the solid volume fraction ϕ_S of the porous solid are linked by the relation

$$2d = t\frac{1 - \phi_S}{\phi_S}. \tag{2.67}$$

The origin $z = 0$ is taken at the midplane so that the planes are located at $z = +d$ and $z = -d$, respectively. The pore space between the planes is filled by an aqueous electrolytic solution made of water, co-ions and counterions. Overall charge neutrality has to be maintained so that the co-ions and the counterions form diffuse concentration profiles away from each plane. Restricting to a 1 : 1 electrolyte, typically a sodium-chloride salt Na^+Cl^-, due to the problem symmetry the concentrations $c_+(z)$ and $c_-(z)$ related to these profiles must therefore satisfy the relation

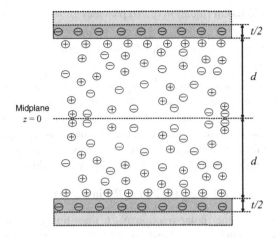

Figure 2.3 The slit-pore and the electric double-layer theory: the pore space between two planes bearing an electric charge density is filled with an aqueous elecrolytic solution whose co-ions and counterions form a diffuse layer away from each plane; this layer ensures the overall charge neutrality

$$q + F \int_0^d [c_+(z) - c_-(z)] \, dz = 0. \tag{2.68}$$

In Equation (2.68) $F \,(= 9.6486 \times 10^4 \,\mathrm{C\,mol^{-1}})$ is the Faraday constant, that is, $F = \mathcal{N}_A e$ where $e = 1.602 \times 10^{-19}$ C is the elementary (protonic) charge. In this simplified standard approach we disregard any molecular interaction between the water molecules forming the solvent and the electrolyte. Accordingly, the mechanical pressure applying on the planar solid walls can be split into the contribution p_W due to water acting as solvent, and the contribution p_{el} due to the presence of the electric charges and of the electrolyte:

$$p = p_W + p_{el}. \tag{2.69}$$

The pressure p_{el} applying to the solid walls, for instance to the one located at $z = +d$, results from three contributions: (i) the electrostatic repulsion $p_{el}^{(1)}$ between the two walls; (ii) the electrostatic interaction $p_{el}^{(2)}$ between the charge q carried by the wall and the co-ions and counterions forming the electrolyte; (iii) the osmotic pressure $p_{el}^{(3)}$ exerted by the electrolyte at $z = d$. Contribution $p_{el}^{(1)}$ is the product of the charge q and the intensity $q/2\varepsilon_r\varepsilon_0$ of the electric field generated by the other plane;[3] contribution $p_{el}^{(2)}$ is the opposite of the electrostatic force exerted by the charged solid wall located at $z = +d$ onto the electrolyte; contribution $p_{el}^{(3)}$ is the sum of the osmotic pressures that can be associated with both the co-ions and counterions with the help of Equation (2.66). This decomposition of p_{el} allows us to write

$$p_{el} = \frac{q^2}{2\varepsilon_r\varepsilon_0} + 2F \int_0^{+d} [c_+(z) - c_-(z)] \frac{q}{2\varepsilon_r\varepsilon_0} dz + RT\,[c_+(d) + c_-(d)]. \tag{2.70}$$

---

[3] ε_0 is the permittivity of free space ($\varepsilon_0 = 8.854 \times 10^{-12}\,\mathrm{C^2\,J\,m^{-1}}$) and ε_r is the relative permittivity of the medium ($\varepsilon_r = 80$, at 293 K for an aqueous solution). The product $\varepsilon = \varepsilon_0\varepsilon_r$ is the dielectric constant of the medium.

Substitution of Equation (2.68) in Equation (2.70) provides

$$p_{el} = -\frac{q^2}{2\varepsilon_r \varepsilon_0} + RT\left[c_+(d) + c_-(d)\right]. \tag{2.71}$$

Interestingly, the electrostatic interaction $p_{el}^{(1)} + p_{el}^{(2)} = -q^2/2\varepsilon_r\varepsilon_0$ is always negative, that is, attractive, whatever the sign of the same electrostatic charge q borne by the solid walls. Their repulsion can therefore only result from the osmotic pressure $RT\left[c_+(d) + c_-(d)\right]$.

In order to determine the latter, let us introduce the chemical potential $\mu_\pm(z)$ of the co-ions and counterions forming the electrolyte. In the absence of any electrostatic effect, the aqueous solution would be ideal and, according to Equation (2.43), we would have

$$\mu_\pm(z) = \mu_\pm^*(p_L, T) + RT \ln \overline{N}_\pm(z). \tag{2.72}$$

The concentration $c_\pm(z)$ and the molar fraction $\overline{N}_\pm(z)$ are linked by the relation $c_\pm(z) = \overline{N}_\pm(z)/\overline{V}(z)$, where $\overline{V}(z)$ stands for the local overall molar volume of the aqueous solution. If the molar volumes of the various components forming the solution do not differ significantly between them, $\overline{V}(z)$ can be assumed to be independent of z. Accounting then for the extra contribution of the electric potential $\psi(z)$ to $\mu_\pm(z)$ and omitting a constant depending only on temperature, Equation (2.72) extends in the form

$$\mu_+(z) = \mu_+^*(p_L, T) + RT \ln c_+(z) + F\psi(z); \tag{2.73}$$
$$\mu_-(z) = \mu_-^*(p_L, T) + RT \ln c_-(z) - F\psi(z).$$

When the equilibrium is established, $\mu_\pm(z)$ must be uniform. Indeed, according to Equation (2.10) the chemical potential $\mu_\pm(z)$ must be equal, whatever the location z, to the uniform chemical potential of the corresponding ion of the electrolyte in a bulk solution of a large reservoir. In this bulk solution no electrostatic effects occur since in such a solution an ion and a counterion are always matched up. Noting c_∞ the concentration of the bulk solution of ion–counterion pairs, we therefore write

$$\mu_\pm(z) = \mu_\pm^*(p_L, T) + RT \ln c_\infty. \tag{2.74}$$

Since no electrostatic effects occur in the bulk solution, the pressure p_L is both the mechanical and the thermodynamic pressure of the bulk solution. When the equilibrium is achieved, because of Equation (2.74) the chemical potential of the electrolyte, both in the bulk solution and between the charged planes, is the same. Since this holds for the solvent too, the pressure p_L can be termed the 'equivalent' thermodynamic pressure of the solution between the charged planes. The term 'equivalent' recalls that the large reservoir acting as a thermodynamic 'pressuremeter' may be fictitious. Yet, because of the electrostatic effects, the equivalent thermodynamic pressure of the solution, p_L, does not merge with the mechanical pressure p defined by Equation (2.69). The purpose of the remaining part of this section is to determine the pressure difference $p - p_L$.

In order to determine the concentration profiles $c_\pm(z)$, from Equation (2.73) and the equilibrium condition (2.74) we first derive the following useful relations:

$$c_\pm(z) = c_\infty \exp\left(\mp\frac{F\psi(z)}{RT}\right); \quad \frac{dc_\pm}{dz} = \mp c_\pm \frac{d}{dz}\left(\frac{F\psi(z)}{RT}\right); \quad c_+(z)c_-(z) = c_\infty^2. \tag{2.75}$$

The factor $a_\pm(z) = \exp(\pm F\psi(z)/RT)$ appears as a local chemical activity coefficient arising from electrostatic effects. Indeed, the two previous equations show that the chemical potential $\mu_\pm(z)$ can be expressed as $\mu_\pm(z) = \mu_\pm^*(p_L, T) + RT \ln[a_\pm(z) c_\pm(z)]$. Now applying the Gauss theorem to this one-dimensional problem, the intensity $E_{el}(z)$ of the electrostatic field satisfies the differential relation

$$\varepsilon_r \varepsilon_0 \frac{dE_{el}}{dz} = F[c_+(z) - c_-(z)]. \tag{2.76}$$

Substitution of the relation linking the electric field E_{el} to the electric potential ψ, namely,

$$E_{el} = -\frac{d\psi}{dz}, \tag{2.77}$$

provides us with the one-dimensional Poisson equation

$$\frac{d^2\psi}{dz^2} = -\frac{F}{\varepsilon_r \varepsilon_0}[c_+(z) - c_-(z)]. \tag{2.78}$$

Integrating Equation (2.78) between $z = 0$ and $z = d$, and using both the electroneutrality Equation (2.68) and the symmetry condition,

$$\frac{d\psi}{dz}\Big|_{z=0} = 0, \tag{2.79}$$

we obtain the boundary condition

$$\frac{d\psi}{dz}\Big|_{z=d} = \frac{q}{\varepsilon_r \varepsilon_0}. \tag{2.80}$$

We now multiply Equation (2.78) by $d\psi/dz$. Then using the second relation of Equation (2.75) and integrating between $z = 0$ and z, we derive

$$\frac{1}{2}\varepsilon_r \varepsilon_0 \left(\frac{d\psi}{dz}\right)^2 = RT[c_+(z) + c_-(z) - c_+(0) - c_-(0)]. \tag{2.81}$$

Applying Equation (2.81) to $z = +d$ and using the boundary condition stated in Equation (2.80), from Equation (2.71) we derive the Donnan equation, namely,

$$p_{el} = RT[c_+(0) + c_-(0)]. \tag{2.82}$$

According to Equation (2.82) p_{el} reduces to the osmotic pressure prevailing at the midplane. This result could have been anticipated: owing to the symmetry Equation (2.79), the intensity of the electric field is zero at the midplane, resulting in the absence of electrostatic interaction there.

According to Equation (2.66), $2RTc_\infty$ is the osmotic pressure that would prevail in the bulk solution if the latter were insulated with a membrane impervious to the electrolyte. The excess of osmotic pressure, ϖ_{EL}, can then be defined as

$$\varpi_{EL} := RT[c_+(0) + c_-(0) - 2c_\infty]. \tag{2.83}$$

Equations (2.82) and (2.83) show that the excess of osmotic pressure, ϖ_{EL}, is the extra mechanical pressure applying to the walls, caused by the presence of electric charges, in addition to the osmotic pressure, $2RTc_\infty$. Because of the last relation of Equation (2.75), the excess of osmotic pressure is always positive, resulting in a repulsion between the solid walls. The excess pressure is induced by the extra thermal agitation offsetting the chemical potential decrease due to electrostatic effects.

As previously pointed out, in the bulk solution there is no electrostatic interaction. According to Equation (2.53) the partial pressure of the electrolyte is then equal to the osmotic pressure, $2RTc_\infty$. In addition, since the solvent is assumed not to interact with the solute, its related partial pressure, p_W, is the same in the bulk solution and in the solution between the charged planes, so that we can write

$$p_L = p_W + 2RTc_\infty. \tag{2.84}$$

Equations (2.69), (2.82)–(2.84) combine into

$$p = p_L + \varpi_{EL}. \tag{2.85}$$

According to Equation (2.85) the excess of osmotic pressure, ϖ_{EL}, represents the difference between the mechanical pressure, p, and the equivalent thermodynamic pressure, p_L, of the solution. In the experiment described at the beginning of Section 2.3.7, the mechanical and thermodynamic pressure of the bulk solution in the reservoir identifies with p_L and does not change during the immersion of the chunk of clay. Differentiating Equation (2.85) and taking $dp_L = 0$, we obtain

$$dp = d\varpi_{EL}, \tag{2.86}$$

showing that the excess of osmotic pressure ϖ_{EL} is eventually identified with the excess of internal pressure causing the swelling.

Substituting the expressions provided by Equation (2.75) for $c_+(0)$ and $c_-(0)$ in Equation (2.83), the excess of osmotic pressure ϖ_{EL} can be expressed as a function of the electric potential $\psi(0)$ at the midplane in the form

$$\varpi_{EL} = 2RTc_\infty \left[\cosh \overline{\psi}(0) - 1\right], \tag{2.87}$$

where $\overline{\psi}$ stands for the dimensionless electric potential $\overline{\psi} = F\psi/RT$. The actual determination of ϖ_{EL} thus requires the determination of the intensity $\overline{\psi}(0)$ which $\overline{\psi}$ takes at the midplane. If we substitute the first part of Equation (2.75) in Equation (2.81), we first have

$$\frac{1}{2}\left(\frac{d\overline{\psi}}{dz}\right)^2 = \frac{1}{\ell^2}\left[\cosh \overline{\psi}(z) - \cosh \overline{\psi}(0)\right], \tag{2.88}$$

where ℓ is the Debye length defined by

$$\ell = \sqrt{\frac{\varepsilon_r \varepsilon_0 RT}{2c_\infty F^2}}. \tag{2.89}$$

The Debye length ℓ is the screening length that captures the order of magnitude of the diffuse layer's thickness. This layer forms away from each plane and is made of co-ions and counterions. At distances much larger than ℓ the intensity of the direct electrostatic interaction between charged objects rapidly falls to zero. For the problem at hand, the electrostatic interactions are scaled by the dimensionless quantity d/ℓ. The smaller the bulk concentration $2c_\infty$, the smaller d/ℓ, and the more the system is affected by electrostatics. For small values

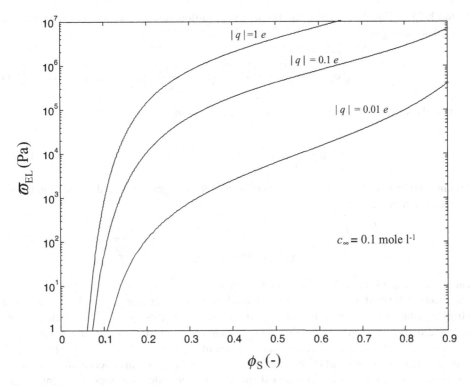

Figure 2.4 Excess of osmotic pressure ϖ_{EL} plotted against the solid volume fraction ϕ_S for three values, namely, $0.01e$, $0.1e$ and $1e$, of the intensity $|q|$ of the electric charge density borne by the solid walls; e is the protonic charge and the value of the thickness t of the solid wall is 1 nm; for any other value t^* of the wall thickness t the same value of ϖ_{EL} will be achieved for a solid volume fraction ϕ_S^* given by $t^*\left(1-\phi_S^*\right)/\phi_S^* = (t = 1\,\text{nm})(1-\phi_S)/\phi_S$

of d/ℓ the overall concentration $c_+(0) + c_-(0)$ at the midplane is then expected to depart significantly from the bulk concentration $2c_\infty$, resulting in a large excess of osmotic pressure ϖ_{EL} for large values of the solid volume fraction ϕ_S. This is actually noticeable in Figure 2.4, where ϖ_{EL} is plotted against the solid volume fraction ϕ_S for various values of the intensity $|q|$ of the electric charge borne by the solid walls.

The value of $\overline{\psi}(0)$ supporting the results reported in Figure 2.4 for the excess of osmotic pressure ϖ_{EL} can be determined as follows: taking into account the symmetry Equation (2.79), together with the boundary condition stated in Equation (2.80) for determining the sign of $d\overline{\psi}/dz$, an integration of Equation (2.88) from $z = 0$ to $z = +d$ provides the relation

$$\frac{d}{\ell} = \text{sgn}(q) \int_{\overline{\psi}(0)}^{\overline{\psi}(d)} \frac{d\overline{\psi}}{\sqrt{2\left(\cosh\overline{\psi} - \cosh\overline{\psi}(0)\right)}}. \tag{2.90}$$

With the help of Equation (2.88), the boundary condition stated in Equation (2.80) can be expressed according to

$$\sqrt{\cosh^2 \frac{\overline{\psi}(d)}{2} - \cosh^2 \frac{\overline{\psi}(0)}{2}} = \frac{|q|}{4c_\infty F \ell}. \tag{2.91}$$

Switching the variable of integration from $\overline{\psi}$ to u defined by

$$u = \sqrt{\cosh^2 \frac{\overline{\psi}}{2} - \cosh^2 \frac{\overline{\psi}(0)}{2}}, \tag{2.92}$$

and taking into account Equation (2.91) to reexpress the integration bounds, Equation (2.90) can be rewritten in the form

$$\frac{d}{\ell} = \int_0^{|q|/4c_\infty F \ell} \frac{du}{\sqrt{u^2 + \cosh^2 \frac{\overline{\psi}(0)}{2}} \sqrt{u^2 + \sinh^2 \frac{\overline{\psi}(0)}{2}}}. \tag{2.93}$$

Equation (2.93) gives us the link that was missing between $\overline{\psi}(0)$, d/ℓ, $|q|$ and c_∞.

The theory described above is instructive but it has severe limitations. Since it addresses the electrolytic solution as a continuum, the co-ions and the counterions have no volumetric extension. As a direct consequence of this the theory does not account for steric effects. Moreover, in Equation (2.73) which gives the electrochemical potential, $\mu_\pm(z)$, the term $\pm c_\pm(z)\psi(z)$ will only be an accurate assessment of the energy cost required for the introduction of the co-ions or counterions at point z of the solution if, in turn, this introduction does not significantly affect the electric potential field $\psi(z)$. In short, this is a mean field theory because none of the correlation effects are accounted for. This mean field theory obviously fails for small values of the ratio d/ℓ as well as large values of the intensity $|q|$ of the electric charge borne by the solid walls which delimit the pore space.

2.3.8 Reactive ideal mixture

Consider now a mixture whose components C_J are subjected to the chemical reaction

$$\sum_J \nu_J C_J \equiv 0, \tag{2.94}$$

where ν_J are the stoichiometric coefficients. These coefficients have a sign, + for the reactants and − for the products. However, since the reaction can occur in both directions, only the sign difference actually matters. Because of the reaction, the change dN_J of the number of moles of component J during the infinitesimal time dt is due both to the external supply and to the internal chemical reaction. We write

$$\frac{dN_J}{dt} = \overset{\circ}{N}_J + \overset{\circ}{N}_{\to J}, \tag{2.95}$$

where $\overset{\circ}{N}_J$ is the external supply rate, and $\overset{\circ}{N}_{\to J} dt$ is the number of moles of component J created by the reaction during the infinitesimal duration dt. Because of the stoichiometry reported in Equation (2.94), the rate $\overset{o}{N}_{\to J}$ relative to component J is linked to the reaction rate $\overset{o}{N}_{\to}$ through the relation

$$\frac{\overset{o}{N}_{\to J}}{\nu_J} = -\overset{o}{N}_{\to}. \tag{2.96}$$

The Clausius–Duhem Equation (2.4) now has to be modified in order to account for the internal reaction. The free energy externally supplied to the mixture related to component J is no longer $\mu_J dN_J$. It is now $\mu_J \overset{o}{N}_J dt$. As a result, if we assume isothermal evolutions, the Clausius–Duhem Equation (2.4) extends to the form

$$-p dV + \sum_J \mu_J \overset{o}{N}_J dt - dF = 0. \tag{2.97}$$

Substitution of Equation (2.95) in Equation (2.97) gives

$$-p\frac{dV}{dt} + \sum_J \mu_J \frac{dN_J}{dt} - \frac{dF}{dt} + \sum_J \nu_J \mu_J \overset{o}{N}_{\to} = 0. \tag{2.98}$$

For infinitely slow reactions, we have $\overset{o}{N}_{\to} \to 0$, so that Equation (2.98) reduces again to the Clausius–Duhem Equation (2.4). The postulate of local equilibrium state extrapolates the Clausius–Duhem Equation (2.4) to any reactive mixture. Substituting Equation (2.4) in Equation (2.98), for isothermal evolutions we then derive

$$\sum_J \nu_J \mu_J = 0. \tag{2.99}$$

The equilibrium Equation (2.99) only applies to nondissipative reactions. For dissipative reactions, the equilibrium relation of Equation (2.99) must be replaced by an inequality and we write

$$A \overset{o}{N}_{\to} \geq 0, \tag{2.100}$$

where A is the affinity of the reaction and is defined by

$$A = \sum_J \nu_J \mu_J. \tag{2.101}$$

A is the driving force of the reaction that measures the difference of potential between the reactants and the products. The sign of the affinity determines in which direction the reaction actually occurs. For instance, if $A > 0$, Equation (2.100) imposes the inequality $\overset{o}{N}_{\to} \geq 0$ so that, according to Equation (2.96), the chemical reaction can then only take place from the reactants ($\nu_J > 0$) towards the products ($\nu_J < 0$). When A becomes zero, the equilibrium Equation (2.99) is satisfied and the reaction stops. Equation (2.99) thereby accounts for the equilibrium towards which the chemical reaction spontaneously tends. 'Elastic' nondissipative reactions ruled by the equilibrium Equation (2.99) are reactions where the achieving equilibrium can be considered as instantaneous compared with the other phenomena involved – molecular diffusion for instance.

If we now assume that the reactive mixture is ideal, substitution of Equation (2.43) in the equilibrium Equation (2.99) provides us with the mass action law, namely,

$$\prod_J \overline{N}_J^{\nu_J} = K(p, T), \qquad (2.102)$$

where the equilibrium constant $K(p, T)$ is given by

$$\Delta G^* = \nu_J \mu_J^*(p, T) = -RT \ln K(p, T). \qquad (2.103)$$

In Equation (2.103), ΔG^* is the Gibbs free energy of formation of the products ($\nu_J < 0$) from the reactants ($\nu_J > 0$). For a reactive gaseous mixture function $K(p, T)$ is explicitly known from Equation (2.28). Besides, in the current applications concerning reactive solutions, the variations of the solution pressure p are small enough for the variations of $\mu_J^*(p, T)$ they induce to be neglected. The mass action law (Equation (2.102)) can finally be rewritten in the form

$$\prod_J \overline{N}_J^{\nu_J} = K(T), \qquad (2.104)$$

where, like $K_H(T)$ in Equation (2.59), $K(T)$ is an equilibrium property depending on temperature only.

2.3.9 Unconditional stability of ideal solutions

Consider a solution consisting of only two components A and B. Let $\overline{N}_A = x$ and $\overline{N}_B = 1 - x$ be the molar fractions of component A, say the solute, and component B, say the solvent, respectively. The solution is stable if the unavoidable local fluctuations of the solute molar fraction x do not spontaneously amplify and ultimately result in the dissociation of the original solution into a composite solution formed of two subsolutions with two different solute molar fractions x_1 and x_2. In other words, the solution is stable if it remains homogeneous with regard to the concentrations of components A and B in the absence of any change of the action of the surroundings upon the solution.

Since there is no change in the action of the surroundings upon the solution, there is no extra external supply of components A or B, nor is there a change in pressure and temperature. The energy balance of Equation (2.6) therefore reduces to $-dG = 0$. However, this equality applies to reversible evolutions only. In order to include the spontaneous irreversible evolutions of a solution as well, instead of the previous equality we must now consider the inequality

$$-dG \geq 0. \qquad (2.105)$$

According to Equation (2.105), the spontaneous dissociation of the original solution in two subsolutions is made possible when this dissociation results in states of lower Gibbs free energy G. Since the Gibbs free energy of mixing (Equation (2.49)) accounts for the free energy change due to mixing, this amounts to determining whether the Gibbs free energy of mixing related to a composite solution made of two subsolutions can be less than that related to the original homogeneous solution.

In the case of homogeneous ideal solutions, the molar variation $\Delta \overline{G} = \Delta G / N$ of free energy while the solute and the solvent are mixed together can be expressed with the help of Equation (2.49) in the form

$$\Delta \overline{G}(x) = -T\Delta \overline{S}(x) = RT\left[x \ln x + (1-x)\ln(1-x)\right]. \tag{2.106}$$

In order to compare $\Delta \overline{G}$ with the free energy of mixing related to the possible composite solution formed from two subsolutions, let the number of moles of the subsolution i, that has x_i for solute molar fraction, be N_i. Since the total number of molecules remains the same, the original solute molar fraction x – and x_1 and x_2 of the two possible subsolutions – must satisfy the inequalities

$$x_1 < x = \alpha_1 x_1 + \alpha_2 x_2 < x_2, \tag{2.107}$$

where α_1 and α_2 are defined by

$$0 < \alpha_i = \frac{N_i}{N} < 1; \quad \alpha_1 + \alpha_2 = 1. \tag{2.108}$$

Owing to the additive character of energy the normalized Gibbs free energy of mixing, $\Delta \overline{\mathcal{G}}$, associated with the composite solution made of the two distinct subsolutions 1 and 2 is given by

$$\Delta \overline{\mathcal{G}}(x_1, x_2) = \alpha_1 \Delta \overline{G}(x_1) + \alpha_2 \Delta \overline{G}(x_2). \tag{2.109}$$

As defined by Equation (2.109), $\Delta \overline{\mathcal{G}}(x_1, x_2)$ is a function of only x_1 and x_2. Indeed, once the molar fraction x of the initial homogeneous solution is given, use of Equations (2.107) and (2.108) provides $\alpha_2 = 1 - \alpha_1$ as a function of x, x_1 and x_2.

In preparation for the stability analysis, let us briefly recall that the function $\Delta \overline{G}(x)$ is convex – concave – for x lying between x_1 and x_2, if in this range the curvature of the representative curve is turned towards the positive – negative – values of $\Delta \overline{G}(x)$. Under Equation (2.109) these definitions imply

$$\Delta \overline{G}(x) \text{ convex:} \quad \Delta \overline{G}(x) \leq \Delta \overline{\mathcal{G}}(x_1, x_2); \tag{2.110a}$$

$$\Delta \overline{G}(x) \text{ concave:} \quad \Delta \overline{G}(x) \geq \Delta \overline{\mathcal{G}}(x_1, x_2). \tag{2.110b}$$

As represented in Figure 2.5 the function $\Delta \overline{G} = -T\Delta \overline{S}$, whose expression is given by Equation (2.106), is convex whatever the value of x so that Equation (2.110a) applies. As a result, since $\Delta \overline{G}$ and $\Delta \overline{\mathcal{G}}$ are both negative, the Gibbs free energy of homogenous solutions is always lower than that of all the possible composite solutions so that an ideal mixture is unconditionally stable. This result could have been anticipated by noting that, whatever the values of α_i, we have

$$\Delta \overline{S}(x) \geq \alpha_1 \Delta \overline{S}(x_1) + \alpha_2 \Delta \overline{S}(x_2), \tag{2.111}$$

stating that the entropy of the homogeneous solution is greater than that of the more ordered composite solution!

In conclusion, a solution can be unstable only if the entropy of mixing of the components A and B and the stabilizing effect it induces are counterbalanced by a destabilizing effect. This destabilizing effect is for instance the interaction energy ΔE existing between components A and B compared with the interaction energy related to the molecules forming the components before mixing. The interaction's destabilizing effect for regular solutions is analyzed in detail in the next section.

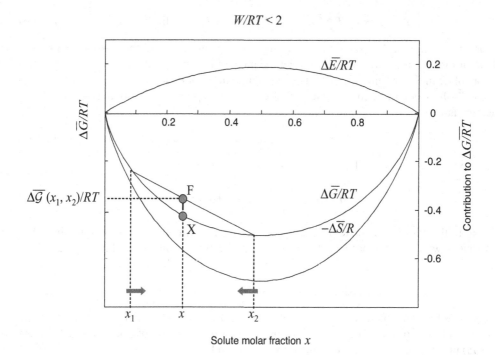

Figure 2.5 Unconditional stability of ideal solutions ($\Delta \overline{G} = -T\Delta \overline{S}$) and of regular solutions ($\Delta \overline{G} = \Delta \overline{E} - T\Delta \overline{S}$) for weak interactions $W/RT < 2$ ($W/RT = 1$ in the case shown); in both cases, the loss in Gibbs free energy due to mixing is always greater for homogeneous solutions than for composite solutions made from two different subsolutions 1 and 2 ($-\Delta \overline{G} > -\Delta \overline{\mathcal{G}}$)

2.4 Regular Solutions

2.4.1 Regular solutions and the interaction between molecules

A solution formed from the mixing of various components is said to be regular if the effects due to mixing are only the increase in entropy, that is, the entropy of mixing ΔS, and the change in the interaction energy between the molecules. In particular, the sum of the volumes occupied by the components before mixing and the volume of the solution resulting from their mixing are the same – there is no volume change. This amounts to assuming that mixing operates through only the substitution of solvent molecules with solute molecules, and conversely. If ΔE denotes the energy variations of molecular interactions due to mixing, instead of Equation (2.106) we now express the Gibbs free energy of mixing ΔG in the form

$$\Delta G = \Delta E - T\Delta S. \tag{2.112}$$

Since there is neither pressure variation nor volume change, ΔG is also the change in the Helmholtz free energy. Under the assumption of a random mixing, the Bragg–Williams model consists of expressing the energy of mixing ΔE in the form

$$\Delta E = NWx(1-x), \tag{2.113}$$

where $2W$ is proportional to the energy associated with the substitution of an AA pair of molecules and a BB pair of molecules by two AB pairs of molecules. This energy of substitution, and consequently the energy of mixing ΔE, is generally positive. More precisely, the energy of substitution ΔE is equal to $\Delta E = N_{AB} w$, where N_{AB} is the number of AB pairs, and $2w$ is the energy required to replace an AA pair and a BB pair by two pairs AB. We therefore have $2w = 2w_{AB} - w_{AA} - w_{BB}$, where w_{XY} is the interaction energy of an XY pair (XY = AA, BB or AB). The interaction energy w_{XY} is negative when it results from attractive intermolecular forces as, for instance, the van der Waals intermolecular forces, that will be addressed more thoroughly in Section 6.1.2 of Chapter 6 which is devoted to capillarity. The interaction energy $w_{XY} = -a_X a_Y$ is then the product, with the opposite sign, of a positive property a_X of an X molecule and a positive property a_Y of a Y molecule. In the end, we obtain: $2w = (a_A - a_B)^2 > 0$. Only N_{AB} remains to be determined. Considering the solution as a network, each site is surrounded by z closest neighbors. If the mixture is assumed to be random, the presence of an A molecule on a given site does not modify the probabilities of having A or B on the neighboring sites. The number N_{AB} of AB pairs is then equal to the number of A molecules, that is, $N \mathcal{N}_A x$, multiplied by the average number of neighboring B molecules around each A molecule, that is, $z(1-x)$. Noting $W = z\mathcal{N}_A w$ we finally obtain Equation (2.113).

Combining Equation (2.106) for $-T\Delta\overline{S}(x)$ with Equations (2.112) and (2.113), we derive the expression $\Delta\overline{G}$ of the molar Gibbs free energy of mixing related to regular solutions:

$$\Delta\overline{G}(x) = x(1-x)W + RT[x \ln x + (1-x)\ln(1-x)]. \tag{2.114}$$

2.4.2 Conditional stability of regular solutions

In Section 2.3.9 ideal solutions were shown to be unconditionally stable since a composite solution formed from two subsolutions would be more ordered, implying that its entropy of mixing is lower. For a regular solution a new element, the molar interaction energy of mixing, $\Delta\overline{E} = x(1-x)W$, comes into play. As illustrated in Figure 2.5, this interaction energy is a concave function of its argument x. According to Equation (2.110b) applied to the interaction energy $\Delta\overline{E}(x)$, whatever the values of α_i defining the number of moles of the two possible subsolutions, the interaction energy $\alpha_1\Delta\overline{E}(x_1) + \alpha_2\Delta\overline{E}(x_2)$ of the composite solution is always lower than the interaction energy $\Delta\overline{E}(x)$ of the homogeneous solution. Since the most stable state corresponds to the lowest Gibbs free energy state, the stability of a regular homogeneous solution now depends on the value of W/RT.

Regarding the homogeneous mixing, the energy W/RT quantifies the unfavorable effect of the interaction energy, $\alpha_1\Delta\overline{E}(x_1) + \alpha_2\Delta\overline{E}(x_2)$, against the favorable effect of the mixing entropy, $\alpha_1\Delta\overline{S}(x_1) + \alpha_2\Delta\overline{S}(x_2)$. For a weak interaction energy corresponding to values of W smaller than $2RT$, the favorable effect of the entropy always wins out over the unfavorable effect of the interaction energy. Despite the contribution of the interaction energy, the mixing Gibbs free energy $\Delta\overline{G}(x)$ now defined by Equation (2.114) remains convex, still satisfying Equation (2.110a). This results in the solution's stability for any proportions of the A and B components. In other words, for a weak interaction energy, the separation of the initial solution into two subsolutions 1 and 2 close to the initial solution is not favorable, since this separation still corresponds to a mixing energy $\Delta\overline{\mathcal{G}}(x_1, x_2) = \alpha_1\Delta\overline{G}(x_1) + \alpha_2\Delta\overline{G}(x_2)$ greater than

$\Delta \overline{G}(x)$ and, thereby, to a Gibbs free energy which is greater for the composite solution than for the homogeneous solution. If, for instance, an unavoidable energy fluctuation results in the separation of the homogeneous solution (point X in Figure 2.5) into two subsolutions 1 and 2 in an energy state close to the initial mixture's (point F in Figure 2.5), the extra Gibbs free energy dissipates, following the second law. This happens because the A and B components of one submixture are not in equilibrium with the A and B components of the other submixture, respectively, and a diffusion process tends to reestablish the initial homogeneous state. In the end, the initial equilibrium corresponding to a lower energy state is recovered (path FX in Figure 2.5).

For larger values of the interaction energy corresponding to values of W greater than $2RT$, the molar Gibbs free energy of mixing $\Delta \overline{G}(x)$ has the aspect shown in Figure 2.6: $\Delta \overline{G}(x)$ is symmetric with regard to the value $x = 0.5$ for which it reaches its maximum. It reaches its minimum $\Delta \overline{G}_{cr}$ for two critical values of the solute molar fraction, x_{cr} and $x = 1 - x_{cr}$, which are the other solutions of $\partial \Delta \overline{G}(x)/\partial x = 0$ and therefore satisfy

$$\frac{x}{1-x} = \exp\left[-\frac{W}{RT}(1-2x)\right]. \qquad (2.115)$$

Even though the Gibbs free energy of mixing can become positive when its maximum which is obtained for $x = 0.5$ is positive, it is not its sign that eventually determines the stability of the mixture. Taking into account Equation (2.107), the same stability analysis as the one

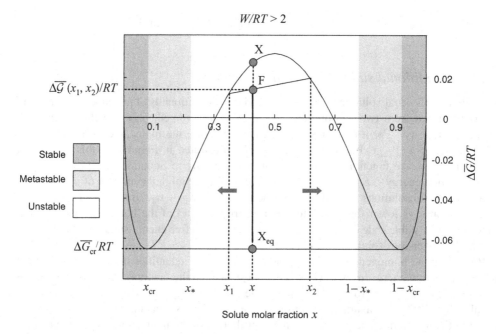

Figure 2.6 Conditional stability of a regular solution for strong energy interactions $W/RT > 2$ ($W/RT = 2.9$ in the case shown). A spontaneous fluctuation, in the form of the separation of an initial homogeneous solution (point X) into two subsolutions (point F) is favorable, since the related free energy of mixing, $\Delta \overline{\mathcal{G}}(x_1, x_2)$, is lower than the initial energy, $\Delta \overline{G}(x)$. This fluctuation amplifies until a stable equilibrium is achieved (point X_{eq})

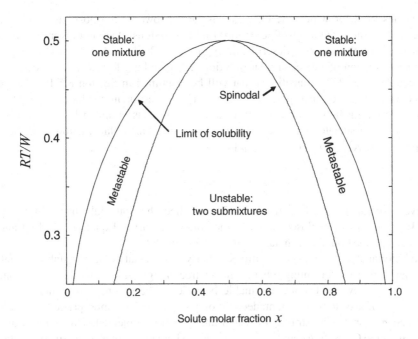

Figure 2.7 Domains of stability, metastability and instability for a regular solution in the plane $(x, RT/W)$

illustrated in Figure 2.5 shows first that the homogeneous mixture remains stable along the convex parts of function $\Delta \overline{G}(x)$, that is, for $0 < x < x_{cr}$ and $1 - x_{cr} < 1$.

In contrast, for $x_{cr} < x < 1 - x_{cr}$, the separation of the homogeneous solution into two subsolutions, of molar fractions $x_1 = x_{cr}$ and $x_2 = 1 - x_{cr}$, respectively, is favorable. Basically, this separation leads to a value of the composite solution's energy of mixing $\Delta \overline{\mathcal{G}}(x_{cr}, 1 - x_{cr})$ consistently lower than the homogeneous state's energy of mixing $\Delta \overline{G}(x)$, that is, defined by the initial molar fraction x:

$$x_{cr} < x < 1 - x_{cr}: \quad \Delta \overline{G}(x) > \Delta \overline{\mathcal{G}}(x_{cr}, 1 - x_{cr}) = \Delta \overline{G}_{cr}. \tag{2.116}$$

For $x_{cr} < x < 1 - x_{cr}$, the stability of the homogeneous solution is thereby no longer ensured. In the plane $(x, RT/W)$ the curve representing the function defined by Equation (2.115) shows the limit of solubility (see Figure 2.7). However, a solution is unstable if the loss of stability is spontaneous. This is the case with the values of x for which the function $\Delta \overline{G}(x)$ has a curvature turned towards the negative values, and is therefore concave. Due to Equation (2.110b), a spontaneous fluctuation is then favorable, since the separation corresponds to an energy of mixing $\Delta \overline{\mathcal{G}}(x_1, x_2)$ lower than the energy $\Delta \overline{G}(x)$ of the initial homogenous solution. This fluctuation will materialize in the separation of the homogeneous solution (point X in Figure 2.6) into two subsolutions 1 and 2, in an energy state close to the initial mixture's (point F in Figure 2.6). This fluctuation increases until a stable equilibrium is reached (path FX$_{eq}$ in Figure 2.6). The final equilibrium corresponds to the lowest possible energy state regarding the mole conservation. It is achieved for a composite solution formed of two subsolutions with molar fractions x_{cr} and $1 - x_{cr}$ (point X$_{eq}$ in Figure 2.6).

With values of x for which the curvature is turned towards the positive values, and is therefore convex, the spontaneous fluctuations (unlike previously) are not favorable, since we find ourselves with the situation described in Figure 2.5. The loss of stability can only occur through the formation of nuclei of a subsolution rich in solute, following a process similar to that occurring for phase transitions, that will be examined in Section 8.5.1 of Chapter 8 which is devoted to unconfined phase transitions. The corresponding solutions are said to be metastable. The limit between metastability and instability is obtained by determining the molar fraction x_* associated with the change of curvature. The molar fraction x_* is therefore the solution of $\partial^2 \left[\Delta \overline{G}(x)\right]/\partial x^2 = 0$, that is,

$$x(1-x) = \frac{2RT}{W}. \qquad (2.117)$$

The curve associated with Equation (2.117) is called the spinodal curve. In the plane $(x, RT/W)$ the domain enclosed between the solubility curve (Equation (2.115)) and the spinodal curve is the domain of metastability (see Figure 2.7).

The stability analysis of regular solutions is fairly representative of any stability analysis. It mainly amounts to determining whether lower free energy states than the current one are available, and if they can become attainable because of the second law, which states that the free energy of a system can spontaneously only decrease. The appropriate free energy to consider depends on the constraints to which the system is subjected. This appropriate free energy is either the Gibbs free energy G, for systems where pressure is maintained constant, or the Helmholtz free energy F, for systems where volume is maintained constant. These states of lower energy are generally related to a possible lower molecular attraction negative energy than the current one, while being in competition with the stabilizing effect of the entropic disorder due to thermal agitation. Metastable states delimited by the spinodal curve only correspond to local minima of free energy. As a result, they are not unconditionally stable, since unavoidable thermal energy fluctuations can allow the system to leave the metastable energy well in order to attain a lower energy state. These situations in particular are those governing confined phase transitions within porous solids, and they will be explored in Chapter 9 devoted to this topic. In this present chapter we have revisited the thermodynamics of fluid mixtures. The various results we have obtained are general and remain valid when the fluid mixture occupies the pore network of a porous solid. This will prove particularly useful in the next chapter when addressing the thermodynamics of fluid–solid mixtures made of a porous solid filled with a fluid mixture.

Further Reading

Atkins, P. W. (1990) *Physical Chemistry*, 4th edn, Oxford University Press, Oxford.
Barrat, J.-L. and Hansen, J.-P. (2003) *Basic Concepts for Simple and Complex Fluids*, Cambridge University Press, Cambridge.
Cabane, B. and Hénon, S. (2003) *Liquids: Solutions, Dispersions, Emulsions, Gels* (in French), Belin.
Gibbs, J. W. (1961) *The Scientific Papers*, Vol. 1, Dover.
Lindley, D. (2001) *Boltzmann's Atom: The Great Debate That Launched A Revolution In Physics*, The Free Press.
Sandler, S. I. (2006) *Chemical, Biochemical, and Engineering Thermodynamics*, 4th edn, John Wiley & Sons, Inc..

3

The Deformable Porous Solid

A fluid is an amorphous substance whose molecules move freely past one another. A fluid at rest is therefore unable to withstand a static shear stress and responds to it with an irrecoverable flow. It eventually tends to assume the shape of its container and to forget its past configurations. As a result, the laws governing the mechanical behavior of a fluid involve its current configuration only. The description is Eulerian, referring to the works of Leohnard Euler (1707–1783) in hydrodynamics. In contrast, a solid is a substance whose molecules remain surrounded by the same molecules. Unlike a fluid, a solid responds to a shear stress in deforming. As a result, the state equations governing the mechanical behavior of a solid always involve an initial reference configuration. The description is Lagrangian, referring to the works in solid mechanics of Joseph-Louis Lagrange (1736–1813).

A porous solid made of a solid matrix and of a connected pore space saturated by a fluid mixture is a fluid–solid mixture. Since neither a Lagrangian description is relevant to formulate the state equations of a fluid, nor an Eulerian description to formulate those of a solid, one may indeed wonder which is the relevant description applying to a porous solid. The answer consists of a two-step approach. The first step is based on the works of the mechanics specialist Maurice Biot (1905–1985): a porous solid can be viewed as a deformable open system exchanging fluid mass with the surroundings. The associated thermodynamic system is the fluid–solid mixture whose material boundaries are those delimiting the porous solid. Since these boundaries always remain the same, a Lagrangian description is the best suited in the search for the state equations of the porous solid. With the help of the thermodynamics of fluid mixtures examined in Chapter 2, the next step consists of removing the fluid mixture from the previous open system: the system left is the porous solid.

In this chapter, in order to work out these two steps progressively, we first revisit the two pillars of continuum solid mechanics, strain and stress, and we show how they combine to produce the strain work. The approach is then extended to the porous solid: the porosity change is introduced as related to the deformation, the pore pressure as related to the stress. In the previous chapter, we stated the free energy balance relative to the fluid mixture. This will enable us to extract the free energy balance restricted to the porous solid from the free energy balance applying to the fluid–solid mixture considered as a whole. The energy balance derived in such a way for the porous solid is valid whatever the actual origin of the pore pressure

---

Mechanics and Physics of Porous Solids Olivier Coussy
© 2010 John Wiley & Sons, Ltd

p being exerted on its internal solid walls. In the next chapter, this energy balance will be the way to the state equations of a poroelastic solid irrespective of those of the fluid mixture saturating its porous space.

3.1 Strain

3.1.1 Deformation gradient and displacement

We need a mathematical tool to describe the deformation of a solid. Let us begin by introducing the deformation gradient, in connection with the displacement. At time $t = 0$ the solid stands in an initial configuration. In this configuration the particles constituting the solid can be located by their position vector $\underline{X}$ with components X_i in a Cartesian coordinate frame of orthonormal basis $(\underline{e}_1, \underline{e}_2, \underline{e}_3)$. As time passes a Lagrangian description of the deformation consists of following the motion of these solid particles and their relative positions (see Figure 3.1). At time t the solid has deformed and stands in the current configuration. In this configuration the particle whose initial position vector was $\underline{X}$ can be located by its current position vector $\underline{x}$ with components $x_i(X_j, t)$. We write

$$\underline{X} = X_i \underline{e}_i; \quad \underline{x} = x_i(X_j, t)\underline{e}_i, \qquad (3.1)$$

with a summation on the repeated subscript i. In the following text this convention is adopted and, providing no further indication is given, the subscripts $i, j, ..$ will refer to a Cartesian coordinate system.

In the initial configuration consider now an infinitesimal vector $d\underline{X}$, which links the solid particle located at $\underline{X}$ to the nearby particle located at $\underline{X} + d\underline{X}$. After deformation $d\underline{X}$ becomes $d\underline{x}$ linking the same solid particles, which are now located by their current position vectors $\underline{x}$ and $\underline{x} + d\underline{x}$ as shown in Figure 3.1. Because they consist of the same solid particles, vectors $d\underline{X}$ and $d\underline{x}$ are material vectors, and the terminology extends to surfaces and volumes. Vector $d\underline{x}$ can be obtained from $d\underline{X}$ by differentiating (3.1). We obtain

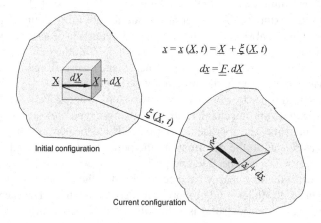

Figure 3.1 The Lagrangian description of the deformation of a body consists of following the motion of its material particles, from the initial configuration to the current configuration, as well as the deformation of the material vectors $d\underline{X}$ linking the nearby particles

$$d\underline{x} = (\partial x_i/\partial X_j) dX_j \underline{e}_i \qquad (3.2)$$

or, equivalently,

$$d\underline{x} = \underline{\underline{F}} \cdot d\underline{X}, \qquad (3.3)$$

where

$$\underline{\underline{F}} = \nabla_{\underline{X}}\, \underline{x}; \quad F_{ij} = \partial x_i/\partial X_j. \qquad (3.4)$$

In Equation (3.4) $\nabla_{\underline{X}}$ stands for the nabla operator relative to the initial configuration. $\underline{\underline{F}}$ is named the deformation gradient. It transports any material vector $d\underline{X}$ onto its deformed $d\underline{x}$. In preparation for transporting any material volume, it is worthwhile noting that the current infinitesimal volume $d\Omega = dx_1 dx_2 dx_3$ is equal to the composed product $(d\underline{x}_1, d\underline{x}_2, d\underline{x}_3)$ of vectors $d\underline{x}_i = dx_i\, \underline{e}_i$. We write

$$d\Omega = (d\underline{x}_1, d\underline{x}_2, d\underline{x}_3) = d\underline{x}_1 \cdot (d\underline{x}_2 \times d\underline{x}_3). \qquad (3.5)$$

The composed product is linear with respect to the vectors it combines. Substituting $d\underline{x}_i = \underline{\underline{F}} \cdot d\underline{X}_i$ in Equation (3.5), we derive that the initial volume $d\Omega_0 = (d\underline{X}_1, d\underline{X}_2, d\underline{X}_3)$ transforms into the volume $d\Omega$ according to the relation

$$d\Omega = \det \underline{\underline{F}}\, d\Omega_0. \qquad (3.6)$$

Now let $\underline{\xi}(\underline{X}, t)$ be the displacement vector of the particle whose initial and current positions are $\underline{X}$ and $\underline{x}$, respectively. We write

$$\underline{x} = \underline{X} + \underline{\xi}. \qquad (3.7)$$

From Equations (3.4) and (3.7) the deformation gradient $\underline{\underline{F}}$ can be expressed as a function of the displacement vector $\underline{\xi}$ according to

$$\underline{\underline{F}} = \underline{\underline{1}} + \nabla_{\underline{X}}\, \underline{\xi}; \quad F_{ij} = \delta_{ij} + \frac{\partial \xi_i}{\partial X_j}, \qquad (3.8)$$

where δ_{ij} is the Kronecker index, namely, $\delta_{ij} = 1$ if $i = j$ and $\delta_{ij} = 0$ if $i \neq j$.

3.1.2 Strain tensor

Deformation induces changes in the length of material vectors and the angle they form. The Green–Lagrange strain tensor $\underline{\underline{\Delta}}$ measures these changes by quantifying the variation of the scalar product of two material vectors $d\underline{X}$ and $d\underline{Y}$, which transform into $d\underline{x}$ and $d\underline{y}$ throughout the deformation. We write

$$d\underline{x} \cdot d\underline{y} - d\underline{X} \cdot d\underline{Y} = 2 d\underline{X} \cdot \underline{\underline{\Delta}} \cdot d\underline{Y}. \qquad (3.9)$$

Substitution of relations $d\underline{x} = \underline{\underline{F}} \cdot d\underline{X}$ and $d\underline{y} = \underline{\underline{F}} \cdot d\underline{Y}$ in Equation (3.9) provides us with the expression for $\underline{\underline{\Delta}}$:

$$\underline{\underline{\Delta}} = \frac{1}{2}\left(\underline{\underline{F}}^t \cdot \underline{\underline{F}} - \underline{\underline{1}}\right), \qquad (3.10)$$

where $\underline{F}^t$ is the transpose of the deformation gradient ($F^t_{ij} = F_{ji}$). By means of Equation (3.8) $\underline{\underline{\Delta}}$ can be expressed explicitly as a function of the displacement vector $\underline{\xi}$:

$$\underline{\underline{\Delta}} = \frac{1}{2}\left(\nabla_{\underline{X}}\,\underline{\xi} + \nabla^t_{\underline{X}}\,\underline{\xi} + \nabla^t_{\underline{X}}\,\underline{\xi}\cdot\nabla_{\underline{X}}\,\underline{\xi}\right); \quad \Delta_{ij} = \frac{1}{2}\left(\frac{\partial \xi_i}{\partial X_j} + \frac{\partial \xi_j}{\partial X_i} + \frac{\partial \xi_k}{\partial X_i}\frac{\partial \xi_k}{\partial X_j}\right). \quad (3.11)$$

The symmetry of the strain tensor entails the existence of real eigenvalues $\Delta_{J=1,2,3}$. They are linked to the set of orthonormal eigen vectors $\underline{u}_J$ according to the relations

$$\underline{\underline{\Delta}}\cdot\underline{u}_J = \Delta_J\,\underline{u}_J \quad \text{(no summation on } J\text{).} \quad (3.12)$$

The eigen values $\Delta_{J=1,2,3}$ and the directions $\underline{u}_J$ are the principal strains and the principal directions of the deformation, respectively. Because of Equation (3.12), the nondiagonal components of the matrix representing the strain tensor $\underline{\underline{\Delta}}$ in the orthonormal basis $(\underline{u}_1, \underline{u}_2, \underline{u}_3)$ are zero. This means that the principal directions $\underline{u}_J$ of the deformation remain orthogonal throughout the deformation.

A first-order approximation to the finite theory can be carried out under the condition of infinitesimal transformation, that is,

$$\left|\partial \xi_i / \partial X_j\right| \ll 1. \quad (3.13)$$

Because of Equation (3.7) linking $\underline{\xi}$ to $\underline{X}$, the successive approximations $\partial/\partial X_i \simeq \partial/\partial x_i$ and $\nabla_{\underline{X}} \simeq \nabla$ ensue from Equation (3.13). As a result, we may omit the subscript $\underline{X}$ when the nabla operator ∇ is used. The current and the initial configurations can eventually be merged, providing only spatial derivations are involved. This does not mean that no difference remains between the Lagrangian and Eulerian descriptions of infinitesimal transformations. Indeed, the very concept of displacement is only relevant in a Lagrangian description, even though the displacement is infinitesimal with respect to some characteristic length related to the structure made of the porous material. Under the condition of infinitesimal transformation (Equation (3.13)), the Green–Lagrange strain tensor $\underline{\underline{\Delta}}$ reduces to the linearized strain tensor $\underline{\underline{\varepsilon}}$, which is defined by

$$\underline{\underline{\Delta}} \simeq \underline{\underline{\varepsilon}}; \quad \underline{\underline{\varepsilon}} = \frac{1}{2}\left(\nabla\underline{\xi} + \nabla^t\underline{\xi}\right); \quad \varepsilon_{ij} = \frac{1}{2}\left(\frac{\partial \xi_i}{\partial x_j} + \frac{\partial \xi_j}{\partial x_i}\right). \quad (3.14)$$

Since Δ_{ij} has the same order of magnitude as $\partial \xi_i/\partial X_j$, an infinitesimal transformation results in an infinitesimal deformation, that is, $|\Delta_{ij}| \ll 1$. In contrast, the deformation may be infinitesimal although the transformation is not. For instance, $\underline{\underline{\Delta}}$ is zero in a rigid body movement, although $\nabla_{\underline{X}}\underline{\xi}$ can have any order of magnitude. Under the approximation of infinitesimal transformation, because of the definition of $\underline{\underline{\Delta}} \simeq \underline{\underline{\varepsilon}}$ (Equation (3.9)) and of the approximation $|\varepsilon_{ij}| \ll 1$, the diagonal term ε_{ii} (no summation) is equal to the linear dilation in the $\underline{e}_i$ direction. Under the approximation of infinitesimal transformation (Equation (3.13)), we also derive

$$\det \underline{\underline{F}} \simeq 1 + \epsilon, \quad (3.15)$$

where ϵ is defined by

$$\epsilon = \varepsilon_{ii}, \quad (3.16)$$

The Deformable Porous Solid

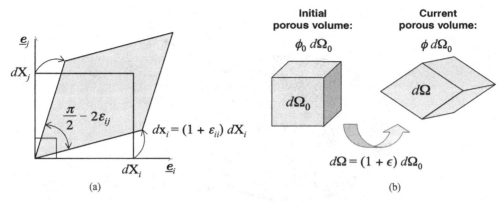

Figure 3.2 (a) Geometrical interpretation of the components ε_{ij} of the strain tensor; (b) definition of Lagrangian porosity ϕ

and is the volumetric strain or dilation – a contraction corresponding to a negative value. Indeed, Equations (3.6) and (3.15) give us the approximation

$$d\Omega \simeq (1+\epsilon)\, d\Omega_0. \tag{3.17}$$

Note that $\epsilon = \varepsilon_1 + \varepsilon_2 + \varepsilon_3$ where $\varepsilon_{J=1,2,3} \simeq \Delta_{J=1,2,3}$. Also, twice the nondiagonal term, that is, $2\varepsilon_{ij}$ where $i \neq j$, is equal to the change undergone by the angle made between the material vectors $\underline{e}_i$ and $\underline{e}_j$, which were normal prior to the deformation. This angle change is named the distortion relative to the material directions $\underline{e}_i$ and $\underline{e}_j$. These geometrical interpretations of the strain tensor are illustrated in Figure 3.2. It is often useful to extract from the deformation tensor $\underline{\underline{\varepsilon}}$ the deviatoric part accounting for the material distortion only. The resulting tensor $\underline{\underline{e}}$ is the deviatoric strain tensor, whose expression is given by

$$\underline{\underline{e}} = \underline{\underline{\varepsilon}} - \frac{\epsilon}{3}\underline{\underline{1}}; \quad e_{ij} = \varepsilon_{ij} - \frac{\epsilon}{3}\delta_{ij}. \tag{3.18}$$

3.2 Stress

3.2.1 The hypothesis of local contact forces

In continuum mechanics, a material domain Ω is subjected to two kinds of forces, namely body and surface forces. This is illustrated in Figure 3.3. The infinitesimal body force $\delta \underline{f}$ acting on the elementary material volume $d\Omega$ is defined by means of a body force density $\underline{f}$:

$$\delta \underline{f} = \rho \underline{f}\left(\underline{x}, t\right) d\Omega, \tag{3.19}$$

where ρ stands for the current mass density of the material volume $d\Omega$. The body force density $\underline{f}$ depends solely on the current position vector $\underline{x}$ and on time t. Accordingly the external body forces applied to the infinitesimal material volume $d\Omega$ are local forces, whose expression holds irrespective of the domain Ω to which the infinitesimal volume $d\Omega$ belongs. Nonlocal forces, such as forces depending on the distance between particles, will not be considered.

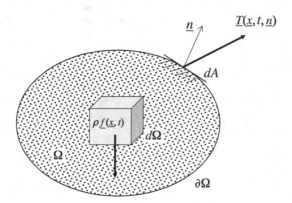

Figure 3.3 Definition of external forces: body forces and surface forces

The surface forces act on the border $\partial\Omega$ of Ω. The infinitesimal surface force $\delta\underline{T}$ acting on the infinitesimal material surface dA is defined through a surface force density $\underline{T}$ according to

$$\delta\underline{T} = \underline{T}\left(\underline{x}, t; \underline{n}\right) dA. \tag{3.20}$$

The surface force density $\underline{T}$ is assumed to depend only on the position vector $\underline{x}$ and time t; and the outward unit normal $\underline{n}$ to dA. This is assuming that the surface forces result only from local contact forces exerted by the immediately adjacent material points. The action of more distant points, action that would at least involve the local curvature of $\partial\Omega$, is excluded here. This hypothesis of 'local contact forces' will soon turn out to be essential to define the stress tensor introduced by Augustin Louis Cauchy (1789–1857).

3.2.2 The action–reaction law

Confining ourselves to a static approach in the following, the mechanical equilibrium of all matter included in any material domain Ω requires the nullity of the overall force to which this domain is subjected. We therefore write

$$\int_\Omega \rho\underline{f}\left(\underline{x}, t\right) d\Omega + \int_{\partial\Omega} \underline{T}\left(\underline{x}, t, \underline{n}\right) dA = 0. \tag{3.21}$$

Since Equation (3.21) must hold for any domain Ω, and since the body force $\underline{f}\left(\underline{x}, t\right)$ acting on the volume $d\Omega$ is irrespective of the choice of the domain Ω to which $d\Omega$ belongs, we may choose the cylinder shown in Figure 3.4(a). The surface $\partial\Omega$ is then formed by the cylinder wall of surface Σ oriented by the unit outward normal $\underline{n}_\Sigma$ and by the two end sides, A^+ and A^- oriented by the unit outward normal vectors $\underline{n}$ and $-\underline{n}$. Applying the equilibrium Equation (3.21) to this cylinder leads to

$$\int_\Omega \rho\underline{f}\left(\underline{x}, t\right) d\Omega + \int_\Sigma \underline{T}(\underline{n}_\Sigma) dA + \int_{A^+} \underline{T}(\underline{n}) dA + \int_{A^-} \underline{T}(-\underline{n}) dA = 0. \tag{3.22}$$

Allowing the cylinder thickness h in the direction normal to A^+ and A^- tend to zero, the integrals over the volume Ω and over the surface Σ in the left-hand side of Equation (3.21) vanish. We derive

The Deformable Porous Solid

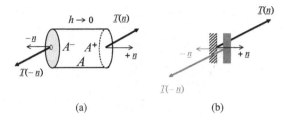

Figure 3.4 (a) Equilibrium of an infinitely thin cylinder; (b) action–reaction law

$$h \to 0: \quad \int_{A^+} \underline{T}(\underline{n}) \, dA + \int_{A^-} \underline{T}(-\underline{n}) \, dA = 0. \tag{3.23}$$

Since this identity must hold whatever the value of dA, we obtain the action–reaction law, namely,

$$\underline{T}(-\underline{n}) = -\underline{T}(\underline{n}). \tag{3.24}$$

As illustrated in Figure 3.4(b) the action–reaction law can be expressed in the following way: the surface force exerted on the domain Ω by the surroundings through the material surface dA is equal in intensity and opposite in direction to the surface force that the domain Ω exerts on the surroundings through the same material surface dA. The action–reaction law, and therefore the hypothesis of contact forces from which the law derives, plays an analogous role in solid mechanics to the role played by the zeroth law in thermodynamics. The zeroth law states that 'if a system A is in thermal equilibrium with a system B, and if the latter is in thermal equilibrium with a system C then C will be in thermal equilibrium with A'. Without the zeroth law thermometers would not exist. Without the action–reaction law mechanical presses would not exist either.

3.2.3 The Cauchy stress tensor

Let us now apply the equilibrium relation (Equation (3.21)) to the infinitesimal small tetrahedron represented in Figure 3.5, whose three faces A_j are parallel to the coordinate planes and oriented by $-\underline{e}_j$. Surfaces A_j are linked to the base surface A oriented by the unit normal $\underline{n}$ according to the relation

$$A_j = A\underline{n} \cdot \underline{e}_j = A n_j. \tag{3.25}$$

Applying Equation (3.21) to this infinitesimally small tetrahedron we obtain

$$\frac{hA}{3} \mathrm{O}\left(\rho \underline{f}\right) + \underline{T}(\underline{n})A + \underline{T}(-\underline{e}_j)A_j \simeq 0, \tag{3.26}$$

where h is the height of the tetrahedron, so that $hA/3$ is its volume, and where $\mathrm{O}\left(\rho \underline{f}\right)$ scales the order of magnitude of $\rho \underline{f}$. Substitution of Equation (3.25) into Equation (3.26), together

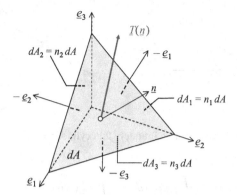

Figure 3.5 Tetrahedron lemma proving the existence of a stress tensor

with the use of the action–reaction law (Equation (3.24)), leads to

$$\underline{T}(\underline{n}) - \underline{T}(\underline{e}_j)n_j \simeq \frac{h}{3}O\left(\rho\underline{f}\right). \tag{3.27}$$

Letting h tend to 0, that is, degenerating the tetrahedron to a point, the right-hand side of Equation (3.27) tends to zero and we derive

$$\underline{T}(\underline{n} = n_j\underline{e}_j) = \underline{T}(\underline{e}_j)n_j. \tag{3.28}$$

This result is known as the tetrahedron lemma, which shows that a linear operator relates the vector $\underline{T}(\underline{x}, t, \underline{n})$ to $\underline{n}$. This linear operator is the stress tensor $\underline{\underline{\sigma}} = \sigma(\underline{x}, t)$ of components σ_{ij}, while vector $\underline{T}(\underline{x}, t, \underline{n})$ is called the stress vector:

$$\underline{T}(\underline{x}, t, \underline{n} = n_j\underline{e}_j) = \underline{\underline{\sigma}} \cdot \underline{n} = \sigma_{ij}n_j\underline{e}_i. \tag{3.29}$$

Equation (3.29) which defines the Cauchy stress $\underline{\underline{\sigma}}$ as a tensorial quantity is the direct consequence of the hypothesis of local contact forces, that is $\underline{T}=\underline{T}(\underline{x},t,\underline{n})$. Figures 3.6(a) and 3.6(b) illustrate the physical significance of the Cauchy stress tensor $\underline{\underline{\sigma}}$ relative to the stress vector: σ_{ij} ($i = 1, 2, 3$) are the components in the orthonormal basis $\underline{e}_i$ ($i = 1, 2, 3$) of the surface force density $\underline{T}(\underline{x}, t, \underline{n} = \underline{e}_j) = \sigma_{ij}\underline{e}_i$ applied to the material surface oriented by the unit normal $\underline{n} = \underline{e}_j$. As a result σ_{jj} (no summation) is the intensity of the tension applied to the material surface in the direction $\underline{e}_j$ normal to it — a pressure corresponding to a negative value of σ_{jj}, whereas σ_{ij} is the intensity of the shear acting in the $\underline{e}_i$ direction along the same surface and is zero for a fluid at rest.

Analogous to the decomposition of the strain tensor $\underline{\underline{\varepsilon}}$ in Equation (3.18) the stress tensor $\underline{\underline{\sigma}}$ can be decomposed between its spherical part $\sigma\underline{\underline{1}}$ and its deviatoric part $\underline{\underline{s}}$, the latter accounting for the shear only. The mean stress σ and the deviatoric stress tensor $\underline{\underline{s}}$ are then defined according to

$$\sigma = \frac{1}{3}\sigma_{ii} \tag{3.30}$$

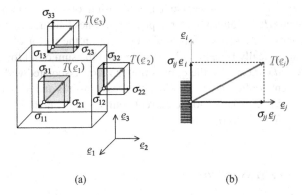

Figure 3.6 Physical significance of the Cauchy stress tensor: (a) significance of the stress tensor components σ_{ij}; (b) significance of the stress vector $\underline{T}(\underline{e}_j)$

and

$$\underline{\underline{s}} = \underline{\underline{\sigma}} - \sigma \underline{\underline{1}}; \quad s_{ij} = \sigma_{ij} - \sigma \delta_{ij}. \tag{3.31}$$

For a fluid at rest note that σ would reduce to the fluid pressure affected by the sign $-$, whereas the deviatoric stress tensor $\underline{\underline{s}}$ would be zero.

3.2.4 Local mechanical equilibrium

Use of Equation (3.29) in the overall mechanical equilibrium relation, that is, Equation (3.21), provides

$$\int_{\Omega} \rho \underline{f} \, d\Omega + \int_{\partial \Omega} \underline{\underline{\sigma}} \cdot \underline{n} \, dA = 0. \tag{3.32}$$

Applying the divergence theorem to the surface integral of Equation (3.32), that is,

$$\int_{\partial \Omega} \underline{\underline{\sigma}} \cdot \underline{n} \, dA \equiv \int_{\partial \Omega} \sigma_{ij} n_j \, dA = \int_{\Omega} \frac{\partial \sigma_{ij}}{\partial x_j} \, d\Omega \equiv \int_{\Omega} \nabla_x \cdot \underline{\underline{\sigma}} \, d\Omega, \tag{3.33}$$

we rewrite Equation (3.32) in the form

$$\int_{\Omega} \left(\nabla_x \cdot \underline{\underline{\sigma}} + \rho \underline{f} \right) d\Omega = 0. \tag{3.34}$$

Since Equation (3.34) must hold irrespective of any particular choice of domain Ω, the integrand must vanish. We obtain

$$\nabla_x \cdot \underline{\underline{\sigma}} + \rho \underline{f} = 0, \tag{3.35}$$

which provides three equations when varying the $\underline{e}_i$ direction, namely,

$$\frac{\partial \sigma_{ij}}{\partial x_j} + \rho f_i = 0. \tag{3.36}$$

Equation (3.35) accounts for the local mechanical equilibrium of the matter contained at time t in the elementary volume $d\Omega$ located at material point $\underline{x}$. It states that the sum of the elementary body forces $\rho \underline{f} d\Omega$ and of the surface forces $\underline{T} dA$ acting on the various faces dA of $d\Omega$ is equal to zero. More specifically Equation (3.36) states that the sum of all forces acting on the material volume $d\Omega = dx_1 dx_2 dx_3$ in the $\underline{e}_i$ direction is zero. Associated with the mechanical equilibrium of the current configuration, the Cauchy stress tensor is an Eulerian quantity. Note also that Equations (3.35)–(3.36) refer to the current configuration and hold irrespective of the hypothesis of infinitesimal transformation.

3.2.5 Symmetry of stress tensor

The mechanical equilibrium of matter also requires that the resulting moment exerted upon any material domain Ω is zero. In the absence of any density of external moment this is expressed by

$$\int_\Omega \underline{x} \times \rho \underline{f} \, d\Omega + \int_{\partial \Omega} \underline{x} \times \underline{\underline{\sigma}} \cdot \underline{n} \, dA = 0. \tag{3.37}$$

The divergence theorem applied to the last term of Equation (3.37) provides

$$\int_{\partial\Omega} \underline{x} \times \underline{\underline{\sigma}} \cdot \underline{n} \, dA = \int_\Omega (\underline{x} \times \nabla_x \cdot \underline{\underline{\sigma}} + 2\Sigma^{as}) \, d\Omega, \tag{3.38}$$

where, in Cartesian coordinates, Σ^{as} is the vector

$$2\Sigma^{as} = (\sigma_{23} - \sigma_{32})\underline{e}_1 + (\sigma_{13} - \sigma_{31})\underline{e}_2 + (\sigma_{12} - \sigma_{21})\underline{e}_3. \tag{3.39}$$

Substitution of Equation (3.38) into Equation (3.37) and use of Equation (3.34) gives

$$\int_\Omega 2\Sigma^{as} \, d\Omega = 0. \tag{3.40}$$

Since Equation (3.40) must hold for any volume Ω, it follows that $\Sigma^{as} = 0$ and, due to the expression for Σ^{as} given in Equation (3.39), it eventually implies the symmetry of the strain tensor $\underline{\underline{\sigma}}$:

$$\sigma_{ij} = \sigma_{ji}. \tag{3.41}$$

The symmetry of the stress tensor results from the absence of any density of external moments, whether volume or surface related. Indeed, in the absence of external couples, Equation (3.41) states that the resulting moment of surface forces $\underline{T} dA$ applied to the various faces dA of $d\Omega$ is zero. For instance, the symmetry $\sigma_{12} = \sigma_{21}$ results from the nullity of the dynamic moment in the $\underline{e}_3$ direction. The symmetry of the stress tensor eventually ensures its eigenvalues $\sigma_{J=1,2,3}$ to be real. These eigenvalues are called the principal stresses, and the associated orthogonal eigenvectors are the principal stress directions.

3.3 Strain Work

3.3.1 Mechanical energy balance

The infinitesimal work dW supplied to the material domain Ω by the surroundings between time t and time $t+dt$ is given by[1]

$$dW = \int_\Omega d\underline{\xi} \cdot \rho \underline{f}\, d\Omega + \int_{\partial\Omega} \left(d\underline{\xi} \cdot \underline{T} \right) dA, \qquad (3.42)$$

where $d\underline{\xi}\left(\underline{x}\left(\underline{X}\right), t\right)$ is the infinitesimal displacement between time t and time $t+dt$ of the particle of porous solid, which is currently located at $\underline{x}$ and was located at $\underline{X}$ in the reference configuration. Using Equation (3.29) we first rewrite the infinitesimal work dW in the equivalent form

$$dW = \int_\Omega d\xi_i\, \rho f_i\, d\Omega + \int_{\partial\Omega} \left(d\xi_i\, \sigma_{ij}\, n_j \right) dA. \qquad (3.43)$$

The divergence theorem and the symmetry of the stress tensor provide the identity

$$\int_{\partial\Omega} d\xi_i\, \sigma_{ij}\, n_j\, dA = \int_\Omega \left[\sigma_{ij} \times \frac{1}{2}\left(\frac{\partial (d\xi_i)}{\partial x_j} + \frac{\partial (d\xi_j)}{\partial x_i} \right) + d\xi_i\, \frac{\partial \sigma_{ij}}{\partial x_j} \right] d\Omega. \qquad (3.44)$$

Substituting Equation (3.44) in Equation (3.43) we obtain

$$dW = \int_\Omega d\xi_i \left(\frac{\partial \sigma_{ij}}{\partial x_j} + \rho f_i \right) d\Omega + \int_\Omega \sigma_{ij} \times \frac{1}{2}\left(\frac{\partial (d\xi_i)}{\partial x_j} + \frac{\partial (d\xi_j)}{\partial x_i} \right) d\Omega. \qquad (3.45)$$

The first integral on the right-hand side of Equation (3.45) is zero because of the local equilibrium Equation (3.36). This allows us to write

$$dW = \int_\Omega dW\, d\Omega; \quad dW = \sigma_{ij} \times \frac{1}{2}\left[\frac{\partial (d\xi_i)}{\partial x_j} + \frac{\partial (d\xi_j)}{\partial x_i} \right], \qquad (3.46)$$

where dW is the infinitesimal strain work density related to the domain $d\Omega$.

3.3.2 Strain work in infinitesimal transformations

Mechanical energy balance (Equation (3.46) looks into the consequences of the local mechanical equilibrium of matter accounted for by Equation (3.36). It does not provide any new physical law. It highlights how the overall work dW, which is supplied to the material domain Ω through its boundary $\partial\Omega$, is transmitted in the form of the local infinitesimal strain work $dW d\Omega$ to the infinitesimal domains $d\Omega$ constituting Ω. The expression for dW in Equation (3.46) is an exact expression. However, when assuming infinitesimal transformations a comparison with Equation (3.14) shows that the term in parenthesis in Equation (3.46) is the infinitesimal variation $d\varepsilon_{ij}$ of the component ε_{ij} of the linearized strain tensor $\underline{\varepsilon}$. In addition,

---

[1] Conforming with the first chapter, in order to make a distinction with the differential operator with respect to space, as in $d\Omega$, the differential operator with respect to time is denoted by the symbol d, as in $d\underline{\xi}$.

for infinitesimal transformations, according to Equation (3.17) we may replace the current volume $d\Omega$ by the initial volume $d\Omega_0$. For infinitesimal transformations the mechanical energy balance stated in Equation (3.46) can then be approximated in the form

$$dW \simeq \int_\Omega dW \, d\Omega_0; \quad dW \simeq \sigma_{ij} d\varepsilon_{ij}. \tag{3.47}$$

Substitution in Equation (3.47) of the decomposition of both the strain tensor (Equation (3.18)) and the stress tensor (Equation (3.31)) in their spherical and deviatoric parts gives

$$dW = \sigma \, d\epsilon + s_{ij} de_{ij}. \tag{3.48}$$

The quantity $dW \, d\Omega_0$ is the infinitesimal strain work produced by the forces applied to the infinitesimal solid having $V = d\Omega$ as current volume and causing its deformation. For a fluid, which cannot sustain shear, we have $\sigma = -p$ and $s_{ij} = 0$ while, due to Equation (3.17), we obtain $d\epsilon \, d\Omega_0 \simeq d(d\Omega) = dV$, so that $dW \, d\Omega_0$ reduces to the familiar expression $-p dV$. As given by Equation (3.48), $dW \, d\Omega_0$ extends this familiar expression to a solid, which can sustain shear.

3.4 From Solids to Porous Solids

A porous solid is a solid having internal walls, which delimit a connected pore space. As a result, the strain tensor $\underline{\varepsilon}$ describes only a part of the deformation undergone by the porous solid because the internal walls of the latter deform too. These internal walls are subjected to the pressure exerted by the fluid mixture filling the pores. Since the pore walls are not subjected to a shear stress, no strain work is associated with the deviatoric deformation of the pore space. As a result, only the pore volumetric deformation needs to be described and is conveniently accounted for through the variation of the porosity. Because of the existence of the internal pore pressure applied on the internal walls, the stress tensor $\underline{\sigma}$ is not sufficient to describe the internal forces acting upon the porous solid. As a result the strain work density dW, defined by Equations (3.46) or (3.48), cannot account for the strain work relative to the porous solid only. This section addresses these various issues, by extending the basic concepts of the mechanics of solids to porous solids.

3.4.1 Porosity and deformation

The porosity relative to a volume of porous solid is the fraction of that volume formed from pores. Let n be the porosity of the *current* volume $d\Omega$ of the solid under concern, so that the pores occupy the volume $n d\Omega$ in the current configuration. Because it refers the current volume to the current configuration, n is termed Eulerian porosity. Since the overall volume $d\Omega$ of the solid changes throughout the deformation, the infinitesimal change dn of the Eulerian porosity n does not account for the actual volume change undergone by the porous space. In contrast to the Eulerian porosity n, the volume change of the porous space is accounted for by the change of the Lagrangian porosity ϕ, which refers the current porous volume to the initial volume $d\Omega_0$ according to

$$\phi \, d\Omega_0 = n \, d\Omega. \tag{3.49}$$

The change in the porous space is then given by $(\phi - \phi_0)\,d\Omega_0$, where ϕ_0 stands for the initial Lagrangian porosity. In the following text we will adopt the notation

$$\varphi = \phi - \phi_0. \tag{3.50}$$

The observable macroscopic volume dilation ϵ undergone by the porous solid is due to both the volumetric dilation φ/ϕ_0 of the porous volume to the volumetric dilation ϵ_S undergone by the solid matrix forming the solid part of the porous solid, albeit the latter is not accessible from purely macroscopic experiments. However, in a way similar to Equation (3.17) the very definition of ϵ_S allows us to write

$$d\Omega^S = (1 + \epsilon_S)\,d\Omega_0^S, \tag{3.51}$$

where

$$d\Omega^S = (1-n)\,d\Omega; \quad d\Omega_0^S = (1-\phi_0)\,d\Omega_0. \tag{3.52}$$

The combination of Equations (3.17) and (3.49)–(3.52) provides the volumetric dilation balance

$$\epsilon = (1 - \phi_0)\epsilon_S + \varphi, \tag{3.53}$$

where the solid volumetric dilation ϵ_S and the pore volumetric dilation φ/ϕ_0 are weighted by the volume fractions $1 - \phi_0$ and ϕ_0, respectively.

The porosity involved later on will always be the connected porosity, that is the porosity relative to the porous space through which the filtration of the saturating fluids can actually occur. In contrast to this connected porosity the porous solid may also contain an occluded porosity, whether saturated or not, but through which no filtration occurs. The deformation of this occluded porosity is included in ϵ_S. When there is no occluded porosity and when, as in the case of soils, the solid part of the porous solid is formed of poorly compressible grains, the solid matrix can be considered as incompressible. This results in $\epsilon_S = 0$ which, substituted in Equation (3.53), leads to

$$\epsilon = \varphi. \tag{3.54}$$

Equation (3.54) means that the volumetric dilation ϵ observed at the macroscopic scale results from the deformation of the pore space only.

In soil mechanics the current degree of compactness of a porous material is suitably accounted for by the void ratio e, which is defined as the ratio of the current pore volume to the current volume of the solid part. Owing to its very definition the void ratio e is an Eulerian variable, with no Lagrangian counterpart, and can be expressed as a function of n in the form

$$e = \frac{n}{1-n}. \tag{3.55}$$

Under the condition $\epsilon_S = 0$ it can be more convenient to use the void ratio e instead of the volumetric dilation ϵ. Combining Equations (3.17), (3.49), (3.54) and (3.55), we derive

$$\epsilon = \frac{e - e_0}{1 + e_0}, \tag{3.56}$$

where e_0 stands for the initial void ratio.

3.4.2 Pore pressure and stress partition

As defined in Section 3.2 the stress tensor σ is associated with the overall mechanical equilibrium of the volume $d\Omega$ centered at the point located by the position vector $\underline{x}$, irrespective of the detailed composition of $d\Omega$. Since $d\Omega$ can be heterogeneous, as it is for a porous solid, the question arises about how this 'macroscopic' stress relates to the averaged stress field prevailing at the lower scale within $d\Omega$. To address this question let $d\Omega_{\underline{x}=0} = \omega(0)$ be an elementary representative volume centered at the origin of coordinates $\underline{x} = 0$ and let $f(\underline{z})$ be a weighting function which has for argument the position vector $\underline{z}$ of points lying in $\omega(0)$. The function $f(\underline{z})$ has continuous derivatives and satisfies

$$f(\underline{z}) = 0 \text{ for } \underline{z} \in \partial\omega(0); \quad \frac{1}{\omega(0)} \int_{\omega(0)} f(\underline{z}) \, d\omega_{\underline{z}} = 1, \tag{3.57}$$

where $\partial\omega(0)$ stands for the border of $\omega(0)$. Since any elementary representative volume $d\Omega_{\underline{x}} = \omega(\underline{x})$ centered at $\underline{x}$ can be obtained by the translation of volume $\omega(0)$ along vector $\underline{x}$, we can then define the 'sliding' average $\langle \mathcal{G} \rangle (\underline{x})$ of any quantity $\mathcal{G}$ related to volume $\omega(\underline{x})$ centered at point $\underline{x}$ according to

$$\langle \mathcal{G} \rangle (\underline{x}) = \frac{1}{\omega(\underline{x})} \int_{\omega(\underline{x})} \mathcal{G}^{\text{micro}}(\underline{z}) f(\underline{z} - \underline{x}) \, d\omega_{\underline{z}}. \tag{3.58}$$

For instance, the sliding average of the stress field $\sigma_{ij}(\underline{z})$ is

$$\langle \sigma_{ij} \rangle (\underline{x}) = \frac{1}{\omega(\underline{x})} \int_{\omega(\underline{x})} \sigma_{ij}^{\text{micro}}(\underline{z}) f(\underline{z} - \underline{x}) \, d\omega_{\underline{z}}. \tag{3.59}$$

From this definition we obtain

$$\frac{\partial \langle \sigma_{ij} \rangle (\underline{x})}{\partial x_j} = -\frac{1}{\omega(\underline{x})} \int_{\omega(\underline{x})} \sigma_{ij}^{\text{micro}}(\underline{z}) \frac{\partial f}{\partial z_j} (\underline{z} - \underline{x}) \, d\omega_{\underline{z}}. \tag{3.60}$$

Integrating by parts Equation (3.60) gives

$$\frac{\partial \langle \sigma_{ij} \rangle (\underline{x})}{\partial x_j} = \frac{1}{\omega(\underline{x})} \int_{\omega(\underline{x})} \left\{ \frac{\partial \sigma_{ij}^{\text{micro}}(\underline{z})}{\partial z_j} f(\underline{z} - \underline{x}) - \frac{\partial}{\partial z_j} \left[\sigma_{ij}^{\text{micro}}(\underline{z}) f(\underline{z} - \underline{x}) \right] \right\} d\omega_{\underline{z}}. \tag{3.61}$$

The action–reaction law stated in Section 3.2.2 ensures the continuity of the stress vector components $\sigma_{ij}^{\text{micro}}(\underline{z}) n_j$ related to any surface lying within $\omega(\underline{x})$ and having $n_j \underline{e}_j$ as normal unit. Since f vanishes on the border $\partial\omega(\underline{x})$ of $\omega(\underline{x})$, this continuity and the divergence theorem combine to show that the last term in Equation (3.61) is zero. As a result we have

$$\frac{\partial \langle \sigma_{ij} \rangle (\underline{x})}{\partial x_j} = \left\langle \frac{\partial \sigma_{ij}^{\text{micro}}(\underline{z})}{\partial z_j} \right\rangle. \tag{3.62}$$

To express the mechanical equilibrium at the macroscopic scale we could have carried out the approach of Section 3.2 at the lower scale, that is, at microscopic points $\underline{z}$. In a similar way to

Equation (3.36), this would have provided

$$\frac{\partial \sigma_{ij}^{\text{micro}}(\underline{z})}{\partial z_j} + \rho(\underline{z}) f_i = 0. \qquad (3.63)$$

Assuming constant body forces, like those associated with gravity, substitution of Equation (3.63) in Equation (3.62) provides

$$\frac{\partial \langle \sigma_{ij} \rangle (\underline{x})}{\partial x_j} + \langle \rho \rangle f_i = 0. \qquad (3.64)$$

A comparison of Equation (3.36) with Equation (3.64) then shows that $\sigma_{ij}(\underline{x}) = \langle \sigma_{ij} \rangle (\underline{x})$ and, therefore, that the macroscopic symmetry $\sigma_{ij}(\underline{x}) = \sigma_{ji}(\underline{x})$ results from the microscopic symmetry $\sigma_{ij}^{\text{micro}}(\underline{z}) = \sigma_{ji}^{\text{micro}}(\underline{z})$. The weighting function $f(\underline{z})$ can be chosen arbitrarily close to the characteristic function $f_{\omega(0)}(\underline{z})$ of $\omega(0)$, that is, the function equal to one for a point lying inside $\omega(0)$ and zero for a point lying on the border $\partial \omega(0)$. In practice, the sliding average therefore reduces to a standard space average. Then, for a porous solid, let $\sigma_{ij}^S(\underline{x})$ be the space averaged stress of the microscopic stress field $\sigma_{ij}^{\text{micro}}(\underline{z})$ prevailing in the solid part only, that is the solid matrix. In addition, in all that follows the internal solid walls of the solid matrix will be subjected to a uniform pressure p so that in the porous space the microscopic stress field $\sigma_{ij}^{\text{micro}}(\underline{z})$ is uniform and reduces to $-p(\underline{x}) \delta_{ij}$. From the equality $\sigma_{ij}(\underline{x}) = \langle \sigma_{ij}(\underline{z}) \rangle$, we then derive the stress partition

$$\sigma_{ij} = (1-n) \sigma_{ij}^S - n p \delta_{ij}, \qquad (3.65)$$

where the macroscopic argument $\underline{x}$ has been omitted.

In a similar way to Equation (3.30) we can introduce the mean stress σ_S prevailing in the solid matrix. In addition, assuming that the transformation is infinitesimal, the current Eulerian porosity n can be replaced in Equation (3.65) by the initial porosity ϕ_0. From Equation (3.65) we then obtain the relation

$$\sigma = (1 - \phi_0) \sigma_S - \phi_0 p. \qquad (3.66)$$

In Chapter 4 Equations (3.53) and (3.66) will turn out to be quite useful linking the macroscopic properties of the porous solid and the properties of the matrix forming its solid part.

3.4.3 Free energy balance for the fluid–solid mixture

Analogously to the representation of deformable solids in continuum mechanics, a deformable porous continuum Ω consists of juxtaposed infinitesimal porous solids $d\Omega(\underline{x})$, whose porous spaces are saturated by a fluid mixture. Any elementary system $d\Omega(\underline{x})$ forming the porous continuum Ω at point $\underline{x}$ is therefore a fluid–solid mixture, which can be addressed as an open thermodynamic system exchanging fluid mixture with the adjacent elementary systems $d\Omega(\underline{x} + d\underline{x})$. Instead of considering overall quantities related to the elementary system $d\Omega$ – like its Helmholtz free energy F, its entropy S or the number of moles N_J of component J it contains – the use of the corresponding volumetric densities f, s or n_J will be more suitable when stating the balance laws that govern the evolutions of $d\Omega$. Furthermore, it will turn out to be more convenient to define these volumetric densities with respect to the *initial* volume $d\Omega_0$

(Lagrangian densities), rather than to the *current* volume $d\Omega$ (Eulerian densities), as concerns the overall free energy F and entropy S; or rather than to the volume $\phi d\Omega_0$ that the fluid mixture actually occupies, as concerns the mole density n_J. Having these density definitions in mind, we write

$$F = f\, d\Omega_0; \quad S = s\, d\Omega_0; \quad N_J = n_J\, d\Omega_0. \tag{3.67}$$

If we first consider the case where the porous solid does not deform, both the overall volume, namely $V = d\Omega_0$, and the volume occupied by the fluid mixture, namely $\phi_0 d\Omega_0$, do not change. The free energy balance stated in Equation (2.4) in Chapter 2 then extends to the fluid–solid mixture in the form of the free energy density balance

$$\mathrm{d}f = \sum_J \mu_J \mathrm{d}n_J - s\mathrm{d}T. \tag{3.68}$$

If the porous solid now deforms, that is, if both $\underline{\varepsilon} \neq 0$ and $\varphi \neq 0$ ($\phi \neq \phi_0$), the current volume of the open system is not $d\Omega_0$ but $d\Omega$. However, according to the definitions given in Equation (3.67), in spite of the deformation of the system, the term $\mu_J \mathrm{d}n_J$ still accounts for the free energy supplied to the open system induced by the increase $\mathrm{d}N_J = \mathrm{d}n_J d\Omega_0$ of the number of moles of component J. In addition, we must now account for the energy density supplied through the deformation of the open system. This energy density is the strain work density $\mathrm{d}W$ introduced in Section 3.3.2. For the fluid–solid mixture with a deformable porous solid, using Equation (3.48), we write

$$\mathrm{d}f = \sigma \mathrm{d}\epsilon + s_{ij}\mathrm{d}e_{ij} + \sum_J \mu_J \mathrm{d}n_J - s\mathrm{d}T. \tag{3.69}$$

3.4.4 Free energy balance for the porous solid

Instead of the free energy balance stated in Equation (3.69) relative to the whole fluid–solid mixture, we now look for the free energy balance relative to the porous solid only. At any time the volume occupied by the fluid mixture is $\phi d\Omega_0$, while its entropy can be written $s_{\mathrm{mix}} d\Omega_0$. As a result the Gibbs–Duhem Equation (2.14), which applies to any fluid mixture, can be rewritten in the equivalent form

$$-\phi \mathrm{d}p + s_{\mathrm{mix}}\mathrm{d}T + \sum_J n_J \mathrm{d}\mu_J = 0. \tag{3.70}$$

Energy and entropy are additive quantities. The Helmholtz free energy a_S and the entropy s_S per unit of the initial volume $d\Omega_0$ of the porous solid can then be obtained by simply removing the free energy and entropy of the fluid mixture from the overall free energy and entropy of the fluid–solid mixture. Using Equation (2.5) relating the Helmholtz free energy of the fluid mixture to its Gibbs free energy, from Equation (2.13) and the definitions given in Equation (3.67) we obtain

$$a_S = f - \left(\sum_J n_J \mu_J - p\phi\right); \quad s_S = s - s_{\mathrm{mix}}. \tag{3.71}$$

Substituting Equation (3.71) in Equation (3.69) and making use of Equation (3.70), we derive

$$da_S = \sigma d\epsilon + pd\phi + s_{ij}de_{ij} - s_S dT. \quad (3.72)$$

Substituting Equation (3.50) in Equation (3.72) we derive the free energy balance related to the porous solid only in the form

$$da_S = \sigma d\epsilon + pd\varphi + s_{ij}de_{ij} - s_S dT. \quad (3.73)$$

The first three terms on the right-hand side of Equation (3.73) account for the infinitesimal strain work density dW_S related to the porous solid only, that is, considered separately from the fluid–solid mixture it constitutes a part of. Shortening $\sigma d\epsilon + s_{ij}de_{ij}$ with the help of Equations (3.18) and (3.31), we write

$$dW_S = \sigma_{ij}d\varepsilon_{ij} + pd\varphi. \quad (3.74)$$

With respect to the expression given in Equation (3.47) of the strain work density relative to an ordinary solid, the expression given in Equation (3.74) relative to a porous solid involves an extra term, $pd\varphi$, which is related to the mechanical work produced by the pore pressure p applying to the internal solid walls of the porous solid. Indeed, between times t and $t+dt$, the initial porous volume $\phi_0 d\Omega_0$ undergoes the exact volume change $dV = d\varphi \, d\Omega_0$. Since this expression holds irrespective of any assumption of infinitesimal transformations, the associated strain work – that is $-pdV = -pd\varphi \, d\Omega_0$ – and the related strain work density per unit of initial volume $d\Omega_0$ of porous solid – that is $-pd\varphi$ – correspond to exact expressions too. This remark will be useful in the next section, when addressing finite transformations. Since no solid matter fills the porous space, Equation (3.74) states that the strain work density related to the porous solid only, namely, dW_S, can be obtained by removing the quantity $-pd\varphi$ from the overall strain work density $\sigma_{ij}d\varepsilon_{ij}$.

3.4.5 Strain work and the effective stress 'principle'

The principle of effective stress is central to soil mechanics. It was intuitively introduced in 1923 by Karl von Terzaghi (1883–1963), the father of modern soil mechanics. Historically based on experimental observations, this principle states that the deformation of the porous solid is only driven by the excess of stress over the pore pressure. The excess of stress is the so-called effective stress, σ'_{ij}, defined by

$$\sigma'_{ij} = \sigma_{ij} + p\delta_{ij}. \quad (3.75)$$

As a result, whenever this 'principle' is relevant, the expression of the strain work dW_S relative to the porous solid given in Equation (3.74) must reduce to

$$dW_S = \sigma'_{ij}d\varepsilon_{ij}. \quad (3.76)$$

The question arises about the validity and the actual meaning of such a principle. According to Equation (3.54), as long as the solid part of the porous solid undergoes negligible volume changes, we have $\delta_{ij}d\varepsilon_{ij} \equiv d\epsilon = d\varphi$, resulting in Equation (3.76). The effective stress 'principle' Equation (3.76) is therefore equivalent to assuming that the solid part of the porous solid is formed of solid grains poorly compressible, and therefore exhibiting no significant volume change. This does in fact hold for soils.

The previous result is not restricted to infinitesimal transformations and can be extended to finite transformations. Indeed, from the previous section we know that the strain work $dW_S\,d\Omega_0$ related to the sole porous solid only is obtained by removing the exact strain work $-p\,d\varphi\,d\Omega_0$ related to the porous space from the overall strain work $dW\,d\Omega$, resulting in

$$dW_S\,d\Omega_0 = dW\,d\Omega + p\,d\varphi\,d\Omega_0. \tag{3.77}$$

To derive from Equation (3.77) a relevant expression for dW_S irrespective of any assumption of infinitesimal transformations, we need to relate the exact expression of $dW\,d\Omega$ given in Equation (3.46) to the initial volume $d\Omega_0$. This can be illustrated through a standard problem encountered in geomechanics, namely, the one-dimensional consolidation or sedimentation of a layer, where body and surface forces (the gravity forces and the applied consolidation pressure) act only in the $\underline{e}_3$ direction. Provided the layer is transversely homogeneous, the displacement ξ_3 in the $\underline{e}_3$ direction is the unique nonzero displacement and depends on the X_3 coordinate only. The exact expression of the overall strain work $dW\,d\Omega$ given in Equation (3.46) then reduces to

$$dW\,d\Omega = \sigma_{33}\left(\frac{\partial d\xi_3}{\partial x_3}\right)d\Omega = \sigma_{33}\left(\frac{\partial d\xi_3}{\partial X_3}\right)\frac{\partial X_3}{\partial x_3}d\Omega. \tag{3.78}$$

In order to relate the overall strain work $dW\,d\Omega$ to the initial volume $d\Omega_0$, we first note the relations

$$x_3 = X_3 + \xi_3; \quad F_{33} = \frac{\partial x_3}{\partial X_3}; \quad \text{or else } F_{ij} = \delta_{ij}, \tag{3.79}$$

resulting in

$$d\Omega = \det \underline{\underline{F}}\,d\Omega_0 = \frac{\partial x_3}{\partial X_3}d\Omega_0 = \left(1 + \frac{\partial \xi_3}{\partial X_3}\right)d\Omega_0. \tag{3.80}$$

If the solid part of the porous solid does not undergo any volume change, as was assumed to derive Equation (3.76), the overall volume change $d(d\Omega)$ of volume $d\Omega$ is only due to the variation of porosity, whereas the volume $(1 - n_0)\,d\Omega_0$ of the solid part remains unchanged. We therefore have

$$d\Omega = (1+\varphi)\,d\Omega_0; \quad (1-n)\,d\Omega = (1-n_0)\,d\Omega_0, \tag{3.81}$$

where n stands for the Eulerian porosity. Equations (3.80) and (3.81) combine to provide

$$\varphi = \frac{\partial \xi_3}{\partial X_3}; \quad \left(\frac{\partial d\xi_3}{\partial X_3}\right)\frac{\partial X_3}{\partial x_3}d\Omega = d\varphi\,d\Omega_0. \tag{3.82}$$

Substituting the second relation of Equation (3.82) in Equation (3.78), and the resulting expression for $dW\,d\Omega$ in Equation (3.77), we obtain

$$dW_S = (\sigma_{33} + p)\,d\varphi = \sigma'_{33}d\varphi. \tag{3.83}$$

In isothermal reversible evolutions, the strain work input is fully stored in the form of Helmholtz free energy. As a result, without assuming infinitesimal transformations, the Helmholtz free energy balance gives

$$da_S = \sigma'_{33} d\varphi, \tag{3.84}$$

where, like f in Equation (3.67), a_S stands for the current Helmholtz free energy density of the porous solid per unit of initial volume $d\Omega_0$. The energy balance stated in Equation (3.84) shows that a_S, and therefore σ'_{33}, are functions of the Lagrangian porosity change φ only, and eventually of the Eulerian porosity n only, because of the relation linking φ to n, which can be derived from Equation (3.81). This allows us to recover the state equation widely used in one-dimensional problems of geophysics, namely,

$$\sigma'_{33} = \sigma'_{33}(n). \tag{3.85}$$

The field Eulerian porosity profile $n(x_3)$, which is an assessment of the state of consolidation of the layer, is therefore directly related to the vertical effective stress.

The constitutive equations of ordinary solids are the relations that associate the strain components ε_{ij} to the stress components σ_{ij}. The effective stress 'principle' constitutes an attractive basis to extend the usual constitutive equations of ordinary solids to porous solids. Indeed, it is tempting to elaborate the constitutive equations of porous solids by consistently replacing the stress components σ_{ij} in the constitutive equations of ordinary solids by the effective stress components σ'_{ij} defined by Equation (3.75). However, the effective stress 'principle' has been shown to amount to assuming the incompressibility of the solid matrix. If this assumption is relevant for soils, whose solid grains are poorly compressible with respect to the compressibility of the soil considered as a whole, it is questionable for the majority of porous solids, such as wood, bones, bricks, concrete, living tissues and so on. In Chapter 4, the free energy balance stated in Equation (3.73) will be the starting point for the operational formulation of the state equations of poroelastic solids, and will allow us to replace the effective stress 'principle' in a more general context of poroelasticity.

Further Reading

Bear, J. and Bachmat, Y. (1990) *Introduction to Modeling of Transport in Porous Media*, Kluwer, Amsterdam.
Coussy, O. (1995) *Mechanics of Porous Continua*, John Wiley & Sons.
Coussy, O. (2004) *Poromechanics*, John Wiley & Sons.
de Boer, R. (2000) *Theory of Porous Media: Highlights in Historical Development and Current State*, Springer-Verlag, Berlin Heidelberg.
Lewis, R. W. and Schrefler, B. A. (1998) *The Finite Element Method in the Static and Dynamic Deformation and Consolidation of Porous Media*, 2nd edn, John Wiley & Sons.
Salençon, J. (2001) *Handbook of Continuum Mechanics: General Concepts, Thermoelasticity*, Springer-Verlag, Berlin Heidelberg.
Ulm, F.-J. and Coussy, O. (2002). *Mechanics and Durability of Solids, I: Mechanics of Solids*, Prentice Hall.

4

The Saturated Poroelastic Solid

The discovery of the linear relationship relating the force to the extension for small strains goes back to the seventeenth century with the early experiments of Robert Hooke (1635–1703). Among others the names of Thomas Young (1703–1829) and Siméon Denis Poisson (1781–1840) are associated with the modern formulation of the constitutive equations of elastic solids. Surprisingly, the natural extension of these constitutive equations to poroelastic solids was only achieved by the middle of the twentieth century by Maurice A. Biot (1905–1985), in particular, in his seminal series of papers devoted to wave propagation applied to seismic exploration. Nowadays the Biot coefficient of elastic porous solids is routinely measured when testing the mechanical behavior of rocks.

In this chapter we first address the constitutive equations of both linear and nonlinear poroelastic solids. The chapter then provides some insights into microporoelasticity, which gives a mean field assessment of the macroscopic poroelastic properties from the porosity and the properties of the solid and fluid components. Later we derive the constitutive equations of poroelastic solids saturated by a compressible fluid and particular attention is given to the presence of gases in the form of both bubbles and solute. The chapter ends by accounting for thermal effects and by considering delayed behaviors induced by viscoelasticity of the solid matrix.

4.1 The Poroelastic Solid

4.1.1 The linear poroelastic solid

In Equation (3.73) of Chapter 3 we derived the free energy balance for the porous solid considered separately from the fluid–mixture it partly constitutes. For isothermal evolutions this balance reads

$$\sigma d\epsilon + p d\varphi + s_{ij} de_{ij} - da_S = 0. \quad (4.1)$$

Instead of considering the solid free energy a_S, it will be more convenient to consider the opposite η_S of its partial Legendre transform with regard to φ, that is,

$$\eta_S = a_S - p\varphi. \quad (4.2)$$

Substitution of Equation (4.2) in Equation (4.1) gives

$$\sigma d\epsilon - \varphi dp + s_{ij}de_{ij} - d\eta_S = 0. \quad (4.3)$$

As a result, $d\eta_S$ is an exact differential such that η_S has ϵ, p and e_{ij} as its only arguments. From Equation (4.3) the state equations of poroelasticity can then be written in the form

$$\sigma = \frac{\partial \eta_S}{\partial \epsilon}; \quad \varphi = -\frac{\partial \eta_S}{\partial p}; \quad s_{ij} = \frac{\partial \eta_S}{\partial e_{ij}}. \quad (4.4)$$

Considering now only isotropic poroelastic solids, no material frame must be preferred to formulate their constitutive equations from Equation (4.4). As a result, in addition to the pore pressure p, the energy function η_S must only depend on the three principal strains, $\varepsilon_J \simeq \Delta_J$, and not on the principal directions of the deformation defined by Equation (3.12) in Section 3.1.2. The three principal strains ε_J can be expressed as functions of the three first invariants of the strain tensor $\underline{\underline{\varepsilon}}$. However, when considering only *linear* isotropic poroelastic solids, only the two first invariants of the strain tensor have to be considered in the expression of the energy η_S. The latter is then a quadratic function of the two first invariants and of the pore pressure p, with no quadratic term involving both the second invariant and the pore pressure. Because it is equivalent to considering the first invariant ϵ of the strain tensor together with the second invariant $e_{ij}e_{ji}$ of the strain deviator $\underline{\underline{e}}$, this quadratic function is conveniently expressed in the form

$$\eta_S = \frac{1}{2}K\epsilon^2 - b\epsilon p - \frac{1}{2}\frac{p^2}{N} + Ge_{ij}e_{ji}. \quad (4.5)$$

Substituting Equation (4.5) in Equation (4.4) we derive the state equations of isotropic linear poroelasticity in the form

$$\sigma = K\epsilon - bp; \quad (4.6a)$$
$$\varphi = b\epsilon + p/N; \quad (4.6b)$$
$$s_{ij} = 2Ge_{ij}. \quad (4.6c)$$

In the constitutive Equations (4.6) of isotropic poroelasticity the deformation and the porosity change are defined with respect to a reference state free of any stress ($\sigma = s_{ij} = 0$) and any pore pressure ($p = 0$). The constitutive equations relative to a zero deformation and porosity reference state, which is now associated with the initial stress components, σ^0 and s_{ij}^0, and the initial pore pressure p_0, are obtained by replacing σ, s_{ij}^0 and p_0 in Equation (4.6) by their variations, $\sigma - \sigma^0$, $s_{ij} - s_{ij}^0$ and $p - p_0$. When dealing with poroelastic materials the existence of nonzero initial stresses and pore pressure can thus easily be addressed by considering the stress and pore pressure variations instead of their net values.

In Equation (4.6) K, b, N and G are the poroelastic properties characterizing the mechanical behavior of the linear elastic porous solid irrespective of the origin of the pore pressure p. In Equation (4.6a) K is the bulk modulus which relates the volumetric dilation ϵ linearly to the mean stress σ providing the pore pressure p is zero. The coupling coefficient b is the Biot coefficient. According to Equation (4.6b) b represents the fraction of the volumetric dilation ϵ due the porosity variation φ when the pore pressure p is zero; in turn, N is the modulus relating the pore pressure p linearly to the porosity variation φ when the volumetric deformation ϵ is zero; G is the shear modulus relating the deviatoric strain component e_{ij} linearly to the corresponding shear s_{ij}. Besides, because of the assumed isotropy it is worthwhile noting that the stress and strain tensors have the same eigen directions.

If the effective stress 'principle' stated in Section 3.4.5 does apply to the porous solid being considered, the volumetric strain ϵ must be governed by only the excess $\sigma' = \sigma + p$ of the mean stress σ over the opposite $-p$ of the pore pressure. For a linear poroelastic solid state Equation (4.6a) must then reduce to

$$\sigma' = \sigma + p = K\epsilon. \qquad (4.7)$$

According to Equation (4.7) a variation $\Delta\sigma'$ of the mean effective stress σ' results in the same variation $\Delta\epsilon$ of the volumetric strain, whether the variation $\Delta\sigma'$ is achieved through a mean stress variation, that is, for $\Delta\sigma' = \Delta\sigma$, or through a pore pressure variation, that is, for $\Delta\sigma' = \Delta p$. This remark provides the experimental means to check the validity of the effective stress 'principle' in tests where the mean pressure $-\sigma$ and the pore pressure p are separately monitored. The mean pressure $-\sigma$ can be imposed on a porous material sample by means of a mechanical triaxial press, whereas the monitoring of the pore pressure is achieved by controlling the pressure of any saturating liquid. According to Equation (4.7) a stress loading path made up of n successive loading elementary steps of intensity $\Delta\sigma = -\Delta\omega$, and performed at zero pore pressure ($p = 0$), will induce a strain history made up of n successive elementary contractions of intensity $\Delta\epsilon = -\Delta\omega/K$. According to Equation (4.7) a pore pressure loading path, performed at the end of the previous stress loading path and made up of n loading elementary steps of intensity $\Delta p = \Delta\omega$, will induce the exact opposite strain history, made up of n successive elementary dilations of intensity $\Delta\epsilon = \Delta\omega/K$. In the plane $(-\epsilon, -\sigma')$ two superimposed lines of equal slope K should thus be observed. This superposition is not observed for the limestone tested using such a procedure in the experiments reported in Figure 4.1. Returning then to the original state Equation (4.6a) the ratio between the slopes of the lines relative to the two previous loading paths is finally identified to be the coefficient b and is for instance found to be 0.63 for the limestone of Figure 4.1.

This departure of b from 1 proves that the solid matrix forming the solid part of the limestone undergoes significant volume changes. Indeed, Section 3.4.5 has shown that the effective stress 'principle' only holds when the solid matrix is poorly compressible, resulting in Equation (3.54), that is, $\epsilon = \varphi$. An inspection of Equation (4.6b) shows that the relation $\epsilon = \varphi$ is met irrespective of the pore pressure value p provided that

$$b = 1; \quad N \to \infty. \qquad (4.8)$$

The value $b = 1$ is then consistent with Equation (4.7), whereas Equation (4.6b), under the second of conditions (4.8), degenerates to $\epsilon = \varphi$. As a result the conditions stated in Equation (4.8) hold whenever the solid matrix does not undergo significant volume changes. When this is not the case the question arises as to how the poroelastic properties K, b, N and G are related to the elastic properties of the solid matrix forming the solid part of the porous solid. This is covered in the next section.

4.1.2 Microporoelasticity

In the previous chapter the macroscopic volumetric strain ϵ and the macroscopic mean stress σ were shown to be the space averages of their microscopic counterparts. This was accounted for by the relations

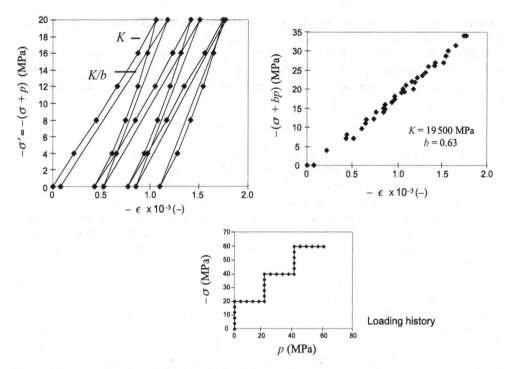

Figure 4.1 Assessing the validity of the effective stress 'principle' for a limestone. If the effective stress 'principle' were valid, two superposed lines of equal slope K would have been observed in the plane $(-\epsilon, -\sigma')$ for the loading path consisting of $n = 5$ elementary loading steps by increasing either the opposite $-\sigma$ of the mean stress or the pore pressure p. In contrast, when plotting $-(\sigma + bp)$ against $-\epsilon$, a unique line is obtained when the Biot coefficient b takes the value 0.63 (Data from Bouteca, M. and Sarda, J.-P. (1995) Experimental measurements of thermoporoelastic coefficients. In *Mechanics of Porous Media*, Charlez, P., ed., Balkema.)

$$\epsilon = (1 - \phi_0)\epsilon_S + \varphi, \tag{4.9}$$

and

$$\sigma = (1 - \phi_0)\sigma_S - \phi_0 p. \tag{4.10}$$

From now on let us assume that the matrix forming the solid part of the porous solid is homogeneous. Then let k_S be the elastic bulk modulus related to the solid matrix, the matrix homogeneity allowing us to write

$$\sigma_S = k_S \epsilon_S. \tag{4.11}$$

The three equations above combine to become

$$\sigma = k_S(\epsilon - \varphi) - \phi_0 p. \tag{4.12}$$

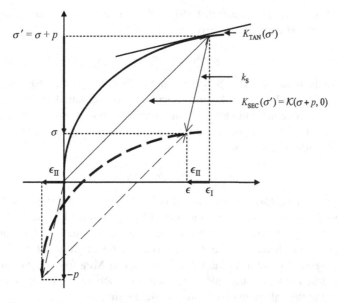

Figure 4.2 Loading steps I = $(\sigma + p, 0)$ and II = $(-p, p)$ used to derive the final volumetric strain ϵ for any value of the loading (σ, p); in the loading path represented by dashed arrows the two steps are reversed, but the secant modulus to be used in step I is not affected by step II because of the linear elasticity of the solid matrix

If we perform a test at zero pore pressure, substituting $p = 0$ in the state Equations (4.6a) and (4.6b) as well as in Equation (4.12), we derive the system

$$\mathcal{M} \cdot \begin{pmatrix} \sigma \\ \epsilon \\ \varphi \end{pmatrix} = 0; \quad \mathcal{M} = \begin{pmatrix} 1 & -K & 0 \\ 0 & b & -1 \\ 1 & -k_S & k_S \end{pmatrix}. \quad (4.13)$$

The variables $(\sigma, \epsilon, \varphi)$ are independent state variables, so that the knowledge of two of them cannot fix the value of the third. Put another way, despite Equation (4.13) the variable set $(\sigma, \epsilon, \varphi)$ must remain unspecified. This results in the condition $\det \mathcal{M} = 0$, providing the relation

$$b = 1 - K/k_S. \quad (4.14)$$

We can also perform a test where $\sigma = -p$, which can be achieved by immersing an unjacketed porous material sample within a liquid whose pressure is monitored. Substituting the condition $\sigma = -p$ in Equation (4.6a) and using Equation (4.14) we first find

$$\epsilon = \sigma/k_S = -p/k_S. \quad (4.15)$$

Substituting the condition $\sigma = -p$ in Equation (4.10) we also obtain

$$\sigma = -p = \sigma_S. \quad (4.16)$$

The volumetric strain average (Equation (4.9)) and the matrix constitutive Equation (4.11) combine with the two last equations to give

$$\varphi/\phi_0 = \epsilon_S = \epsilon. \quad (4.17)$$

Indeed, because of the relation $\sigma = -p$, the internal walls of the porous space are subjected to the same pressure p as the border of the porous sample. As accounted for by Equation (4.17) this results in a homogeneous volumetric strain within the porous solid, causing the matrix and the porous space to deform in the same way. Substitution of Equations (4.15) and (4.9) in Equation (4.6b) provides the relation

$$1/N = (b - \phi_0)/k_S. \quad (4.18)$$

Provided the porosity ϕ_0 and the bulk modulus k_S of the solid matrix are both known, the poroelastic properties b and N are then determined by Equations (4.14) and (4.18) from the knowledge of K alone. Conversely it is worthwhile noting that Equation (4.14) provides k_S as a function of b and K. For instance, from the values $b = 0.63$ and $K = 19\,500$ MPa found for the limestone of Figure 4.1, we infer the value $k_S = 52\,700$ MPa from Equation (4.14). This high value of k_S, close to that of pure calcite, leads us to conclude that no significant occluded porosity exists within the calcite matrix of this particular limestone.

The matrix bulk modulus k_S can also be measured directly by means of the test where the confining pressure is set equal to the pore pressure, resulting in the condition $\sigma = -p$. According to Equation (4.15), in this experiment the confining pressure $\sigma = -p$ plotted against ϵ is a line, whose slope is the matrix bulk modulus k_S. (This is illustrated for a sandstone specimen in Figure 4.3(a).) The measurement of k_S proves to be particularly useful when analyzing the cracking pattern which a porous solid can undergo beyond some stress

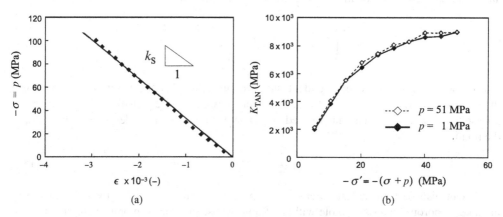

Figure 4.3 Experimental confirmation of nonlinear constitutive Equations (4.34)–(4.37) for a sandstone specimen: (a) confirmation of the linear elastic behavior of the matrix; (b) confirmation of the dependence of the tangent bulk modulus K_{TAN} on Terzaghi's effective stress $\sigma' = \sigma + p$. Because of the progressive closure of cracks, as the intensity of the applied compressive confining stress increases, the tangent bulk modulus K_{TAN} increases and the sandstone specimen progressively stiffens (Data from Bemer E. *et al* (2001) Poromechanics: from linear poroelasticity to non-linear poroelasticity and poroviscoelasticity. *Oil & Gaz Science and Technology, Rev. IFP*, **56**, 531–544).

threshold. The cracks can initiate either within the solid matrix or from the existing porous space. In the latter case the cracking will result in the increase of the connected porosity and, in turn, will affect only the bulk modulus K, while the matrix bulk modulus k_S will remain unaltered in contrast to the former case.

Provided the solid matrix is homogeneous, Equations (4.14) and (4.18) hold irrespective of the morphology of the pore space. In contrast, there is no general expression for K as a function of the matrix elastic properties and the porosity ϕ_0, because of the dependency of K upon the pore space morphology. Nonetheless, an assessment of K can be obtained when the porous solid can be likened to a solid matrix embedding spherical voids as porous space. From the standard theory of elastic solids, a spherical void with no internal pressure, embedded in a linearly elastic solid matrix having k_S and g_S as bulk and shear moduli respectively, and whose imposed volumetric strain ϵ_S is some distance from the void, undergoes a volumetric dilation φ/ϕ_0 given by

$$\varphi/\phi_0 = \left(1 + \frac{3k_S}{4g_S}\right)\epsilon_S. \tag{4.19}$$

Substitution of Equation (4.19) in Equation (4.9) gives

$$\varphi/\phi_0 = \frac{4g_S + 3k_S}{3\phi_0 k_S + 4g_S}\epsilon; \quad \epsilon_S = \frac{4g_S}{3\phi_0 k_S + 4g_S}\epsilon. \tag{4.20}$$

The factors affecting ϵ in Equation (4.20) are termed the volumetric strain localization factors. Because of the linear elasticity of the matrix, they linearly localize the overall volumetric strain ϵ both in the void (the first relation in Equation (4.20)), and in the solid matrix (the second relation in Equation (4.20)). Equation (4.10), where we take $p = 0$, together with Equations (4.11) and (4.20), combine to provide

$$\sigma = (1 - \phi_0)\frac{4k_S g_S}{3\phi_0 k_S + 4g_S}\epsilon, \tag{4.21}$$

leading to the assessment

$$K = (1 - \phi_0)\frac{4k_S g_S}{3\phi_0 k_S + 4g_S}. \tag{4.22}$$

An assessment of the shear modulus G involved in Equation (4.6c) can be similarly obtained by applying the deviatoric strain e_{ij}^S some distance from the void. The procedure gives[1]

$$G = (1 - \phi_0)\frac{(9k_S + 8g_S)g_S}{9k_S\left(1 + \frac{2}{3}\phi_0\right) + 8g_S\left(1 + \frac{3}{2}\phi_0\right)}. \tag{4.23}$$

Expressions given in Equations (4.22) and (4.23) for the bulk and shear moduli K and G are those related to a perfect isotropic poroelastic solid, whose porous space is made up of spherical voids. These expressions have both been derived by imposing the solid strain ε_{ij}^S some distance from the voids. This approach may mislead us into thinking that they are accurate only for dilute systems where $\phi_0 \ll 1$. Let us then note that the original Equations (4.22) and

---

[1] For the solution of the general problem of an ellipsoidal inclusion embedded in a solid matrix, see the celebrated paper of Eshelby, J. D. (1957) Determination of the Elastic Field of an Ellipsoidal Inclusion, and Related Problems, *Proceedings of the Royal Society of London. Series A, Mathematical and Physical Sciences*, **241**, 376–396.

(4.23) are not identical to the approximate ones that can be derived by using the approximation $\phi_0 \ll 1$ in Equations (4.22) and (4.23). Basically, the same expressions as the ones given by Equations (4.22) and (4.23) are found for the apparent bulk and shear moduli of an elastic hollow sphere whose external radius R_{ext} and internal radius R_{int} would meet the condition $R_{\text{int}}^3/R_{\text{ext}}^3 = \phi_0$. This means that the expressions given in Equations (4.22) and (4.23) for K and G relate to a mean field theory: the porous material is approached as an ideal isotropic porous solid formed from noninteracting hollow spheres, which may have different sizes but always meet the requirement $R_{\text{int}}^3/R_{\text{ext}}^3 = \phi_0$. Indeed, it can be shown that Equations (4.22) and (4.23) provide[2] upper bounds for the values of the bulk and shear moduli of any isotropic poroelastic solid with the same properties k_S, g_S and ϕ_0.

4.1.3 The nonlinear poroelastic solid

Even though the solid matrix forming the solid part of the porous solid is linearly elastic, and therefore obeys Equation (4.11), the overall mechanical behavior of porous solids is often nonlinear because of the softening or stiffening induced by the progressive opening or closure of pores and cracks induced by a tensile or compressive confining stress. The bulk modulus K is then no longer a constant property of the porous solid, as observed in Figure 4.3(b) reporting the experimental data relative to a sandstone specimen. In order to capture this nonlinearity let h_S be the enthalpy of the porous solid defined as the Legendre transform of η_S according to

$$h_S = \sigma\epsilon + s_{ij}e_{ij} - \eta_S. \qquad (4.24)$$

Substituting Equation (4.24) in Equation (4.3) we obtain

$$\epsilon d\sigma + \varphi dp + e_{ij}ds_{ij} - dh_S = 0, \qquad (4.25)$$

which gives the inversion of the state Equations (4.4) in the form

$$\epsilon = \frac{\partial h_S}{\partial \sigma}; \quad \varphi = \frac{\partial h_S}{\partial p}; \quad e_{ij} = \frac{\partial h_S}{\partial s_{ij}}. \qquad (4.26)$$

The comments made for η_S, when looking into the consequences of Equation (4.4) for isotropic porous solids, extend to h_S, providing the role of the strain tensor $\underline{\epsilon}$ is now played by the stress tensor $\underline{\sigma}$. Therefore, for isotropic poroelastic solids, in addition to the pore pressure p, the energy function h_S depends only on the first three invariants of the stress. In practice, it is usually sufficient to use only the first two invariants. It is then equivalent to considering one third of the first invariant σ of the stress tensor, together with the square root τ of the second invariant of the stress deviator $\underline{s}$, that is,

$$\tau = \sqrt{\frac{1}{2}s_{ij}s_{ji}}. \qquad (4.27)$$

Because of the nonlinearity induced by the pores, unlike a linear poroelastic solid, the expression for h_S is no longer a quadratic expression of its arguments σ, p and τ. However, most

---

[2] They are known as Hashin–Shrikman bounds. See Hashin, Z. (1983) Analysis of composite materials – a survey, *Journal of Applied Mechanics*, **50**, 481–504.

of the experimental results can be accounted for splitting h_S into a volumetric term and a deviatoric term according to

$$h_S = h_S^{\text{vol}}(\sigma, p) + h_S^{\text{dev}}(\tau). \tag{4.28}$$

Use of Equation (4.28) in Equation (4.26) provides the nonlinear poroelastic constitutive equations in the form

$$\epsilon = \frac{1}{\mathcal{K}(\sigma, p)} \sigma + \frac{\beta(\sigma, p)}{\mathcal{K}(\sigma, p)} p; \tag{4.29a}$$

$$\varphi = \frac{1}{\mathcal{P}(\sigma, p)} p + \frac{\beta(\sigma, p)}{\mathcal{K}(\sigma, p)} \sigma; \tag{4.29b}$$

$$e_{ij} = \frac{1}{2\mathcal{G}(\tau)} s_{ij}, \tag{4.29c}$$

where $\mathcal{K}(\sigma, p)$, $\mathcal{G}(\tau)$ and $\mathcal{P}(\sigma, p)$ are secant moduli, while $\beta(\sigma, p)$ is the secant Biot coefficient.

Consider now a porous material whose matrix is both homogeneous and linearly elastic, so that Equation (4.11) holds. The current volumetric loading (σ, p) can be achieved through two successive loading phases I and II, according to the loading decomposition $(\sigma, p) = (\sigma + p, 0) + (-p, p)$:

1. In phase I the loading $(\sigma + p, 0)$ is applied. According to Equations (4.29a) and (4.29b) the loading phase I generates the volumetric strain ϵ_I and the porosity variation φ_I given by

$$\epsilon_I = \frac{1}{\mathcal{K}(\sigma + p, 0)} (\sigma + p); \quad \varphi_I = \frac{\beta(\sigma + p, 0)}{\mathcal{K}(\sigma + p, 0)} (\sigma + p). \tag{4.30}$$

2. In phase II the loading $(-p, p)$ is added to the previous loading in order to finally achieve the current loading (σ, p). The loading phase II corresponds to the loading process we examined in Section 4.1.2 to derive Equation (4.18). According to Equations (4.15)–(4.17), the macroscopic volumetric strain ϵ_{II} and the porosity variation φ_{II} caused by the second loading step are given by

$$\epsilon_{II} = -\frac{p}{k_S}; \quad \varphi_{II} = -\phi_0 \frac{p}{k_S}. \tag{4.31}$$

Since the matrix is linearly elastic the bulk modulus k_S does not depend on the stress state. Hence, the nonlinear phase I does not affect the volumetric strain and the porosity variation Equation (4.31) caused by the loading phase II. Adding the volumetric strains and porosity variations given by Equations (4.30) and (4.31), we obtain

$$\epsilon = \frac{1}{\mathcal{K}(\sigma + p, 0)} \sigma + \left(\frac{1}{\mathcal{K}(\sigma + p, 0)} - \frac{1}{k_S} \right) p; \tag{4.32a}$$

$$\varphi = \left(\frac{\beta(\sigma + p, 0)}{\mathcal{K}(\sigma + p, 0)} - \frac{\phi_0}{k_S} \right) p + \frac{\beta(\sigma + p, 0)}{\mathcal{K}(\sigma + p, 0)} \sigma. \tag{4.32b}$$

Comparing Equation (4.29) with Equation (4.32) we derive

$$\mathcal{K}(\sigma, p) = \mathcal{K}(\sigma + p, 0); \quad \frac{1}{\mathcal{P}(\sigma, p)} = \frac{\beta(\sigma + p, 0)}{\mathcal{K}(\sigma + p, 0)} - \frac{\phi_0}{k_S}; \quad (4.33a)$$

$$\beta(\sigma, p) = \beta(\sigma + p, 0) = 1 - \frac{\mathcal{K}(\sigma + p, 0)}{k_S}. \quad (4.33b)$$

It is instructive to note that these identifications could not have been achieved in the case of a nonlinear elastic matrix where the matrix modulus k_S would have depended on the mean stress σ_S in Equation (4.11). The dependence of k_S on p, in the right-hand side of Equation (4.33b), would then not have allowed the function β to depend only on $\sigma + p$, although this is required by the left-hand side of Equation (4.33b). Indeed, in the case of a nonlinear elastic matrix the loading phase I would definitely affect the volumetric strain and porosity variations resulting from the loading phase II. In contrast, because of the poroelastic behavior of the porous material, either linear or not, its final deformation state does not depend upon the specific loading path achieving the final loading (σ, p). As a result, the loading steps I and II can be reversed, with no change in the final volumetric strain ϵ and change φ of porosity. Since there is no change in the volumetric deformation $\epsilon_{II} = -p/k_S$ induced by the loading step II, $(-p, p)$, when it is applied first, there is also no change in the deformation $\epsilon_I = \epsilon - \epsilon_{II}$ induced by the loading step I, $(\sigma + p, 0)$, when it is subsequently applied. From this remark it stems that the expression given by Equation (4.30) for ϵ_I remains unchanged and, as illustrated in Figure 4.2, the secant modulus $\mathcal{K}$ is not affected by any loading step of type II.

Reorganization of Equation (4.32) gives

$$\sigma = K_{\text{SEC}}(\sigma') \epsilon - b_{\text{SEC}}(\sigma') p; \quad (4.34a)$$

$$\varphi = b_{\text{SEC}}(\sigma') \epsilon + \frac{p}{N_{\text{SEC}}(\sigma')}, \quad (4.34b)$$

where $\sigma' = \sigma + p$, $K_{\text{SEC}}(\sigma') = \mathcal{K}(\sigma', 0)$ and $b_{\text{SEC}}(\sigma') = \beta(\sigma', 0)$ are the secant bulk modulus and the secant Biot coefficient satisfying the relation

$$b_{\text{SEC}}(\sigma') = 1 - \frac{K_{\text{SEC}}(\sigma')}{k_S}; \quad \frac{1}{N_{\text{SEC}}(\sigma')} = \frac{b_{\text{SEC}}(\sigma') - \phi_0}{k_S}. \quad (4.35)$$

According to Equation (4.34), as illustrated in Figure 4.2 the secant bulk modulus $K_{\text{SEC}}(\sigma')$ can be measured in a test performed on the dry porous material where $p = 0$. In the plane (σ, p), the measurement of $K_{\text{SEC}}(\sigma')$ and the knowledge of the solid matrix bulk modulus k_S provide the means to determine the volumetric strain ϵ for any value of the loading (σ, p) (see Figure 4.2). Alternatively, using Equations (4.34) and (4.35) we can write

$$d\sigma = K_{\text{TAN}}(\sigma') d\epsilon - b_{\text{TAN}}(\sigma') dp; \quad (4.36a)$$

$$d\varphi = b_{\text{TAN}}(\sigma') d\epsilon + \frac{dp}{N_{\text{TAN}}(\sigma')}, \quad (4.36b)$$

where $K_{TAN}(\sigma')$ (see Figure 4.2), $b_{TAN}(\sigma')$ and $N_{TAN}(\sigma')$ are tangent poroelastic properties

$$K_{TAN}(\sigma') = \frac{K_{SEC}(\sigma')}{1 - (\sigma'/K_{SEC})(dK_{SEC}/d\sigma')}$$
$$b_{TAN}(\sigma') = 1 - \frac{K_{TAN}(\sigma')}{k_S} \quad (4.37)$$
$$\frac{1}{N_{TAN}(\sigma')} = \frac{b_{TAN}(\sigma') - \phi_0}{k_S}.$$

Based solely on the assumption of a linearly elastic homogeneous matrix, Equations (4.34)–(4.37) constitute a convenient nonlinear extension of linear isotropic poroelasticity. In particular, Equations (4.37) extend Equations (4.14) and (4.18). It is noteworthy that the secant and the tangent poroelastic properties depend on Terzaghi's effective stress, namely, $\sigma' = \sigma + p$, in spite of the matrix elastic compressibility resulting in a value of Biot's coefficient departing from unity.

Such a nonlinear extension of poroelasticity can adequately account for the nonlinear behavior of porous materials such as rocks whose porous network is often formed from connected pores and cracks. The source of nonlinearities is then due to the progressive opening of the cracks under the pressurization of their faces by the saturating fluid. As illustrated in Figure 4.3 for a sandstone specimen, the validity of the nonlinear constitutive Equations (4.34)–(4.37) can successfully be tested experimentally. An experiment where the loading varies according to phase II, that is such that $(\sigma, p) = (-p, p)$, is performed first, in order to check the validity of Equation (4.15), that is, the linear elastic behavior of the sandstone matrix. This validity is ensured since the solid matrix modulus k_S is shown to be insensitive to p (see Figure 4.3(a)). The tangent modulus $K_{TAN}(\sigma')$ is then plotted against Terzaghi's effective stress, $\sigma' = \sigma + p$, from experiments where the overall volumetric strain ϵ is measured as a function of σ, whereas the pore pressure p is held constant, and takes different values. As indicated in Figure 4.3(b), the two functions $K_{TAN}(\sigma')$ obtained for two distinct values 1 MPa and 51 MPa of the pore pressure p merge in the limit of experimental accuracy, confirming the validity of the nonlinear constitutive Equations (4.34)–(4.37).

4.2 Filling the Porous Solid

The constitutive equations we have derived to this point are related to the poroelastic solid considered alone, irrespective of the source of the pressure acting on its solid internal walls. We are now going to address the formulation of the constitutive equations of the fluid–solid mixture examined in Section 3.4.3, that is the constitutive equations of the open system formed by the porous solid, whose porous space is filled by a fluid mixture which can be exchanged with the surroundings. The saturating fluid mixture can be either a poorly compressible liquid solution or a gas mixture. However, in many situations the gas can also exist in the form of a solute within the liquid solution and, when appropriate pressure conditions are met, the solute gas can form bubbles, significantly affecting the compressibility of the liquid solution. In the following sections we progressively derive the constitutive equations related to these various fluid–solid mixtures.

4.2.1 Filling by a compressible fluid

Let us first address the case where the fluid mixture is a compressible fluid, whose mass density ρ_F is therefore governed by the constitutive equation

$$d\rho_F/\rho_F = dp/K_F, \qquad (4.38)$$

where K_F is the fluid bulk modulus. Noting that $\rho_F \phi$ is the mass of fluid currently contained within the porous space, per unit of initial volume $d\Omega_0$, and recalling that $(\phi - \phi_0)/\phi_0 = \varphi/\phi_0 \ll 1$, Equations (4.6b) and (4.38) combine to provide

$$d(\rho_F \phi)/\rho_F = b\,d\epsilon + dp/M, \qquad (4.39)$$

where M is the modulus defined by

$$1/M = \phi_0/K_F + 1/N. \qquad (4.40)$$

The constitutive equations of the open system are then made up of Equations (4.6a), (4.38) and (4.39). Basically, Equation (4.6a) remains unchanged providing p is identified as the fluid mixture pressure.

When the fluid mixture is a liquid, identified by subscript F = L, in the context of infinitesimal transformations for the liquid we write

$$|(\rho_L - \rho_{L0})/\rho_{L0}| \ll 1, \qquad (4.41)$$

while $1/\rho_F$ can be replaced by $1/\rho_{L0}$ in Equation (4.39). In addition, the bulk compressibility $K_F = K_L$ involved in

$$d\rho_L/\rho_{L0} \simeq dp/K_L \qquad (4.42)$$

can be considered as being insensitive to the pressure.

When the fluid mixture is an ideal gas, identified by subscript F = G, the ideal gas Equation (2.23) gives

$$p = \rho_G RT/M_G, \qquad (4.43)$$

where M_G and ρ_G are respectively the molar mass and the current mass density of the gas. As a result, we obtain

$$d\rho_G/\rho_G = dp/p; \quad K_G = p. \qquad (4.44)$$

4.2.2 Filling by a mixture containing gas bubbles – bubble pressure

Let us now consider the case where the porous space is saturated by a fluid mixture (subscript mix) consisting of both a solution (subscript L) and gas bubbles (subscript G). Neglecting surface energy effects associated with the bubble–solution interface,[3] the gas pressure and

---

[3] According to the Laplace Equation (6.52) examined in Chapter 6 devoted to capillarity, the pressure difference between the liquid and the gas enclosed in a bubble of radius r is $2\gamma_{GL}/r$, where γ_{GL} is the liquid–gas interface energy. When the liquid is water, and the gas is air, we have $\gamma_{GL} = 73$ mJ m^{-2}. The order of magnitude of the pressure difference is then that of the atmospheric pressure for $r = 1$ μm. As a result, surface energy effects become negligible for air bubbles having a radius larger than 1 μm, and the pressure p can be considered as the same within the gas bubble and the surrounding liquid solution. The formation of gas bubbles is analyzed in Sections 8.4 and 8.5.1.

that of the solution are assumed to be equal in the following text. In the initial state, referred to by a subscript 0, this common pressure value is noted by p_0, while it is p in the current state. Also let v_0 and v be respectively the initial and current volume fractions of gas related to the mixture consisting of both the solution and the gas bubbles. The expressions of the initial and the current mass density of the mixture, ρ_{mix0} and ρ_{mix}, are given by

$$\rho_{mix0} = \rho_{L0}(1-v_0) + \rho_{G0}v_0; \quad \rho_{mix} = \rho_L(1-v) + \rho_G v. \tag{4.45}$$

After deformation of the porous solid, under the effect of the fluid pressure variations an initial volume V_0 of the mixture transforms into V. The mass conservation successively applied to the mixture, the gas and the solution, gives

$$\rho_{mix} V = \rho_{mix0} V_0; \quad \rho_G v V = \rho_{G0} v_0 V_0; \quad \rho_L(1-v) V = \rho_{L0}(1-v_0) V_0. \tag{4.46}$$

Eliminating the ratio V/V_0 and v, from the above equations we obtain

$$\frac{\rho_{mix0}}{\rho_{mix}} = v_0 \frac{\rho_{G0}}{\rho_G} + (1-v_0) \frac{\rho_{L0}}{\rho_L}. \tag{4.47}$$

Differentiation of this relation and use of the liquid and gas constitutive Equations (4.42) and (4.44), together with the approximation given in Equation (4.41), give

$$\frac{d\rho_{mix}}{\rho_{mix}} \simeq \frac{\rho_{mix}}{\rho_{mix0}} \left(\frac{v_0 p_0}{p^2} + \frac{1-v_0}{K_L} \right) dp. \tag{4.48}$$

As expected, Equation (4.48) reduces to Equation (4.42) when $v_0 = 0$ and $\rho_{mix} = \rho_L \simeq \rho_{mix0}$, and to Equation (4.44) when $v_0 = 1$ and $\rho_{mix}/\rho_{mix0} = \rho_G/\rho_{G0} = p/p_0$. If, in addition, we assume that both $v_0 \ll 1$ and $v \ll 1$, we have $\rho_{mix} \simeq \rho_{mix0}$, so that the previous relations can be approximated in the form

$$\frac{d\rho_{mix}}{\rho_{mix}} \simeq \left(\frac{v_0 p_0}{p^2} + \frac{1-v_0}{K_L} \right) dp. \tag{4.49}$$

The above expression for ρ_{mix} does not account for the presence of the substance constituting the gas bubbles in the form of a solute within the solution. In order to account for this presence, let us now introduce the molar fraction x related to the solute in the solution and defined by

$$x = \frac{n_G}{n_S + n_G} \Longrightarrow n_G = n_S \frac{x}{1-x}, \tag{4.50}$$

where n_G and n_S are the number of moles per unit volume of the solution of respectively the solute and the solvent.[4] According to Henry's law, as the fluid pressure varies the molar fraction x varies. Instead of Equation (4.46) we now write the mass conservation of the various components in the form

$$\rho_{mix} V = \rho_{mix0} V_0; \quad \rho_G V = \rho_{G0} V_0; \tag{4.51}$$
$$\rho_G^{bubble} v V + \rho_L(1-v) V = \rho_{G0}^{bubble} v_0 V_0 + \rho_{L0}(1-v_0) V_0,$$

---

[4] In the following the pressure p is assumed to be great enough to neglect the possible presence of the solvent in vapor form in the bubbles.

where, in contrast to Equation (4.46), ρ_G stands for the overall mass density of the gas, including both the solute form and the gas within the bubbles, whose mass density is ρ_G^{bubble}. Then let $\mathcal{M}_G^{\text{solute}}$ be the mass of gas in the solute in any volume V of the mixture. Use of Equation (4.50) gives

$$\mathcal{M}_G^{\text{solute}} = n_S (1 - \upsilon) V M_G \frac{x}{1 - x}. \qquad (4.52)$$

In Equation (4.52) the factor $n_S (1 - \upsilon) V$ represents the number of moles of the solvent. Therefore it remains constant and equal to $\rho_{S0} (1 - \upsilon_0) V_0 / M_S$, where M_S is the solvent molar mass. In the dilute approximation we have $x \ll 1$ and $\rho_{S0} \simeq \rho_{L0}$, so that we rewrite $\mathcal{M}_G^{\text{solute}}$ in the form

$$\mathcal{M}_G^{\text{solute}} \simeq \frac{\rho_{L0} (1 - \upsilon_0) V_0 M_G}{M_S} x. \qquad (4.53)$$

The overall mass of gas $\rho_G V$ can therefore be expressed in the form

$$\rho_G V = \rho_G^{\text{bubble}} \upsilon V + \frac{\rho_{L0} (1 - \upsilon_0) V_0 M_G}{M_S} x. \qquad (4.54)$$

Equation (4.54) provides expressions for $\rho_G V$ and for $\rho_{G0} V_0$ which can now be substituted in the second relation of Equation (4.52) expressing the conservation of the overall mass of gas. We obtain the relation

$$\upsilon = \left[\upsilon_0 \frac{\rho_{G0}^{\text{bubble}}}{\rho_G^{\text{bubble}}} + \frac{\rho_{L0} (1 - \upsilon_0)}{\rho_G^{\text{bubble}}} \frac{M_G}{M_S} (x_0 - x) \right] \frac{V_0}{V}. \qquad (4.55)$$

From the mass conservation of the solution expressed through the last relation of Equation (4.52), making use of the approximation $\rho_G^{\text{bubble}} \ll \rho_L$ we also have

$$1 - \upsilon = (1 - \upsilon_0) \frac{\rho_{L0} V_0}{\rho_L V}. \qquad (4.56)$$

Since $V/V_0 = \rho_{\text{mix}0}/\rho_{\text{mix}}$, eliminating υ between the two previous equations we obtain

$$\frac{\rho_{\text{mix}0}}{\rho_{\text{mix}}} = \left[\upsilon_0 \frac{\rho_{G0}^{\text{bubble}}}{\rho_G^{\text{bubble}}} + \frac{\rho_{L0} (1 - \upsilon_0)}{\rho_G^{\text{bubble}}} \frac{M_G}{M_S} (x_0 - x) \right] + (1 - \upsilon_0) \frac{\rho_{L0}}{\rho_L}. \qquad (4.57)$$

Differentiating this relation and using the liquid and gas constitutive Equations (4.42) and (4.44), together with the approximation of Equation (4.41), while assuming $\upsilon_0 \ll 1$ as well as $\upsilon \ll 1$ so that $\rho_{\text{mix}} \simeq \rho_{\text{mix}0}$, we finally derive

$$\frac{d\rho_{\text{mix}}}{\rho_{\text{mix}}} = \left(\upsilon_0 \frac{p_0}{p^2} + \frac{1 - \upsilon_0}{K_L} \right) dp + (1 - \upsilon_0) \frac{RT \rho_{L0}}{M_S} d\left(\frac{x - x_0}{p} \right). \qquad (4.58)$$

With regard to Equation (4.49) the term involving $x - x_0$ accounts for the change of the gas mass contained in the bubbles owing to further possible gas dissolution or exsolution.

The previous relation holds irrespective of any consideration about the thermodynamic equilibrium between the gas contained within the bubbles and the solute within the solution. Gas bubbles do exist in the reference state, that is, for $\upsilon_0 > 0$, if a thermodynamic equilibrium can be achieved between the solute gas and the gas present within the bubbles. This equilibrium is governed by Henry's law stated in Equation (2.59). Since here p_J identifies with the gas pressure p within the bubbles, this law leads to the relation $p = K_H x$. Let p_{bubble} be the bubble pressure, that is, the pressure above which no gas bubbles can exist. This bubble pressure is given by

$$p_{\text{bubble}} = K_H x_{\max}; \quad x_{\max} := \frac{\rho_{G0}}{M_G} \bigg/ \frac{\rho_{L0}}{M_S}. \tag{4.59}$$

In Equation (4.59) the molar fraction $x_{\max}$ is the molar fraction related to a reference state where no gas bubbles exist. It is not an intrinsic gas property. It is simply the total number of moles of gas present in the pore space divided by the number of moles of solvent. In a consistent way, the expression in Equation (4.59) for $x_{\max}$ is obtained by taking $v_0 = v = 0$, $V = V_0$, $\rho_G = \rho_{G0}$ and $x = x_{\max}$ in Equation (4.54). A reference state with no gas bubbles present ($x_0 = x_{\max}$) will be achieved if the initial pressure p_0 is larger than the bubble pressure p_{bubble} associated by Henry's law in Equation (4.59) with the molar fraction $x_{\max}$. We therefore write

$$p_0 \geq p_{\text{bubble}}, \text{ then } v_0 = 0, \ x_0 = x_{\max}; \quad p_0 < p_{\text{bubble}}, \text{ then } v_0 \neq 0, \ x_0 < x_{\max}, \ p_0 = K_H x_0. \tag{4.60}$$

We will see a direct application of these results in the next section, when examining the detrimental effects of successive unloadings of a gassy sediment sample.

4.2.3 Undrained poroelasticity

Undrained poroelasticity consists of looking into the evolutions of the closed system made of both the porous solid and the saturating fluid, when the latter is not allowed to escape from the former. As a result there is no variation of the fluid mass content. So taking $d(\rho_F \phi) = 0$ in Equation (4.39) and combining the resulting equation with Equation (4.6a), the constitutive equations of undrained poroelasticity are governed by

$$d\sigma = K_u d\epsilon; \quad K_u = K + b^2 M, \tag{4.61}$$

where K_u is the so-called undrained bulk modulus. Because M is positive, the undrained property K_u is larger than the bulk modulus K related to drained evolutions where $p = 0$. We have $K_u > K$ because the saturating fluid, which is not allowed to leave the porous space in undrained evolutions, makes the closed system stiffer than the porous solid.

As an illustration of undrained poroelasticity, let us now consider the case where, as explored in Section 4.2.2, the saturating fluid is a mixture (subscript F = mix) containing gas bubbles. In the context of infinitesimal transformations, where the porosity ϕ varies only slightly, the condition $d(\rho_{\text{mix}}\phi) = 0$ governing undrained evolutions can be approximated in the form

$$\phi_0 \frac{d\rho_{\text{mix}}}{\rho_{\text{mix}}} + d\phi = 0. \tag{4.62}$$

Combining Equations (4.6a), (4.6b) and (4.18), we also obtain

$$d\phi = \frac{b}{K} d\sigma + \left(\frac{b - \phi_0}{k_S} + \frac{b^2}{K} \right) dp. \tag{4.63}$$

Substituting Equations (4.58) and (4.63) in Equation (4.62) and integrating, we finally derive

$$(p - p_0)\left(1 + v_0\frac{\kappa}{p}\right) + \lambda(1 - v_0)\frac{x - x_0}{p} = B_{v_0}(P - P_0), \qquad (4.64)$$

where $P = -\sigma$ is the confining pressure, and where we have introduced the notation

$$\kappa = \frac{\phi_0 K M_{v_0}}{K + b^2 M_{v_0}}; \quad \lambda = \kappa \frac{RT\rho_{L0}}{\phi_0 M_S}; \quad B_{v_0} = \frac{bM_{v_0}}{K + b^2 M_{v_0}}; \quad \frac{1}{M_{v_0}} = \frac{b - \phi_0}{k_S} + \phi_0 \frac{1 - v_0}{K_L}. \qquad (4.65)$$

Equation (4.64) holds at any stage of an undrained process.

Consider now an instantaneous drop of the applied confining pressure $-\sigma$, from the initial value P_0 to the current value P. Simultaneously the pore pressure drops from p_0 to p_{0+}, with no change in the molar concentration of the solute. Indeed, a change in the molar concentration requires a diffusion process, and any diffusion process cannot take place instantaneously. As a result, the value of the pore pressure p_{0+}, which prevails immediately after the drop of the confining pressure, is obtained by taking $x = x_0$ in Equation (4.64), and we obtain[5]

$$(p_{0+} - p_0)\left(1 + \frac{v_0 \kappa}{p_{0+}}\right) = B_{v_0}(P - P_0). \qquad (4.66)$$

Since $P < P_0$, Equation (4.66) shows that $p_{0+} < p_0$. If p_0 is lower than the bubble pressure p_{bubble} defined at the end of Section 4.2.2, the instantaneous pressure drop from p_0 to p_{0+} will result in a thermodynamic imbalance between the solute, at molar concentration x_0, and the gas forming the bubbles at the new pressure $p_{0+} < K_H x_0$. After the instantaneous unloading, in order to restore the thermodynamic equilibrium the molar fraction starts decreasing. This is achieved by a diffusion process of the solute towards the gas bubbles. In turn, this diffusion process results in a simultaneous increase of the pore pressure in the bubbles, which slows down the diffusion process, and weakens the demand for the molar fraction to decrease. The thermodynamic equilibrium is restored when the asymptotic pore pressure $p_\infty > p_{0+}$ is linked to the molar fraction by Henry's law. Substitution of $x = x_\infty = p_\infty/K_H$ in Equation (4.64) requires the asymptotic pressure p_∞ to satisfy

$$(p_\infty - p_0)\left(1 + \frac{v_0 \kappa}{p_\infty}\right) + \frac{(1 - v_0)\lambda(p_\infty - K_H x_0)}{K_H p_\infty} = B_{v_0}(P - P_0). \qquad (4.67)$$

The experiments reported in Figure 4.4 reproduce the successive unloadings that a gassy sediment sample undergoes during its extraction from the deep seabed. In most situations the presence of gas, predominantly methane or carbon dioxide, is known to be detrimental during this extraction. As shown in Figure 4.4 successive drops of the current confining pressure $P^{(n)}$, referred to by the subscript $n = 1, 2, \ldots$, induce successive drops in the pore pressure $p_0^{(n)}$ prevailing before the nth drop down to the instantaneous value $p_{0+}^{(n)}$, followed by a delayed increase in the pore pressure up to the asymptotic value $p_\infty^{(n)} = p_0^{(n+1)}$. However, the delayed increase in the pore pressure is observed only after the second drop in the applied pressure. It can therefore be inferred that both the initial pore pressure $p_0^{(1)}$ and the pore pressure $p_{0+}^{(1)}$ after

---

[5] Irrespective of the presence of gas, property $B = bM/K_u$ is the so-called Skempton coefficient of soil mechanics, which relates the pore pressure variations $p - p_0$ to those of the confining pressure $P - P_0$, in any undrained evolution.

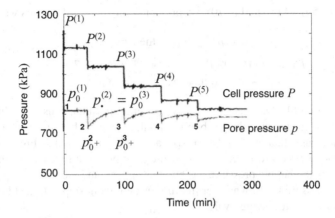

Figure 4.4 Experimental results reproducing the successive unloadings of a gassy sediment sample. A delayed increase of the pore pressure is only observed after the second drop of the applied pressure P. It can therefore be inferred that both the initial pore pressure and the pore pressure immediately after the first drop occurring at $t = 0^+$ were greater than the bubble pressure p_{bubble}. The latter corresponds to the plateau asymptotically observed whatever the further unloadings (Data from Amaratunga, A. S. (2006) *On the undrained unloading behaviour of gassy soils*, MSc Thesis, University of Calgary, Calgary, Alberta, Canada.)

the first drop were larger than the bubble pressure p_{bubble}, resulting in $v_0 = 0$ and $x_0 = x_{\max}$ according to Equation (4.60).

The detrimental presence of gas can then be understood as follows. The solid grains forming the solid part of gassy sediments are poorly compressible compared with the overall compressibility of the sediment softened by a large porosity. This results in values of k_S and K_L/ϕ_0 much larger than those of K. From Equations (4.14) and (4.65), and since $v_0 = 0$, we then obtain

$$\kappa \simeq \phi_0 K; \quad \lambda/K_H \simeq KRT\rho_{L0}/M_S K_H; \quad B_{v_0} \simeq 1. \tag{4.68}$$

The order of the bulk compressibility is reported to be $K = 10^5$ kPa for the sediment tested in Figure 4.4. The value of the Henry's constant related to CO_2 is $K_H = 1.65 \times 10^2$ MPa, while for CH_4 we have $K_H = 0.415 \times 10^2$ MPa. Recalling that $R = 8.314$ J K^{-1} mol^{-1}, and retaining for K_H the largest value related to CO_2, and thereby providing the lowest value for λ/K_H, if we adopt $T = 293$ K we derive

$$\lambda/K_H = 82\,020 \text{ kPa} = 809.47 \text{ atm}. \tag{4.69}$$

In practice, we therefore have both $\lambda/K_H \gg p_\infty$ and $\lambda/K_H \gg P - P_0$, so that Equation (4.67) reduces to

$$(p_\infty - p_0)\left(1 + \frac{v_0 \kappa}{p_\infty}\right) + \frac{(1 - v_0)\lambda(p_\infty - K_H x_0)}{K_H p_\infty} \simeq 0. \tag{4.70}$$

From Equations (4.59), (4.60) and (4.70), keeping only the main order terms we obtain

$$p_0 \text{ and } p_0^+ \geq p_{\text{bubble}} \text{ then } p_\infty = p_0^+;$$
$$p_0 \geq p_{\text{bubble}} \text{ and } p_0^+ < p_{\text{bubble}} \text{ then } p_\infty = p_{\text{bubble}}; \quad (4.71)$$
$$p_0 < p_{\text{bubble}} \text{ then } p_\infty \simeq p_0 = K_H x_0.$$

According to Equation (4.71), when $p_0 < p_{\text{bubble}}$, after an instantaneous unloading the diffusion of the dissolved gas towards the gas bubbles finally reestablishes the initial pore pressure p_0. In the opposite case, that is for $p_0 > p_{\text{bubble}}$, as actually observed in Figure 4.4 after the second drop in the applied pressure, as soon as the pore pressure (because of the unloading process) falls below the bubble pressure, gas bubbles appear, and the diffusion process establishes the bubble pressure itself, whatever the further unloading steps. It must be pointed out that these results are due to the high values of λ / K_H.

That $p_\infty = p_0$ or p_{bubble} explains the detrimental character of gas bubbles as follows. Since the solid matrix making up the solid part of gassy sediments is poorly compressible, resulting in $b \simeq 1$, the associated effective stress defined by Equation (4.7) reads $\sigma' = -P + p_\infty$. A sediment, which is mainly formed of particles in contact, cannot sustain a tensile effective stress. As a result, its collapse will occur during its extraction from the deep seabed when $P = p_\infty$ and, therefore, when $P \simeq p_0$ or p_{bubble}. Accordingly a sediment will fail when the intensity of the external load P applied after extraction falls below the initial pore pressure p_0 or the bubble presure p_{bubble}, according to the case under consideration.

4.3 The Thermoporoelastic Solid

4.3.1 The linear thermoporoelastic solid

Up to now we have only considered isothermal evolutions, that is, the temperature is held constant. For nonisothermal evolutions, considering the thermal contribution in the free energy balance Equation (3.73), in addition to the state Equations (4.4) we obtain

$$s_S = -\frac{\partial \eta_S}{\partial T}, \quad (4.72)$$

where s_S is the entropy of the porous solid defined by Equation (3.71). Restricting ourselves to isotropic linear thermoporoelastic solids, in addition to Equation (4.6c) which remains unchanged, from Equations (4.4) and (4.72) we now derive

$$\sigma = K\epsilon - bp - 3\alpha K (T - T_0); \quad (4.73a)$$
$$\varphi = b\epsilon + p/N - 3\alpha_\varphi (T - T_0); \quad (4.73b)$$
$$s_S = s_{S0} + 3\alpha K\epsilon - 3\alpha_\varphi p + C(T - T_0)/T_0, \quad (4.73c)$$

where T_0 is the initial reference temperature; 3α and $3\alpha_\varphi$ are the volumetric thermal dilation coefficients of the porous solid and the porous space, respectively, while C is a heat capacity. According to Equation (4.73) their measurement requires experiments where the strain is held constant. Experiments where the stress is held constant – in practice at atmospheric pressure – are obviously easier to carry out. Combing Equations (4.73a) and (4.73c) allows us to express the entropy variation as a function of the volumetric stress σ instead of the volumetric strain ϵ according to

$$s_S = s_{S0} + 3\alpha\sigma + 3(\alpha b - \alpha_\varphi)p + C_\sigma(T - T_0)/T_0. \tag{4.74}$$

In Equation (4.74) $C_\sigma = C + 9T_0\alpha^2 K$ is the volumetric heat capacity at constant stress and is eventually the heat capacity measured in practice, albeit in most cases the difference $C_\sigma - C = 9T_0\alpha^2 K$ turns out to be negligible.

In order to state the compatibility relations concerning the thermal properties, analogously to Equation (4.74), we first write

$$\Sigma_S - \Sigma_{S0} = 3\alpha_S\sigma_S + C_{\sigma_S}(T - T_0)/T_0, \tag{4.75}$$

where Σ_S is the solid matrix entropy, while C_{σ_S} is the solid matrix volumetric heat capacity at constant stress. The additive character of entropy also allows us to write

$$s_S - s_{S0} = (1 - \phi_0)(\Sigma_S - \Sigma_{S0}). \tag{4.76}$$

Combining Equations (3.66), (4.75) and (4.76), we derive

$$s_S = s_{S0} + 3\alpha_S\sigma + 3\alpha_S\phi_0 p + (1 - \phi_0)C_{\sigma_S}(T - T_0)/T_0. \tag{4.77}$$

A comparison between Equations (4.74) and (4.77) finally provides the relations

$$\alpha = \alpha_S; \quad \alpha_\varphi = \alpha_S(b - \phi_0); \quad C_\sigma = (1 - \phi_0)C_{\sigma_S}. \tag{4.78}$$

With the help of Equations (3.53) and (4.73), in accordance with physical evidence, relations (4.78) ensure that, whenever $\sigma = p = 0$ while the infinitesimal heat supply is equal to δQ, the thermal strains are uniform within the porous solid, and are given by $\epsilon = \varphi/\phi_0 = \epsilon_S = 3\alpha_S(T - T_0)$, while the temperature change is $T - T_0 = \delta Q/(1 - \phi_0)C_{\sigma_S}$.

4.3.2 The linear thermoporoelastic fluid–solid mixture

Let us now fill the porous solid with a compressible fluid. Extending Equation (4.38) to nonisothermal evolutions, we write

$$d\rho_F/\rho_F = dp/K_F - 3\alpha_F dT; \quad ds_F = -3\alpha_F dp/\rho_F + C_p dT/T, \tag{4.79}$$

where $3\alpha_F$ is the fluid tangent coefficient of volumetric thermal dilation, while C_p is the fluid tangent volumetric specific heat capacity at constant pressure. If the saturating fluid is an ideal gas, a comparison of the equation resulting from the differentiation of Equation (4.43) with the first relation in Equation (4.79) provides us with the identification

$$3\alpha_G = 1/T, \tag{4.80}$$

which completes Equation (4.44), and ensures the compatibility of the second relation in Equation (4.79) with Equation (2.27). Equation (4.73b) and the first relation in Equation (4.79) allow us to extend the isothermal Equation (4.39) to nonisothermal evolutions in the form

$$d(\rho_F\phi)/\rho_F = bd\epsilon + dp/M - 3\alpha_{\rho_F\phi}dT, \tag{4.81}$$

where $\alpha_{\rho_F\phi}$ is a thermal dilation coefficient related to the fluid–solid mixture which can be expressed as

$$\alpha_{\rho_F\phi} = \alpha_\varphi + \phi\alpha_F. \tag{4.82}$$

In addition, combining Equation (4.73c) and the second relation in Equation (4.79), we derive the expression for the entropy related to the fluid–solid mixture as a whole, namely, $s = s_S + \rho_F \phi s_F$, in the form

$$ds = s_F d(\rho_F\phi) + 3\alpha K d\epsilon - 3\alpha_{\rho_F\phi} dp + C_d dT/T, \tag{4.83}$$

where C_d is a heat capacity related to the fluid–solid mixture which can be expressed as

$$C_d = C + \rho_F \phi C_p. \tag{4.84}$$

4.4 The Poroviscoelastic Solid

In the absence of any loading a solid remains at rest in its state of lowest energy. Any loading raises the energy of the solid, which deforms under the internal stress induced by the loading. In a perfect crystalline solid, which exhibits no defect, there is no possibility for the internal stress to relax to a lower energy state. This idealization corresponds to elastic behaviors where the response to any change in the loading is instantaneous. It results in a one-to-one relationship between stress and strain for a solid, and between the stress couple (σ_{ij}, p) and the strain couple $(\varepsilon_{ij}, \varphi)$ for a poroelastic solid. Nonetheless, in solids there are always local defects of various origins. Having been loaded, the solid tends to recover a lower energy state, through local rearrangments of matter made possible by the defects. These local rearrangments are thermally activated, with a characteristic time to escape from the local configuration associated with the defect. At constant loading a delayed or viscous behavior is then observed. For clay-like soils, the local rearrangments are due to the relative viscous microslidings of the platelets forming the solid matrix. These viscous slidings result from the lubricating action of water between the ends of adjoining platelets. This water is electrically bound to the platelets and does not participate in the fluid flow. In the following section we show how the constitutive equations of a porous solid, whose solid matrix exhibits such a delayed behavior, can be addressed in the context of poroviscoelastic solids.

4.4.1 The linear viscoelastic solid matrix

Consider then the infinitesimal volumetric strain $d\epsilon_S$ imposed on the solid matrix between time τ and time $\tau + d\tau$. Owing to viscous effects the infinitesimal mean stress $d\sigma_S$ required to maintain $d\epsilon_S$ is no longer constant and evolves with time. Assuming the linearity of the viscoelastic behavior, instead of Equation (4.11) we now write

$$d\sigma_S = k_S(t \geq \tau, \tau) d\epsilon_S. \tag{4.85}$$

The stress is said to relax, k_S being the related bulk relaxation function. Also, owing to linearity the mean stress history $\sigma_S(t)$ corresponding now to a time history $\epsilon_S(t)$ of the strain is obtained by summing the above relation with regard to τ, resulting in

$$\sigma_S(t) = \int_0^t k_S(t \geq \tau, \tau) \, d\epsilon_S(\tau). \tag{4.86}$$

If the solid matrix is not aging, the relaxation function will only depend upon the time $t - \tau$ separating the current time t from the application time τ of the infinitesimal volumetric strain $d\epsilon_S$, resulting in

$$k_S(t \geq \tau, \tau) = k_S(t - \tau \geq 0). \tag{4.87}$$

For nonaging materials whose bulk relaxation function satisfies Equation (4.87), the constitutive Equation (4.86) can be rewritten in the concise form

$$\sigma_S = k_S \odot \epsilon_S, \tag{4.88}$$

where $f \odot g$ stands for the Stieltjes convolution product of functions $f(t)$ and $g(t)$. Considering here functions of zero value for negative time but including possible discontinuities of function g at time t_i, in particular the one occurring at the origin of time caused by the possible discontinuity in loading, the expression of the Stieltjes convolution product $f \odot g$ is

$$(f \odot g)(t) = \int_0^t f(t-u) \, dg(u) + \sum_i f(t - t_i) \left[g(t_i^+) - g(t_i^-) \right]. \tag{4.89}$$

The convolution product $\odot$ has the same properties as an ordinary product, that is, commutativity, associativity and distributivity with regard to addition. Besides, the inverse f^{-1} of f in the sense of the convolution product satisfies

$$f^{-1} \odot f = H; \quad H(t < 0) = 0, \, H(t > 0) = 1, \tag{4.90}$$

where H is the Heaviside step function, and plays the role of the neutral element. Accordingly, the constitutive Equation (4.88) can be inverted in the form

$$\epsilon_S = k_S^{-1} \odot \sigma_S, \tag{4.91}$$

where k_S^{-1} is the function such that the history of the infinitesimal volumetric strain $d\epsilon_S$ produced by the application of the infinitesimal mean stress $d\sigma_S$ at time τ is given by

$$d\epsilon_S = k_S^{-1}(t - \tau) \, d\sigma_S. \tag{4.92}$$

Owing to viscous effects the infinitesimal volumetric $d\epsilon_S$ associated with the infinitesimal constant stress $d\sigma_S$ is no longer constant and evolves with time. The material is said to creep, k_S^{-1} being the related bulk creep function.

A standard relaxation function is given by the expression

$$k_S(t) = \left[k_S^\infty + (k_S^\infty - k_S^0) \exp(-t/\tau_r) \right] H(t), \tag{4.93}$$

where τ_r is the characteristic relaxation time. The related creep function k_S^{-1} is obtained by inverting $k_S(t)$ in the form

$$k_S^{-1}(t) = \left[1/k_S^\infty + (1/k_S^\infty - 1/k_S^0) \exp(-t/\tau_c) \right] H(t); \quad \tau_c k_S^\infty = \tau_r k_S^0, \tag{4.94}$$

where τ_c is the characteristic creep time. The strain response to a step loading, that has the form

$$\sigma_S = \Delta \sigma_S H(t),$$

is

$$\epsilon_S(t) = \left[1/k_S^\infty + \left(1/k_S^\infty - 1/k_S^0\right)\exp(-t/\tau_c)\right]\Delta\sigma_S. \tag{4.95}$$

Moduli k_S^0 and k_S^∞ are therefore identified with the instantaneous bulk modulus and the asymptotic delayed bulk modulus, respectively, since we have

$$\epsilon_S(t = 0^+) = \Delta\sigma_S/k_S^0; \quad \epsilon_S(t \to \infty) = \Delta\sigma_S/k_S^\infty. \tag{4.96}$$

The elastic energy which is ultimately stored in the solid matrix, that is, $(\Delta\sigma_S)^2/2k_S^\infty$, is lower than the elastic energy which is instantaneously stored, namely, $(\Delta\sigma_S)^2/2k_S^0$, since part of this latter energy is progressively dissipated into heat through the viscous phenomena at work in the solid matrix during the creep process. We therefore derive the conditions $k_S^\infty < k_S^0$ and $\tau_r < \tau_c$.

4.4.2 The linear poroviscoelastic solid

A comparison between Equations (4.11) and (4.88) shows that the viscoelastic constitutive equations can be derived formally from the elastic constitutive equations by simply replacing the ordinary product by the convolution product $\odot$. In a similar way, the poroviscoelastic constitutive equations can be derived from the poroelastic constitutive Equations (4.6) by simply replacing the ordinary product by the convolution product $\odot$. The procedure provides

$$\sigma = K \odot \epsilon - b \odot p; \tag{4.97a}$$

$$\varphi = b \odot \epsilon + N^{-1} \odot p; \tag{4.97b}$$

$$s_{ij} = 2G \odot e_{ij}. \tag{4.97c}$$

In addition, Equations (4.14) and (4.18) provided by microporoelastic considerations extend to poroviscoelasticity in the form

$$b = H - K \odot k_S^{-1}; \quad N^{-1} = (b - \phi_0 H) \odot k_S^{-1}. \tag{4.98}$$

This chapter has addressed in detail the constitutive equations of porous solids in the context of thermoporoelasticity, as well as poroviscoelasticity. These constitutive equations capture the coupling existing between the deformation of the porous solid and the pore pressure of the fluid mixture saturating the porous space at the macroscopic material scale. The related macroscopic poroelastic properties depend on the microscopic elastic properties of the solid matrix and on the porosity when adopting a mean field theory. At both the macroscopic spacescale and the timescale of observation the pore pressure is assumed to be homogeneous within the fluid mixture. At the larger scale of the porous structure made of such juxtaposed fluid–solid systems as, for instance, a soil layer, a bone, a tree, an oil reservoir and so on, this is no longer the case. Pore pressure gradients exist, which induce a fluid flow through the porous structure. The analysis of the coupling occurring between the fluid flow and the deformation of the porous structure is the subject of the next chapter.

Further Reading

Bourbié, T., Coussy, O. and Zinszner, B. (1987) *Acoustics of Porous Media*, Technip.
Charlez, P. (ed.) (1995) *Mechanics of Porous Media*, Balkema.
Coussy, O. (2004) *Poromechanics*, John Wiley & Sons, Ltd.
Detournay, E. and Cheng, A. H.-D (1994) Fundamentals of Poroelasticity, *Comprehensive Rock Engineering* (ed. J. Hudson), Pergamon Press.
Dormieux, L., Kondo, J. and Ulm, F.-J. (2006) *Microporomechanics*, John Wiley & Sons, Ltd.
Wang, H. F. (2000) *Theory of Linear Poroelasticity*, Princeton series in geophysics, Princeton University Press.
Zimmerman, R. W. (1991) *Compressibility of Sandstones*. Elsevier.

5

Fluid Transport and Deformation

Throughout the nineteenth century, motivated by potential industrial and civil engineering applications, many scientists have contributed significantly to the theoretical development of continuum mechanics of elastic solids. They elaborated various methods to determine the deformation field of an elastic solid continuum, having any shape and subjected to any body forces and tractions. The name of Claude Louis M. H. Navier (1785–1836) is associated with the displacement method consisting of taking the displacement components as principal unknowns. Later on named after him, the equations to be solved are derived by substituting the constitutive equations of elastic solids in the equilibrium Equations (3.36). Navier originally derived them by invoking the existence of underlying intermolecular forces responsible for the elasticity of solids. In 1823 he also published the first successful attempt to derive the equations governing the flow of a viscous fluid. The 'lack of slipperiness' of a fluid was attributed to the relative molecular motion occurring between the various portions of the flowing fluid. In contrast to Euler equations the differential equations derived by Navier were second-order equations with regard to the velocity components, so that the fluid velocity could be held equal to zero on the solid walls enclosing the flow and exerting friction upon the fluid. Nowadays these equations are named the Navier–Stokes equations, giving also the credit to George Gabriel Stokes (1819–1903), who established two decades later the same equations albeit in a different form, and gave them their definitive experimental support.

Navier's concern for the derivation of the equations governing the deformation of elastic bodies, or that governing the flow of a viscous fluid, was mainly theoretical. In contrast the laws governing the flow of a fluid within a porous solid, and how the latter deforms when it is subjected to an internal pressure, were discovered experimentally by engineers, whose immediate motivations were mainly practical. In 1856, in the famous Appendix D of a memoir entitled *Les Fontaines Publiques de la Ville de Dijon* Henry P. G. Darcy (1803–1858) reported his experimental investigations on the water flow through columns of natural sands, which lead him to formulate the law later on named after him. Nowadays Darcy's law can be macroscopically found by thermodynamic restrictions. The mechanics and the physics of porous solids often involve the molecular diffusion of a solute within the solution saturating the porous space. The law governing the molecular diffusion is Fick's law. In 1855, Adolf Fick (1829–1901), a physiologist, showed experimentally that the diffusion rate of a gas through a fluid membrane was proportional to the pressure difference through the membrane divided by the membrane thickness. Later on he was able to give some theoretical support to his law by inspecting the basic diffusion process from layer to layer and making some considerations on the motion of the molecules. Fick is also usually credited with the invention of contact lenses.

The discovery that the deformation of soils is governed by the effective stress introduced in Sections 3.4.5 and 4.1 was made by Karl von Terzaghi (1883–1963), and is associated with the one-dimensional theory of the consolidation of soils which he developed from his experimental observations. Later on, with the help of an elegant and operational energy approach, Maurice A. Biot (1905–1985) extended the picture by accounting for the compressibility of the solid part of the porous solid in consolidation theories. Since the beginning of the 1970s, the coupling between the deformation of a porous elastic solid and the flow of a linear viscous fluid have been extensively revisited with the help of modern upscaling homogenization procedures. Continuum poroelasticity and Darcy's law have finally been mathematically shown to derive from the microscopic Navier equations governing the deformation of the elastic solid forming the solid part of the porous solid, and from the Navier–Stokes equations governing the flow at the pore scale.

In this chapter, we first address the derivation of transport laws in the context of continuum thermodynamics, and we show how they can receive some support by combining thermodynamic restrictions with an upscaling dimensional analysis. These results associated with those of the previous chapters allow us to address the coupling of the deformation of the porous solid with the viscous flow of the fluid saturating the porous space. This coupling is finally illustrated by exploring the one-dimensional theory of the consolidation of a soil layer.

5.1 Transport Laws

5.1.1 Mole and mass conservation

In the search for the laws governing the transport of a fluid mixture through a deformable porous medium the starting point is to express the mole conservation of each of the components J forming the mixture. As defined in Section 2.2.5, let $\underline{u}_J$ be the molar flux vector related to component J. Also, as defined by Equation (3.67), let n_J be its molar density with regard to the overall volume of porous solid. Owing to the remarks made in Section 2.2.5 the mole balance relative to component J can be expressed in the form

$$\frac{\partial n_J}{\partial t} + \nabla \cdot \underline{u}_J = 0, \tag{5.1}$$

where infinitesimal transformations have been assumed so that Equation (3.13) is satisfied, resulting in $\nabla \equiv \nabla_X \equiv \nabla_x$.

5.1.2 Dissipation associated with transport

The transport of each component with regard to the averaged motion of the fluid mixture involves molecular diffusion. The averaged motion of the fluid mixture involves viscous friction at the internal solid walls. These molecular diffusion and viscous flow are both dissipative, and the second step in the search for the transport laws is to express this dissipation. Due to the entropy production associated with this dissipation, instead of an equality the balance of free energy provides an inequality. According to Section 2.2.5 the volumetric rate of free energy supplied to the fluid–solid mixture through the molar flow of component J is the quantity $-\nabla \cdot (\mu_J \underline{u}_J)$. As a result, assuming that the temperature is uniform within the porous solid, instead of Equation (3.69) we now write

$$\sigma \frac{d\epsilon}{dt} + s_{ij} \frac{de_{ij}}{dt} - \sum_J \nabla \cdot (\mu_J \underline{u}_J) - s \frac{dT}{dt} - \frac{df}{dt} \geq 0. \tag{5.2}$$

Substituting Equation (5.1) in Equation (5.2) we obtain

$$\sigma \frac{d\epsilon}{dt} + s_{ij} \frac{de_{ij}}{dt} + \sum_J \mu_J \frac{dn_J}{dt} - s \frac{dT}{dt} - \frac{df}{dt} - \sum_J \nabla \mu_J \cdot \underline{u}_J \geq 0. \tag{5.3}$$

Making use of the Gibbs–Duhem Equation (3.70), and substituting Equation (3.71) in Equation (5.3), we obtain

$$\sigma \frac{d\epsilon}{dt} + p \frac{d\varphi}{dt} + s_{ij} \frac{de_{ij}}{dt} - s_S \frac{dT}{dt} - \frac{da_S}{dt} - \sum_J \nabla \mu_J \cdot \underline{u}_J \geq 0. \tag{5.4}$$

The dissipation related to the deformation of a poroelastic solid is zero so that Equation (3.73) applies, and Equation (5.4) reduces to state the positivity of the dissipation related to the transport processes only:

$$-\sum_J \nabla \mu_J \cdot \underline{u}_J \geq 0. \tag{5.5}$$

Restricting Equation (2.8) to the case where, in addition to the mixture pressure and temperature, the chemical potential of a component J depends only on its own molar fraction $\overline{N}_J$ we write

$$\mu_J = \mu_J (p, T, \overline{N}_J). \tag{5.6}$$

Substituting Equation (5.6) in Equation (5.5), while taking $\nabla T = 0$ in order to address the dissipation associated with mass transfer only, we obtain

$$-\nabla p \cdot \underline{q} + p \sum_J -\nabla \overline{N}_J \cdot \underline{v}_J \geq 0, \tag{5.7}$$

where $\underline{q}$ is the overall velocity defined by

$$\underline{q} = \sum_J \overline{N}_J \underline{q}_J; \quad \underline{q}_J := \frac{1}{\overline{N}_J} \frac{\partial \mu_J}{\partial p} \underline{u}_J, \tag{5.8}$$

while $\underline{v}_J$ is another velocity related to component J defined by

$$\underline{v}_J := \frac{1}{p} \frac{\partial \mu_J}{\partial \overline{N}_J} \underline{u}_J. \tag{5.9}$$

In Equation (5.7) the first term is the dissipation associated with the viscous flow of the fluid mixture as a whole. The second term is the dissipation associated with the molecular diffusion of component J within the fluid mixture. We rewrite $\underline{v}_J$ in the form

$$\underline{v}_J = \underline{v} + \underline{V}_J; \quad \underline{V}_J := \sum_K \overline{N}_K \left(\underline{v}_J - \underline{v}_K\right). \tag{5.10}$$

In Equation (5.10), as $\underline{q}$ stands for the molar average of $\underline{q}_J$, $\underline{v}$ stands for the molar average of $\underline{v}_J$ according to

$$\underline{v} = \sum_J \overline{N}_J \underline{v}_J, \tag{5.11}$$

while $\underline{V}_J$ is the relative velocity of component J with regard to the molar average $\underline{v}$ of the velocity $\underline{v}_J$. With the help of Equations (5.10) and (5.11), the dissipation associated with the molecular diffusion can now be rewritten in the form

$$p \sum_J -\nabla \overline{N}_J \cdot \underline{v}_J = p \sum_J -\nabla \overline{N}_J \cdot \underline{V}_J. \tag{5.12}$$

According to the definitions given in Equations (5.8) and (5.9), a more specific expression for velocities $\underline{q}_J$ and $\underline{u}_J$, and eventually their very physical meaning, require further information on the molar Gibbs energy μ_J. An instructive example is the case of an ideal gas mixture for which Equations (2.28) and (2.43) combine to provide us with the following expression for μ_J:

$$\mu_J \left(p, T, \overline{N}_J\right) = f_J(T) + RT \ln \left(\overline{N}_J p\right). \tag{5.13}$$

From Equations (2.37) and (5.13) we then obtain

$$\frac{1}{\overline{N}_J} \frac{\partial \mu_J}{\partial p} = \frac{1}{p} \frac{\partial \mu_J}{\partial \overline{N}_J} = \frac{V}{N_J} = \frac{\phi}{n_J}. \tag{5.14}$$

Substituting Equations (5.14) in Equations (5.8) and (5.9) we obtain

$$\underline{q}_J = \underline{v}_J = \frac{\phi}{n_J} \underline{u}_J; \quad \underline{q} = \underline{v}. \tag{5.15}$$

As a result, $\underline{q}_J = \underline{v}_J$ and $\underline{V}_J$ are now identified with the volume flow vector and the relative volume flow vector of moles of the component J, respectively. Accordingly the vector $\underline{q}$ is the overall volume flow vector of the fluid mixture, and is usually termed the filtration vector, while $\underline{q}_J$ is the partial molar filtration vector. When applied to the diffusion within an ideal gas mixture, the previous results can be used as a first approach to the diffusion within ideal solutions.

Regarding some applications it can be more convenient to consider mass conservation instead of mole conservation. Since $n_J = \phi \rho_J / M_J$, where ρ_J and M_J are respectively the mass density and the molar mass of component J, from Equations (5.1) and (5.15) we alternatively derive the so-called continuity equation

$$\frac{\partial (\phi \rho_J)}{\partial t} + \nabla \cdot \left(\rho_J \underline{q}_J\right) = 0. \tag{5.16}$$

5.1.3 Fick's law

The advective flow of the fluid mixture as a whole addressed in the previous section, and the molecular diffusion of a component through that mixture, are quite distinct phenomena. As a result, in Equation (5.7) we can assume that the dissipations related to these two phenomena are positive irrespective of each other. The separate positivity of the dissipation related to molecular diffusion, whose expression is given in Equation (5.12), can then be written in the form

$$\sum_J -\nabla \overline{N}_J \cdot \underline{V}_J \geq 0. \tag{5.17}$$

The molecular diffusion is governed by the law relating the relative molar flow $\underline{V}_J$ to the driving force $-\nabla \overline{N}_J$ at the origin of it. According to the kinetic theory of ideal gases, this is Fick's law, given by

$$\overline{N}_J \underline{V}_J = -\phi \tau D_J \nabla \overline{N}_J, \tag{5.18}$$

where D_J is the diffusion coefficient corresponding to a free space, while the factor $\phi \tau$ accounts both for the reduction of the space offered to the diffusion with the help of porosity ϕ, and for the tortuosity of the porous space with the help of the so-called tortuosity τ. Strictly speaking Fick's law corresponds to the law governing the molecular diffusion in a pure fluid mixture, that is for $\phi = 1$ and therefore $\tau = 1$. Substitution of Equation (5.18) in Equation (5.10) and use of Equation (5.15) produce

$$\underline{v}_J = \underline{q} - \phi \tau D_J \nabla \ln \overline{N}_J. \tag{5.19}$$

An important case is the situation where the fluid mixture is wet air made of dry air (subscript $J = A$) and water vapor (subscript $J = V$). This is, for instance, the two-component mixture involved in the drying of porous materials. For such a two-component mixture, from Equation (5.10) we have $\overline{N}_V \underline{V}_V = -\overline{N}_A \underline{V}_A = \overline{N}_A \overline{N}_V (\underline{v}_V - \underline{v}_A)$ and Equation (5.18) reduces to the law

$$\overline{N}_A \overline{N}_V (\underline{v}_V - \underline{v}_A) = -\phi \tau D \nabla \overline{N}_V, \tag{5.20}$$

where D is the free air–vapor diffusion coefficient which can be experimentally shown to be well accounted for through the following expression[1]

$$D = D_0 \frac{p_{\text{atm}}}{p_G}; \quad D_0 = \delta_0 \left(\frac{T}{T_0}\right)^{1.88}, \tag{5.21}$$

where p_G is the wet air pressure; $p_{\text{atm}} = 101\,325\,\text{Pa}$ is the atmospheric pressure, while $\delta_0 = 2.17 \times 10^{-5}\,\text{m}^2\,\text{s}^{-1}$ and $T_0 = 273\,\text{K}$. Vapor and dry air can be accurately assumed to be ideal gases. According to Equation (2.37) we then have $\overline{N}_V = p_V/p_G$. Use of Equations (5.20) and (5.21), with $\underline{q} = \underline{q}_G$ allows us to rewrite Equation (5.19) in the more operational form

---

[1] See de Vries, D. A. and Kruger, A. J. (1966) On the value of the diffusion coefficient of water vapour in air. In *Phénomènes de transport avec changement de phase dans les milieux poreux ou colloïdaux*, CNRS ed., 561–572.

$$\underline{v}_V = \underline{q}_G - \phi\tau D_0 \frac{p_{atm}}{p_V}\nabla\left(\frac{p_V}{p_G}\right); \qquad (5.22a)$$

$$\underline{v}_A = \underline{q}_G + \phi\tau D_0 \frac{p_{atm}}{p_A}\nabla\left(\frac{p_V}{p_G}\right). \qquad (5.22b)$$

5.1.4 Darcy's law

Permeability

The separate positivity of the dissipation associated with the viscous flow of the fluid mixture as a whole through the porous solid can be written in the form

$$-\nabla p \cdot \underline{q} \geq 0. \qquad (5.23)$$

The fluid mixture filtration is governed by the law relating the filtration vector $\underline{q}$ to the driving force $-\nabla p$ producing the flow. The simplest form that this law can take is Darcy's law, which linearly relates $\underline{q}$ to $-\nabla p$. In the isotropic case Darcy's law is expressed in the form

$$\underline{q} = -k\nabla p, \qquad (5.24)$$

where k is the permeability of the porous solid with respect to the fluid. The permeability must be positive in order to ensure the positivity of the dissipation associated with the fluid flow (Equation (5.23)).

Intrinsic permeability

Darcy's law is a linear law. Using dimensional analysis this law can be shown to apply whenever the macroscopic filtration law derives from the microscopic viscous flow of a Newtonian fluid through the porous network. For this purpose let η_{mix} be the dynamic (shear) viscosity of the fluid mixture considered as a Newtonian fluid. Also let ℓ be the characteristic length associated with the geometry of the porous network where the flow takes place. Accounting for the viscous force resisting the flow, the isotropic filtration law relating the component q_i of $\underline{q}$ to the component $-\partial p/\partial x_i$ of ∇p can be formally expressed in the form

$$q_i = f(-\partial p/\partial x_i, \eta_{mix}, \ell, \phi). \qquad (5.25)$$

The previous relation will be physically relevant only if it is dimensionally consistent, that is, if q_i and f both have the physical dimension of a velocity, namely LT^{-1} in the LMT dimension basis, with L standing for length, M for mass and T for time. In this dimension basis any physical quantity Q can be expressed through its dimension function $[Q] = L^\alpha M^\beta T^\gamma$, that is, as a power function of the fundamental dimensions. For instance, since viscosity η_{mix} linearly relates a (shear) stress (having the dimension of pressure) to a strain rate (having the inverse dimension of a time) its dimension function is $[\eta_{mix}] = L^{-1}MT^{-1}$. The physical dimension of function f results from the physical dimensions of its arguments, $-\partial p/\partial x_i$, η_{mix}, ℓ and ϕ. The dimension functions are conveniently summarized in the form of the exponent matrix of dimensions, in which the exponents, α, β and γ, form the columns

	q_i	$-\partial p/\partial x_i$	η_{mix}	ℓ	ϕ
L	1	-2	-1	1	0
M	0	1	1	0	0
T	-1	-2	-1	0	0

(5.26)

The number of dimensionally independent quantities among q_i, $-\partial p/\partial x_i$, η_{mix}, ℓ and ϕ, is three, which turns out to be the rank of their exponent matrix of dimensions, that is, the maximum number of linearly independent columns in the exponent matrix above. Except for the porosity ϕ which is dimensionless, only one dimensionless quantity Π based on the dimensionally independent set $-\partial p/\partial x_i$, η_{mix} and ℓ remains to be formed from q_i, reading

$$\Pi = \frac{\eta_{mix} q_i}{\ell^2 (-\partial p/\partial x_i)}. \tag{5.27}$$

Consequently, dimensional analysis requires for the relation linking q_i to $-\partial p/\partial x_i$, η_f, ℓ and ϕ, to be expressed in the form

$$\Pi = \delta(\phi), \tag{5.28}$$

resulting in

$$q_i = -\frac{\varkappa}{\eta_{mix}} \partial p/\partial x_i; \quad \varkappa = \ell^2 \delta(\phi), \tag{5.29}$$

and providing us with the isotropic Darcy's law

$$\underline{q} = -\frac{\varkappa}{\eta_{mix}} \nabla p, \tag{5.30}$$

where the permeability k is identified as

$$k = \frac{\varkappa}{\eta_{mix}} = \frac{\ell^2}{\eta_{mix}} \delta(\phi). \tag{5.31}$$

Equation (5.31) for k relies essentially on the physical dimension of η_{mix}, and therefore on the assumed linear viscous behavior of the Newtonian fluid. The material property $\varkappa = \ell^2 \delta(\phi)$ represents the square of a length scaling the geometry of the flow. It is called the *intrinsic permeability* of the porous solid, since it depends only on the geometry of the porous network irrespective of the fluid. The usual unit used for intrinsic permeability $\varkappa$ is the Darcy — equal to 10^{-12} m^2. In addition to porosity ϕ, a single length ℓ has been considered to characterize the geometry of the porous network where the flow occurs. Actually, the other lengths ℓ_1, ℓ_2, ... required to characterize the geometry are present in the function $\delta(\phi)$ through the hidden arguments ℓ_1/ℓ, ℓ_2/ℓ, ... which the dimensional analysis would reveal. For simple geometries involving a single length ℓ various expressions for $\delta(\phi)$ have been derived. An expression often referred to is the Kozeny–Carman formula, which relates to a solid matrix formed by the packing of regular spheres and reads

$$\delta(\phi) = \frac{\phi^3}{1 - \phi^2}. \tag{5.32}$$

Table 5.1 Order of magnitude of intrinsic permeability for various materials

Material	$\varkappa$ [m² (= 10^{12} Darcy)]
Concrete	10^{-16}—10^{-21}
Clays	10^{-16}—10^{-20}
Bone	10^{-20}
Granites, gneiss, compact basalts	10^{-16}—10^{-20}
Marble	10^{-19}
Sandstones	10^{-11}—10^{-17}
Limestone	10^{-12}—10^{-16}
Fine sands, silts and loess	10^{-12}—10^{-16}
Gravels and sands	10^{-9}—10^{-12}

For more complex geometries, or when the filtration process does not result from a viscous flow, the intrinsic permeability $\varkappa$ must be experimentally determined. Typical values of $\varkappa$ are given in Table 5.1.

Hydraulic permeability

When considering the existence of body forces $\underline{b}$, $-\nabla p$ must be replaced by $-\nabla p + \rho_{mix}\underline{b}$. In practical applications, when not considering dynamic body forces due to inertia effects, body forces often reduce to gravitational forces $-\rho_{mix} g \underline{e}_z$ acting in the Oz vertical direction. The driving force producing the flow becomes $-\nabla(p + \rho_{mix}gz)$ and Equation (5.30) extends to the form

$$\underline{q} = -\frac{\varkappa}{\eta_{mix}} \nabla(p + \rho_{mix}gz). \qquad (5.33)$$

Equation (5.33) allows us to state Darcy's law in the form commonly used in hydrology and soil mechanics, namely,

$$\underline{q} = -\lambda \nabla H; \quad \lambda = \frac{\rho_{mix} g \varkappa}{\eta_{mix}}, \qquad (5.34)$$

where H is the fluid particle head defined by

$$H = \frac{p}{\rho_{mix} g} + z. \qquad (5.35)$$

At zero pressure H is the liquid height, whose field determination can be achieved by means of piezometric measurements. Property λ is a characteristic filtration velocity and is commonly termed the hydraulic permeability. The filtration velocity λ is the vertical one that would result from gravitational forces in the absence of any pressure gradient whatever the origin of it. For liquid water we have $\eta_{mix} \simeq 10^{-3}$ kg m^{-1} s^{-1}, and a useful numerical relation between the dimensioned properties $\varkappa$ and λ is

$$\varkappa\,(\mathrm{m}^2) \simeq 10^{-7} \lambda\,(\mathrm{m\,s^{-1}}). \qquad (5.36)$$

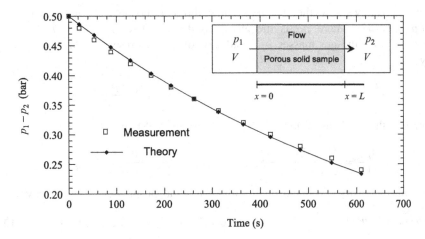

Figure 5.1 Validation of the pulse test for a mortar sample (Data from Skoczylas, F., Coussy, O. and Lafhaj, Z. (2003) On the reliability of heterogenous permeability values measured by gas injection (in French). *Revue Française de Génie Civil*, 7, 4)

The pulse test

Among the many experimental procedures aimed at the determination of the intrinsic permeability $\varkappa$, an interesting test is the pulse test. In this test, sketched out in Figure 5.1, the two end faces of a bar, formed from the porous solid whose permeability has to be determined, are connected to two large reservoirs of volume V. The length of the bar is L along the Ox direction, and the area of the face normal to this direction is A. The initial fluid pressure within the system is uniform and equal to p_0. The upstream reservoir – subscript 1 – is momentarily disconnected from the system, in order to separately raise its pressure $p_1(t)$ up to the initial pressure $p_1(0) = p_0 + \omega$, where $\omega \ll p_0$ is the pulse intensity. The upstream reservoir is then connected again to the system, resulting in the diffusion of the fluid within the permeable bar from the upstream reservoir to the downstream reservoir – subscript 2. The fluid diffusion tends to make the pressure $p_2(t)$ of the downstream reservoir become closer to the pressure of the upstream reservoir. As shown in detail below, the decay of the pressure difference $p_1(t) - p_2(t)$, which is slowed down by the fluid diffusion, occurs in an exponential form, whose recording finally provides the means for determining the permeability k. The more permeable to the fluid the bar is, the faster the pressure difference $p_1(t) - p_2(t)$ vanishes. Since, according to Equation (5.31), the smaller the fluid mixture viscosity, the larger the permeability k, the test duration will be shortened by using poorly viscous fluids so that gases are good candidates. So let $p(x, t)$ be the gas pressure within the bar. For sufficiently low values of the pulse intensity ω, the porosity ϕ does not change significantly during the test and will be kept equal to its initial value ϕ_0. Substitution of the ideal gas law (Equation (4.43)) and of Darcy's law (Equation (5.24)) in the mass conservation Equation (5.16) gives us the diffusion equation

$$\phi_0 \frac{\partial p}{\partial t} = \frac{1}{2} k \frac{\partial^2 p^2}{\partial x^2}. \qquad (5.37)$$

According to Darcy's law the gas mass flow through the area A of the bar face is $-Ak\rho_G \partial p(x,t)/\partial x$, where ρ_G is the gas mass density. Taking $\rho_G(x=0,t) = \rho_1$ and $\rho_G(x=L,t) = \rho_2$, the gas masses $\rho_1 V$ and $\rho_2 V$ of the two reservoirs are therefore governed by

$$\frac{d(\rho_1 V)}{dt} = \rho_1 Ak \frac{\partial p(x,t)}{\partial x}\bigg|_{x=0}; \quad \frac{d(\rho_2 V)}{dt} = -\rho_2 Ak \frac{\partial p(x,t)}{\partial x}\bigg|_{x=L}. \tag{5.38}$$

Substitution of the ideal gas Equation (4.43) in Equation (5.38) gives

$$\frac{dp_1}{dt} = \frac{1}{2}\frac{Ak}{V}\frac{\partial p^2(x,t)}{\partial x}\bigg|_{x=0}; \quad \frac{dp_2}{dt} = -\frac{1}{2}\frac{Ak}{V}\frac{\partial p^2(x,t)}{\partial x}\bigg|_{x=L}. \tag{5.39}$$

Since the pulse intensity ω is much smaller than the initial gas pressure p_0, we can write

$$p(x,t) = p_0(1 + \Delta\overline{p}(x,t)); \quad \Delta\overline{p}(x,t) = (p(x,t) - p_0)/p_0 \ll 1. \tag{5.40}$$

Substituting Equation (5.40) in Equation (5.39), and only keeping the leading order terms, we obtain

$$\frac{d\Delta\overline{p}_1}{d\bar{t}} = \frac{\partial \Delta\overline{p}}{\partial \bar{x}}\bigg|_{\bar{x}=0}; \quad \frac{d\Delta\overline{p}_2}{d\bar{t}} = -\frac{\partial \Delta\overline{p}}{\partial \bar{x}}\bigg|_{\bar{x}=1}, \tag{5.41}$$

where

$$\bar{x} = x/L; \quad \bar{t} = t/T; \quad T = \frac{VL}{p_0 Ak}. \tag{5.42}$$

Besides, substitution of Equation (5.40) in Equation (5.37) and use of Equation (5.42) provide us with the dimensionless diffusion equation

$$\varepsilon \frac{\partial \Delta\overline{p}}{\partial \bar{t}} = \frac{\partial^2 \Delta\overline{p}}{\partial \bar{x}^2}; \quad \varepsilon = \frac{\phi_0 AL}{V}. \tag{5.43}$$

We have $\varepsilon \ll 1$, since ε represents the ratio of the porous volume of the bar to the volume of the large reservoirs. Letting ε tend to zero in Equation (5.43), the solution of the resulting diffusion equation $\partial^2 \Delta\overline{p}/\partial \bar{x}^2 = 0$ is

$$\Delta\overline{p}(x,t) = (1-\bar{x})\Delta\overline{p}_1(t) + \bar{x}\Delta\overline{p}_2(t). \tag{5.44}$$

The linearized solution given in Equation (5.44) corresponds to the main order term of the exact solution of the stationary nonlinear diffusion equation $\partial^2 p^2/\partial \bar{x}^2 = 0$, namely,

$$p^2(x,t) = (1-\bar{x})p_1^2(t) + \bar{x}p_2^2(t). \tag{5.45}$$

An inspection of the diffusion Equation (5.43) shows that the characteristic time associated with the diffusion within the bar itself is εT, so that the diffusion within the bar, because of the large volume V of the upstream and downstream reservoirs, occurs much faster than the diffusion at the end faces. It results in spatially linear profiles of Equation (5.45) for the square $p^2(x,t)$ of the gas pressure, whose decreasing slope, namely, $p_2^2(t) - p_1^2(t)$, depends only on time. These profiles correspond to successive stationary diffusion processes within the bar, with time-dependent boundary conditions $p_1(t)$ and $p_2(t)$. Under assumption (5.40) these time-depending boundary conditions can be determined by substituting Equation (5.44) in Equation (5.41), resulting in

$$\frac{d\Delta\overline{p}_1}{d\overline{t}} + \Delta\overline{p}_1(t) - \Delta\overline{p}_2(t) = 0; \quad \frac{d\Delta\overline{p}_2}{d\overline{t}} + \Delta\overline{p}_2(t) - \Delta\overline{p}_1(t) = 0. \quad (5.46)$$

Solving Equation (5.46) with initial conditions $\Delta\overline{p}_1(0^+) = \omega$ and $\Delta\overline{p}_2(0^+) = 0$, we derive the pressure difference $p_1(t) - p_2(t) = p_0\left[\Delta\overline{p}_1(t) - \Delta\overline{p}_2(t)\right]$ in the form of the exponential decay

$$p_1(t) - p_2(t) = \omega\exp(-2t/T). \quad (5.47)$$

The experimental determination of the characteristic time $T = VL/p_0Ak$ leads to the determination of k. When the two reservoirs have different volumes, say V_1 and V_2, the expression to consider for $1/T$ in Equation (5.47) becomes $1/T = p_0Ak(1/V_1 + 1/V_2)/2L$.

An illustration of the results provided by the pulse test is given in Figure 5.1. The gas is argon, whose dynamic viscosity value is assessed to be 2.2×10^{-5} Pa s at 20°C, while the values of the initial gas pressure p_0 and the pulse intensity ω are respectively 1.85 MPa and 0.05 MPa. The reservoir volumes are different, with $V_1 = 2V_2 = 2 \times 10^{-3}$ m^3. In a first step the value of the intrinsic permeability is found equal to 10^{-16} m^2 from a stationary test. In a stationary test the pressures p_1 and p_2 are held constant, and the value of q_x is measured in the direction of the flow. The intrinsic permeability $\varkappa$ is then determined from Darcy's law (Equation (5.29)) and the pressure profile of Equation (5.45). The theoretical decay for the pressure difference $p_1(t) - p_2(t)$, represented by black lozenges and obtained by adopting $\varkappa = 10^{-16}$ m^2 in the assessment of time T governing the exponential term in Equation (5.47), compares well with the pulse test measurements represented by empty squares.

The Klinkenberg effect

Although the intrinsic permeability $\varkappa$ should depend only on the current porous network geometry, the experimental values of $\varkappa$ estimated from gaseous flows are generally found to be higher than the ones estimated from liquid flows. This difference is commonly attributed to the Klinkenberg effect, namely, the possible sliding of the gas molecules along the internal walls of the porous network. The sliding does occur when the mean free path λ of a molecule has the same order of magnitude as the pore diameter d. The kinetic theory of ideal gases gives the expression

$$\lambda = \frac{2\eta_G}{\rho_G \langle v \rangle}, \quad (5.48)$$

where the subscript G refers to the gas and where $\langle v \rangle$ is the mean molecular velocity

$$\langle v \rangle = \sqrt{\frac{8RT}{\pi M_G}}. \quad (5.49)$$

Substitution of the ideal gas Equation (4.43) and of Equation (5.49) in Equation (5.48) allows us to express the mean free path λ in the form

$$\lambda = \frac{\eta_G}{p}\sqrt{\frac{\pi RT}{2M_G}}. \tag{5.50}$$

The lower the gas pressure p, the more rarefied the gas and the higher the mean free path λ. The sliding phenomenon becomes significant when the value of the Knudsen number, namely $\text{Kn} = \lambda/d$, is close to unity. In order to capture the 'slip flow' when Kn becomes close to unity, according to statistical mechanics the permeability k involved in Darcy's law (Equation (5.24)) must be modified according to the expression

$$k = \frac{\varkappa}{\eta_G} + \frac{4}{3}l\frac{\langle v \rangle}{p}, \tag{5.51}$$

where l is a characteristic length depending on the geometry of the porous network only. Equation (5.51) is conveniently rewritten according to the Klinkenberg formula[2]

$$k = \frac{\varkappa}{\eta_G}\left(1 + \frac{\varpi}{p}\right), \tag{5.52}$$

where ϖ is a characteristic pressure depending on both the gas and the porous network geometry. According to Equation (5.52) the intrinsic permeability, $\varkappa = \eta_G k$, is then obtained by looking in the plane$(1/p, \varkappa = \eta_G k)$ for the value of $\eta_G k$ as $1/p$ tends to zero, along the line extrapolating the experimental results obtained for different values of $1/p$. In this regime of large pressures with respect to ϖ the viscosity effects are then dominant and the sliding phenomenon is negligible. In contrast, when the Knudsen number Kn becomes larger and larger as p becomes smaller and smaller with respect to ϖ, the viscosity effects vanish and the flow reduces to free molecular diffusion – the Knudsen flow. The procedure of such a determination of $\varkappa = \eta_G k$ is illustrated in Figure 5.2 for a limestone sample, the injected gas being argon. The experimental results also show that the intrinsic permeability $\varkappa$ of the tested limestone is poorly sensitive to the confining pressure $-\sigma$, and thereby to the change of porosity it induces, in the range 4 MPa to 28 MPa.

5.2 Coupling the Deformation and the Flow

In Chapter 4 we detailed how the deformation $\underline{\varepsilon}$ and the change of porosity φ of a poroelastic solid, through its constitutive equations, are governed by the applied stress $\underline{\sigma}$ and pore pressure p. In the previous sections we have shown how the gradient of the pore pressure ∇p governs the flow of the fluid through the porous network. As a result, in any problem of continuum poromechanics the deformation and the flow are coupled. The purpose of this section is to investigate how this coupling can be quantitatively addressed by combining the results of Chapter 4 and the laws governing the flow.

---

[2] See Klinkenberg, L. J. (1941) The permeability of porous media to liquids and gases. *Drilling and production practices*, American Petroleum Institute, New York, 200–214.

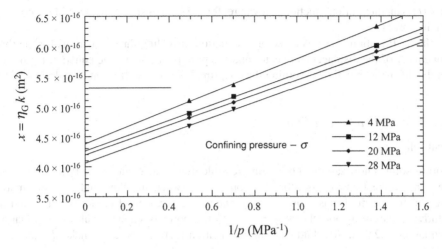

Figure 5.2 Determination of the intrinsic permeability $\varkappa$ of a limestone sample from the Klinkenberg formula, that is Equation (5.52) (Data from Lion, M., Skoczylas, F. and Ledésert, B. (2004) Determination of the main hydraulic and poro-elastic properties of a limestone from Bourgogne, France, *International Journal of Rock Mechanics & Mining Sciences*, **41**, 915–925)

5.2.1 The Navier equation

We will restrict ourselves to the displacement method, which consists of choosing the three displacement components $\xi_{i=1,2,3}$ as the main unknowns. Since there are three equations accounting for the mechanical equilibrium in the three directions, these three displacement components are governed by three equations which we are now going to derive.

Substituting Equation (3.14) into the poroelastic constitutive Equations (4.6a) and (4.6c), we can express the stress components σ_{ij} as functions of the derivatives of the displacement components ξ_i. Substituting these expressions in Equations (3.35) or (3.36) expressing the mechanical equilibrium and assuming that the body forces are those due to gravity, we derive

$$\left(K + \frac{4}{3}G\right)\nabla\left(\nabla \cdot \underline{\xi}\right) - G\nabla \times \left(\nabla \times \underline{\xi}\right) - b\nabla p - \rho g \mathbf{e}_z = 0, \tag{5.53}$$

which is known as the Navier equation in the elastic case when p is not present. Recalling that $\nabla \cdot \underline{\xi} = \epsilon$ and applying the divergence operator, $\nabla \cdot$, to Equation (5.53), we derive the so-called dilation equation,

$$\left(K + \frac{4}{3}G\right)\nabla^2 \epsilon - b\nabla^2 p = 0, \tag{5.54}$$

which relates the volumetric dilation ϵ to the fluid pressure p. A special case of interest is the case of irrotational displacements where $\nabla \times \underline{\xi} = 0$. Substituting $\nabla \times \underline{\xi} = 0$ in Equation (5.53), the Navier equation integrates to become

$$\epsilon = \frac{b}{K + 4G/3} p + f(t), \tag{5.55}$$

where $f(t)$ is an integration function and where, for the sake of simplicity, we did not consider the gravity forces.

The Navier Equation (5.53) is a vectorial equation providing three equations for the three components ξ_i of the displacement vector and the pore pressure p. One equation is missing. This is the diffusion equation governing the coupling between the flow and the deformation.

5.2.2 The diffusion equation

General relations

The porous solid is now saturated by a compressible fluid whose mass density ρ_F is governed by the constitutive Equation (4.38). The coupling between the flow and the deformation is governed by the fluid mass conservation, Darcy's law, the fluid compressibility and the constitutive equation of poroelasticity related to the porosity change. Substituting Darcy's law (Equation (5.24)) in the fluid continuity Equation (5.16), where we make J equal to F, we derive

$$\frac{\partial (\phi \rho_F)}{\partial t} = \nabla \cdot (\rho_F k \nabla p). \tag{5.56}$$

Substituting the fluid constitutive Equation (4.38) in Equation (5.56), while assuming the permeability k to be homogeneous within the medium, we obtain

$$\frac{\partial (\phi \rho_F)}{\partial t} = \rho_F k \left[\nabla^2 p + \frac{1}{K_F} (\nabla p)^2 \right]. \tag{5.57}$$

Poorly compressible fluid

If the fluid is a poorly compressible liquid (subscript F = L), Equation (4.38) can be approximated by Equation (4.42). Bearing in mind that $\varphi = \phi - \phi_0 \ll \phi_0$, this allows us to write

$$\phi \rho_L = (\phi_0 + \varphi)(\rho_{L0} + \rho_L - \rho_{L0}) \simeq \rho_{L0} \left[\phi_0 + \varphi + \frac{\phi_0}{K_L} (p - p_0) \right]. \tag{5.58}$$

In addition, owing to the poor compressibility of the fluid resulting in small values of $1/K_F$, the second term on the right-hand side of Equation (5.57) can be neglected. Substitution of Equation (5.58) in Equation (5.57) and the use of the constitutive Equation (4.6b) governing the porosity change φ allow us to rewrite Equation (5.57) in the form

$$b \frac{\partial \epsilon}{\partial t} + \frac{1}{M} \frac{\partial p}{\partial t} = k \nabla^2 p, \tag{5.59}$$

where M is given by Equation (4.40). Any problem of poroelasticity is then governed by the Navier Equation (5.53) and the coupled diffusion Equation (5.59) governing the pore pressure p, with the appropriate boundary and initial conditions.

Combining Equations (5.54), (5.58) and (4.6b) we also derive

$$\nabla^2 (\phi \rho_L) = \frac{1}{M} \frac{K_u + 4G/3}{K + 4G/3} \rho_{L0} \nabla^2 p, \tag{5.60}$$

where K_u is the undrained bulk modulus defined by Equation (4.61). Neglecting again the second term in the right-hand side of Equation (5.57), substitution of Equation (5.60) in Equation (5.57) provides us with the diffusion equation

$$\frac{\partial (\phi \rho_L)}{\partial t} = c\nabla^2 (\phi \rho_L); \quad c := kM \frac{K + 4G/3}{K_u + 4G/3}. \tag{5.61}$$

Whatever the poroelastic problem at hand involving a poorly compressible fluid, the liquid mass content $\phi \rho_L$ is therefore governed by the single uncoupled diffusion Equation (5.61), where c is thus identified with the poroelastic intrinsic diffusion coefficient irrespective of the problem at hand. Indeed, this is the extensive mass variable $\phi \rho_L$ which is subjected to a pure uncoupled diffusion process, and not the intensive pressure variable p.[3] However, whenever the displacement field is irrotational, Equation (5.55) applies and from Equation (5.59) we then obtain that the pore pressure p is governed by the diffusion equation

$$\frac{bc}{k}\frac{df}{dt} + \frac{\partial p}{\partial t} = c\nabla^2 p. \tag{5.62}$$

The coefficient of the diffusion Equation (5.62) governing the pore pressure remains, as expected, equal to the intrinsic diffusion coefficient c. However, the porous solid deformation induces a pressure source which is accounted for by the first term in the left-hand side of Equation (5.62).

According to Darcy's law (Equation (5.24)) the pore pressure gradient ∇p is proportional to the filtration vector $\underline{q}$, which is spatially continuous except at possible instantaneous injection points or interfaces between porous media. As a result $\nabla^2 p$, and therefore $\nabla^2 (\phi \rho_L)$ because of Equation (5.60), remain finite everywhere. Equation (5.61) then shows that the time derivative of $\phi \rho_L$ must remain finite too, so that the liquid mass content $\phi \rho_L$ cannot undergo discontinuities with regard to time. Indeed, the resistant viscous forces associated with the fluid prevent any instantaneous displacement of the fluid. As a result, the instantaneous response of the porous medium to any external loading is undrained, with no possible instantaneous variation of $\phi \rho_L$ between the time $t = 0^-$ prior to instantaneous loading and time $t = 0^+$ just after loading. Therefore, from Equation (4.39) the instantaneous undrained volumetric deformation can be expressed in the form

$$\epsilon (t = 0^+) = -\frac{1}{bM} p (t = 0^+). \tag{5.63}$$

Gaseous fluid

Now let the fluid be a gas (subscript F = G). Use of the ideal gas Equation (4.43), Equation (4.44) and the relation $\nabla^2 p^2 = 2p\nabla^2 p + 2(\nabla p)^2$ allows us to express Equation (5.57) in the form

$$\frac{\partial (\phi p)}{\partial t} = \frac{1}{2}k\nabla^2 p^2. \tag{5.64}$$

---

[3] For thermal diffusion within theromoelastic solids it can be shown in a similar way that, because of the latent heat of deformation, this is not the intensive variable involved in the process, namely, the temperature, which is governed by an uncoupled diffusion equation, but the extensive variable, namely, the solid entropy.

Substituting the poroelastic constitutive Equation (4.6b) in Equation (5.64) we obtain

$$\phi \frac{\partial p}{\partial t} + bp \frac{\partial \epsilon}{\partial t} + \frac{1}{2N} \frac{\partial p^2}{\partial t} = \frac{1}{2} k \nabla^2 p^2. \tag{5.65}$$

In the specific case of a poorly compressible porous solid, where ϵ, p/N and $(\phi - \phi_0)/\phi_0$ are much smaller than one and become negligible, Equation (5.65) reduces to Equation (5.37). In the general case of compressible porous solids, no universal uncoupled diffusion equation, such as Equation (5.61) which we derived for the poorly compressible liquid, exists for the compressible gas. In order to encompass both cases it is interesting to address the intermediary case where the saturating fluid is a solution formed from a poorly compressible liquid and compressible gas bubbles. The solution mass density ρ_{mix} is then governed by Equation (4.49) and similar developments to those that lead us to Equation (5.65) give the diffusion equation

$$b \frac{\partial \epsilon}{\partial t} + \left[\frac{1}{M} + \phi_0 v_0 \left(\frac{p_0}{p^2} - \frac{1}{K_L} \right) \right] \frac{\partial p}{\partial t} = k \left[\nabla^2 p + \left(\frac{v_0 p_0}{p^2} + \frac{1 - v_0}{K_L} \right) (\nabla p)^2 \right]. \tag{5.66}$$

The diffusion Equation (5.66) reduces to Equation (5.59) for $v_0 = 0$. However, to recover Equation (5.65) from Equation (5.66) the condition $v_0 = 1$ is not sufficient. We also have to assume $p_0/p^2 \simeq 1/p$, because, in the derivation of Equation (5.66), we used the approximation of Equation (4.49) instead of the more accurate Equation (4.48).

5.3 Consolidation of a Soil Layer

5.3.1 Consolidation equation

When a water-saturated soil layer is subjected to an extra loading with regard to gravity, the liquid water undergoes an excess of pressure. Subsequently this excess of pressure progressively vanishes, owing to the diffusion process of the fluid towards the boundary of the soil layer that remains drained. In turn, the porous solid progressively has to sustain alone the extra loading, and a delayed settlement of the soil layer occurs. The whole process is the celebrated consolidation problem of soil mechanics, principally associated with the name of Terzaghi.[4] It is a cornerstone problem since an unexpected consolidation can endanger the overlying structures. Similar problems in petroleum geophysics are the subsidence of reservoirs induced by oil extraction or by the sudden release of an abnormal fluid overpressure following the breakthrough of a reservoir-layer by a drilling tool. In biomechanics the bone of a leg is subjected to repeated consolidation processes, as the blood is flowing in and out the core of the bone at each step as you walk.

Consider the infinite soil layer represented in Figure 5.3. It rests on a rigid impervious base at depth $z = h$, while its upper surface at $z = 0$ remains drained, that is, at zero pressure if we take the hydrostatic pressure induced by the atmospheric pressure and the gravitational forces as the reference pressure. The hydraulic boundary conditions read

$$z = 0: \quad p = 0; \quad z = h: \quad \frac{\partial p}{\partial z} = 0, \tag{5.67}$$

---

[4] Terzaghi, K. (1923) Die berechnung der durchlassigkeitsziffer des tones aus dem verlauf der hydrodynamischen spannungserscheinungen. *Stizungsber. Akad. Wissen, Wien Math. Naturwiss. Kl., Abt. IIa*, **132**, 105–124.

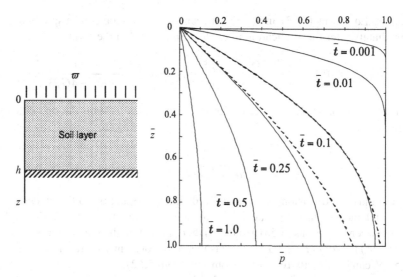

Figure 5.3 Normalized profiles of the overpressure $\bar{p}$ during the consolidation process of the soil layer sketched out in the left-hand side of the figure. The profiles are plotted against the normalized depth $\bar{z} = z/h$ for various normalized times $\bar{t} = ct/h^2$. For $\bar{t} = 0.1, 0.25$ the dashed lines represent the early time solution which cannot be distinguished from the exact solution for $\bar{t} = 0.001, 0.01$

where, according to Darcy's law, the last condition ensures a zero flow at the boundary $z = h$. After the application of a vertical constant load $\sigma_{zz} = -\varpi$ at the upper surface $z = 0$, the mechanical equilibrium in the vertical direction requires $\partial \sigma_{zz}/\partial z = 0$, yielding

$$\sigma_{zz} = -\varpi. \tag{5.68}$$

According to Equation (3.31) the vertical stress σ_{zz} can be expressed as $\sigma_{zz} = s_{zz} + \sigma$ so that

$$-\varpi = s_{zz} + \sigma. \tag{5.69}$$

Substitution of the constitutive Equations (4.6a) and (4.6c) in Equation (5.69) then gives

$$-\varpi = 2Ge_{zz} + K\epsilon - bp. \tag{5.70}$$

Furthermore the displacement in the soil layer is vertical, reading $\xi = \xi(z,t)\underline{e}_z$ so that the only nonzero strain component is ε_{zz}. Using Equations (3.14) and (3.16) we write

$$\varepsilon_{zz} = \epsilon = \frac{\partial \xi}{\partial z}. \tag{5.71}$$

According to Equations (3.18) and (5.71) the deviatoric strain component e_{zz} is given by

$$e_{zz} = \frac{2}{3}\frac{\partial \xi}{\partial z}. \tag{5.72}$$

The three relations above combine to provide us with the relations

$$\frac{\partial \xi}{\partial z} = \epsilon = \frac{b}{K + 4G/3}p - \frac{\varpi}{K + 4G/3}. \tag{5.73}$$

Now combining Equations (5.63) and (5.73), the soil layer instantaneous response at $t = 0^+$, that is, immediately after the application of the load, is derived in the form

$$p(z, t = 0^+) = \frac{bM\varpi}{K_u + 4G/3}; \quad \frac{\partial \xi}{\partial z}(z, t = 0^+) = -\frac{\varpi}{K_u + 4G/3}. \quad (5.74)$$

Substitution of Equation (5.73) into Equation (5.59) provides us with the one-dimensional consolidation equation

$$\frac{\partial p}{\partial t} = c \frac{\partial^2 p}{\partial z^2}, \quad (5.75)$$

where c is the intrinsic diffusion coefficient we defined in Equation (5.61). Indeed, for the problem at hand the displacement field is irrotational so that Equation (5.55) applies. A comparison between Equations (5.55) and (5.73) then shows that the integration function $f(t)$ in Equation (5.55) is $-\varpi/(K + 4G/3)$, and is therefore constant with regard to time, so that Equation (5.75) can be directly recovered from Equation (5.62).

The liquid pressure p is appropriately scaled by the pressure $p(z, t = 0^+)$ of Equation (5.74) induced by the instantaneous loading. The depth z is scaled by the finite thickness h of the soil layer, which is the relevant diffusion length to define the overall diffusion characteristic time τ. We write

$$p = \frac{bM\varpi}{K_u + 4G/3} \bar{p}; \quad z = h\bar{z}; \quad t = \tau\bar{t}; \quad \tau = h^2/c. \quad (5.76)$$

As a result we rewrite Equation (5.75) in the dimensionless form

$$\frac{\partial \bar{p}}{\partial \bar{t}} = \frac{\partial^2 \bar{p}}{\partial \bar{z}^2}, \quad (5.77)$$

with the boundary and initial conditions given in Equations (5.67) and (5.74) now reading

$$\bar{z} = 0: \bar{p} = 0; \quad \bar{z} = 1: \frac{\partial \bar{p}}{\partial \bar{z}} = 0; \quad \bar{t} = 0: \bar{p} = 1. \quad (5.78)$$

Substituting Equation (5.76) into Equation (5.73) and integrating over the layer thickness, while taking into account the boundary condition $\xi(z = h, t) = 0$, we derive the settlement $s(t) = \xi(z = 0, t)$ in the form

$$s(t) = s_\infty + (s_0 - s_\infty) \int_0^1 \bar{p}(\bar{z}, \bar{t}) \, d\bar{z}. \quad (5.79)$$

In the above equation s_0 stands for the instantaneous undrained settlement ($\bar{p} = 1$), while s_∞ stands for the drained settlement achieved as time goes to infinity so that the initial liquid overpressure has vanished ($\bar{p} = 0$). Their respective expressions are:

$$s_0 = \frac{\varpi h}{K_u + 4G/3}; \quad s_\infty = \frac{\varpi h}{K + 4G/3}. \quad (5.80)$$

5.3.2 Early time solution

The early time solution can be determined by exploring the times which are much smaller than the overall characteristic diffusion time τ. We set

$$\vartheta \ll 1: \quad \bar{t} = \vartheta t^*; \quad \bar{z} = \sqrt{\vartheta} z^*, \tag{5.81}$$

so that Equations (5.77) and (5.78) can be rewritten in the form

$$\frac{\partial \bar{p}}{\partial t^*} = \frac{\partial^2 \bar{p}}{\partial z^{*2}} \tag{5.82}$$

and

$$z^* = 0: \bar{p} = 0; \quad z^* = \frac{1}{\sqrt{\vartheta}} \gg 1: \frac{\partial \bar{p}}{\partial z^*} = 0; \quad t^* = 0: \bar{p} = 1, \tag{5.83}$$

respectively. For early times such as $\sqrt{\vartheta} \ll 1$, Equation (5.83) shows that the impermeability condition of the rigid base is repelled at $z^* \to \infty$. At early times the diffusion process involves only the region close to the drained upper surface where it tends to restore a zero liquid pressure. Indeed, the early time solution behaves as if the soil layer was semi-infinite. The search for the invariants of Equations (5.82) and (5.83) shows that the early time solution can be looked for in the form $\bar{p} = \bar{p}\left(z^*/\sqrt{t^*} = \bar{z}/\sqrt{\bar{t}}\right)$. Substituting this form for $\bar{p}$ in Equations (5.82) and (5.83), we derive an ordinary differential equation, whose solution is

$$\bar{t} \ll 1: \bar{p} = \text{erf}\left(\frac{\bar{z}}{2\sqrt{\bar{t}}}\right), \tag{5.84}$$

where erf stands for the error function defined by

$$\text{erf}(u) = \frac{2}{\sqrt{\pi}} \int_0^u \exp\left(-\lambda^2\right) d\lambda. \tag{5.85}$$

5.3.3 Any time solution

A general solution to Equations (5.77) and (5.78) can be investigated in the form of the infinite series

$$\bar{p}(\bar{z}, \bar{t}) = \sum_{n=0}^{n=\infty} \bar{p}_n \sin\left[\frac{(2n+1)\pi}{2} \bar{z}\right] \times \exp\left[-\frac{(2n+1)^2 \pi^2}{4} \bar{t}\right], \tag{5.86}$$

where each term of the series satisfies the diffusion Equation (5.77) and the boundary conditions of Equation (5.78) at $\bar{z} = 0$ and $\bar{z} = 1$. The initial condition requires $\bar{p}(\bar{z}, 0) = 1$, reading

$$\sum_{n=0}^{n=\infty} \bar{p}_n \sin\left[\frac{(2n+1)\pi}{2} \bar{z}\right] = 1. \tag{5.87}$$

When multiplying Equation (5.87) by $\sin((2m+1)\pi\bar{z}/2)$ and integrating from 0 to 1, the left-hand side of the resulting equation is nonzero only for $n = m$, providing us with the values

$$\overline{p}_n = \frac{4}{\pi(2n+1)}, \qquad (5.88)$$

so that

$$\overline{p}(\bar{z},\bar{t}) = \sum_{n=0}^{n=\infty} \frac{4}{\pi(2n+1)} \sin\left[\frac{(2n+1)\pi}{2}\bar{z}\right] \times \exp\left[-\frac{(2n+1)^2\pi^2}{4}\bar{t}\right]. \qquad (5.89)$$

In Figure 5.3 the normalized liquid pressure $\overline{p}$ is plotted against the normalized depth $\bar{z}$ for various normalized times $\bar{t}$, with a comparison with the early time solution of Equation (5.84).

5.3.4 Layer apparent creep

Substitution of Equation (5.89) into Equation (5.79) gives

$$s(t) = s_\infty + (s_0 - s_\infty) \sum_{n=0}^{n=\infty} \frac{8}{\pi^2(2n+1)^2} \times \exp\left[-\frac{(2n+1)^2\pi^2}{4}\frac{t}{\tau}\right], \qquad (5.90)$$

so that the slope at the origin of the consolidation process is

$$\frac{ds}{dt}(t = 0^+) = 2(s_\infty - s_0)\frac{c}{h^2}. \qquad (5.91)$$

Like Equation (5.47) in the pulse test, Equation (5.91) provides us with an experimental mean of the experimental assessment of the diffusion coefficient c.

At the layer scale the consolidation process appears as a creep process under constant load. Indeed, invoking the linearity of the whole set of equations governing the consolidation process, we can derive the settlement due to any history $\varpi(t)$ of the overload in the form

$$s(t) = \int_0^t s_{\varpi=1}(t-u)\,d\varpi(u), \qquad (5.92)$$

where $s_{\varpi=1}(t)$ is the settlement history obtained by taking $\varpi = 1$ in Equation (5.90), and stands for the creep function which can be associated with the layer consolidation. Therefore, similarly to the constitutive Equation (4.91) of viscoelasticity, we can rewrite Equation (5.92) in the concise form

$$s = s_{\varpi=1} \odot \varpi, \qquad (5.93)$$

where $\odot$ is the convolution product defined by Equation (4.89). However, it must be pointed out that the layer creep, as accounted for by Equations (5.92) and (5.93), is only due to the coupling between the deformation of the poroelastic solid and the viscous flow of the saturating fluid caused by the initial pore pressure gradient consecutive to the instantaneous loading of the layer. The layer creep considered above is not due to viscous phenomena associated with the viscoelastic behavior of the solid matrix itself, which would then involve another characteristic creep time distinct from the consolidation characteristic time τ we defined in (5.76). In soil mechanics the intrinsic layer creep which the viscoelasticity of the solid matrix would induce is commonly termed 'secondary' consolidation, because it occurs as the 'primary' consolidation caused by the sole diffusion of the saturating fluid is already achieved. Indeed, the characteristic time τ_c, which scales the intrinsic creep of the solid matrix in Equation (4.94), is generally much greater than the characteristic time τ which scales the fluid diffusion in Equation (5.76).

The consolidation process is due to the progressive dissipation by means of liquid diffusion of the overpressure induced by the instantaneously applied load. Under the applied constant overload ϖ, the energy required for the liquid to diffuse against the resistant viscous forces is supplied by a part of the elastic free energy instantaneously stored in the porous solid and the compressible liquid. Indeed, according to Equation (5.23) the energy D, which is dissipated through the liquid conduction throughout the whole consolidation process, can be written in the form

$$D = \int_0^\infty dt \int_0^h k \left(\frac{\partial p}{\partial z}\right)^2 dz = \left(\frac{3bM\varpi}{3K_u + 4G}\right)^2 \frac{k\tau}{h} \mathcal{I}; \quad \mathcal{I} = \int_0^\infty d\bar{t} \int_0^1 \left(\frac{\partial \bar{p}}{\partial \bar{z}}\right)^2 d\bar{z}. \quad (5.94)$$

Integrating by parts with respect to space, while making use of Equations (5.77) and (5.78) in Equation (5.94), and reversing the integration order, we successively derive

$$\mathcal{I} = \int_0^\infty d\bar{t} \left(\left[\frac{\partial \bar{p}}{\partial \bar{z}} \bar{p}\right]_0^1 - \int_0^1 \frac{\partial^2 \bar{p}}{\partial \bar{z}^2} \bar{p} \, d\bar{z}\right) = -\int_0^\infty d\bar{t} \int_0^1 \frac{\partial \bar{p}}{\partial \bar{t}} \bar{p} \, d\bar{z} = \frac{1}{2}. \quad (5.95)$$

With the help of Equation (5.80) the factor affecting $\mathcal{I}$ in Equation (5.94) for D can be identified as $\varpi s_\infty - \varpi s_0$. Collecting the above results, we obtain

$$D = \frac{1}{2}\varpi s_\infty - \frac{1}{2}\varpi s_0. \quad (5.96)$$

The interpretation of dissipation D can be understood more easily through the decomposition of the dissipation in the form $D = \varpi s_0/2 + (\varpi s_\infty - \varpi s_0) - \varpi s_\infty/2$. The first term, $\varpi s_0/2$, accounts for the elastic energy stored by both the porous solid and the compressible liquid during the instantaneous application of the load, that is, between times $t = 0^-$ and $t = 0^+$. The second term, $\varpi s_\infty - \varpi s_0$, accounts for the external mechanical work supplied to the layer from $t = 0^+$ to $t \to \infty$. The expression of the dissipation is eventually obtained by subtracting the last term, $\varpi s_\infty/2$, which represents the elastic energy finally stored by the sole porous solid at the end of the consolidation process. This is sketched out in Figure 5.4, where the path with an arrow represents a drained unloading process that would be performed infinitely slowly, in order to keep the liquid at zero pressure and to avoid any dissipation.

Up to now we have addressed only saturated conditions, where the porous volume is filled up with a single fluid. This is not usually the case in many processes, such as for instance imbibition, drying, freezing, fluid injection and so on. The porous space is then filled with at least two fluids. A new player then comes into the world of porous solids, namely, the interface energy between the fluids and the solid matrix. The variations of this interface energy govern the invasion of the porous network by the fluids or their recession from it. Before addressing, in the forthcoming chapters, the complex situations where the porous solid deforms under unsaturated conditions, in the next chapter we will first explore for undeformable porous solids how surface energy arises from the molecular scale to determine ultimately at the macroscopic scale how the porous network can be simultaneously filled with several fluids.

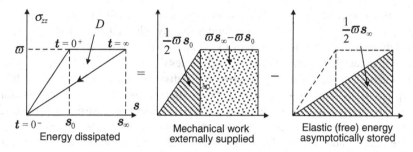

Figure 5.4 Analysis of the dissipated energy in the consolidation process. The path with an arrow represents a drained unloading process that would be performed infinitely slowly in order to keep the fluid at zero pressure and that no dissipation occurs. The energy picture so sketched out is analogous to the one relative to a creep process

Further Reading

Adler, P. (1992) *Porous Media: Geometry and Transports*, Butterworth-Heinemann.
Bear, J. (1988) *Dynamics of Fluids in Porous Media*, Dover Publications, New York (reprint of 1972 edition published by Elsevier, New York).
Bear, J. and Cheng, A. H.-D. (2010) *Modeling Groundwater Flow and Contaminant Transport*, Series: Theory and Applications of Transport in Porous Media, Springer.
Carman, P. C. (1956) *Flow of Gases Through Porous Media*, Butterworths, London.
Coussy, O. (1995) *Mechanics of Porous Continua*, John Wiley & Sons Ltd.
Coussy, O. (2004) *Poromechanics*, John Wiley & Sons Ltd.
de Boer, R. (2000) *Theory of Porous Media: Highlights in Historical Development and Current State*, Springer-Verlag Berlin Heidelberg.
de Groot, S. R. and Mazur, P. (1983) *Non-equilibrium Thermodynamics*, Dover Publications, New York.
de Marsily, J. (1986) *Quantitative Hydrogeology. Groundwater Hydrology for Engineers*, Academic Press, New York.
Detournay, E. and Cheng, A. H.-D (1994) Fundamentals of Poroelasticity, *Comprehensive Rock Engineering* (ed. Hudson J.), Pergamon Press.
Dullien, F. A. L. (1979) *Porous Media: Fluid Transport and Pore Structure*, Academic Press, New York.
Lewis, R. W. and Shrefler, B. A. (1998) *The Finite Element Method in the Static and Dynamic Deformation and Consolidation of Porous Media*, 2nd edn, John Wiley & Sons.
Selvadurai, A. P. S. (ed.) (2008) *Mechanics of Poroelastic Media*, Kluwer Academic Publishers.
Tokaty, G. A. (1994) *A History and Philosophy of Fluid Mechanics*, Dover Publications, Inc., New York.
Wang, H. F. (2000) *Theory of Linear Poroelasticity*, Princeton series in geophysics, Princeton University Press.

6

Surface Energy and Capillarity

As early as the beginning of the eighteenth century the height of the capillary rise of water occurring in a narrow tube against gravity was recognized to be inversely proportional to the diameter of the tube. Although no satisfactory theory was truly available at that time, the physical origin was rightly attributed to intermolecular forces causing the adhesion of water onto the walls of the tube. At the beginning of the nineteenth century Pierre Simon Laplace (1749–1827) elaborated a mathematical approach to capillarity. Involving short-range attractive forces, Laplace found the equation governing the mechanical equilibrium of the meniscus separating the water from the gaseous region above the surface. At the same time Thomas Young (1773–1829), ignoring the complex mathematical manipulations of Laplace, derived an alternative theory based on the surface tension of water. Introducing the concept of the angle of contact, he was able to derive the equation governing the mechanical equilibrium of the triple line. While recognizing in later works the independent research of Young, Laplace criticized him for not relating the capillarity to the existence of intermolecular forces. These intermolecular forces are nowadays termed the van der Waals forces. In 1873, Johannes Diderik van der Waals (1837–1923) gave them their first firm theoretical support by deriving from them the law governing the behavior of real gases. Van der Waals attractive intermolecular forces have various causes. They even exist between nonpolar molecules because of the electric interaction between the instantaneous dipoles of each atom - a dipole made up of the nucleus and the electrons forming the atoms. These forces are nowadays termed dispersive van der Waals forces or London forces, after Fritz Wolfgang London (1900–1954), who derived their expression from quantum mechanics in the late twenties of the twentieth century.

In this chapter, a new energy comes into the world of porous solids, namely, surface energy. This chapter examines in detail how, because of the creation of new solid–fluid interfaces, surface energy governs the intrusion of a fluid phase within a porous solid. Except for the effect of the surface deformation on surface energy, the analysis is restricted to undeformable porous solids, leaving until Chapter 7 the exploration of the mechanics of deformable porous solids in unsaturated conditions. After addressing in a simple way the intermolecular origin of the interface energy existing within two substances, special attention is given to making the distinction between surface energy and surface stress. This distinction turns out to be crucial when examining the equilibrium of a solid–fluid interface. The chapter continues by

---

Mechanics and Physics of Porous Solids   Olivier Coussy
© 2010 John Wiley & Sons, Ltd

successively deriving the Young–Dupré equation and the Laplace equation, which govern the equilibrium of the triple line between three substances and the equilibrium of the interface between two substances, respectively. Further analysis looks at how interfacial energy can be modified by adsorption and how intermolecular forces can induce a disjoining pressure in thin liquid films lying on a solid substrate. The Laplace equation and the surface energy balance combine to afford the microscopic interpretation of the capillary pressure curve, which provides the saturation of the wetting fluid in a porous solid. In relation to the pore size distribution, this capillary curve turns out to be the key state function to describe the macroscopic effects due to capillarity, possibly hysteretic, upon porous solids. After addressing transport laws in unsaturated conditions, the chapter ends by analyzing how advection and diffusion compete in the injection of a wetting phase in a porous medium.

6.1 Physics and Mechanics of Interfaces

6.1.1 Origin of surface energy

When a volume of liquid is separated in two parts, the attractive intermolecular forces which the two parts exert upon each other tend to put them back together. When one of the two parts is removed, the molecules of the remaining part are no longer subjected to the attractive forces that the molecules of the removed part exerted before the separation. This is the origin of surface energy. While the separation has little effect on the bulk molecules some distance from the surface, the molecules near the surface experience a different environment from the one which prevailed before the separation. They are pulled inwards by other molecules deeper inside the liquid whilst they are no longer attracted by the molecules of the removed part. Therefore, all of the molecules at the surface are subject to an inward force of molecular attraction. This force is balanced by the liquid's resistance to compression, so that there is eventually no net inward force. However, the molecules belonging to the surface layer have lost half the energy related to the attraction they were subjected to from the neighboring molecules prior to the separation (see Figure 6.1). Since an attraction energy is negative, this corresponds to an actual gain in energy for the molecules belonging to the surface layer. This gain has the effect of reducing the density in the region close to the surface. In order to lower their state of energy, and thereby to achieve the most stable state, liquids spontaneously tend to

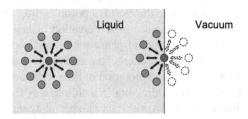

Figure 6.1 Origin of surface energy: compared with the bulk molecules some distance from the liquid surface, the molecules belonging to the surface layer of a liquid lose half the attractive energy from the neighboring molecules. Since an attractive energy is negative, bringing new molecules to the surface of a liquid requires a positive energy proportional to the area of the surface created. The proportionality factor is the surface energy

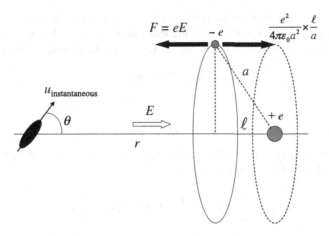

Figure 6.2 Origin of dispersive van der Waals (London) forces for nonpolar molecules: a nonpolar molecule is polarized by the electric field related to the instantaneous dipole associated with a neighboring molecule. An attractive intermolecular force derives from the electrostatic interaction energy between the dipole so induced and the dipole at the origin of it

minimize the surface which delimits them. Indeed, since the increase of the delimiting surface needs to bring new molecules to the surface, a surface increase results in an energy increase proportional to the surface increase. The surface energy is the proportionality factor. The next sections quantitatively address the effects of the surface energy described above, and of the interface energy related to the surface separating two distinct substances, either fluid or solid.

6.1.2 Basic approach to van der Waals forces

The behavior of a gas is governed by the collisions between the molecules due to thermal agitation. In contrast, a liquid is in a condensed state where, in addition to the collisions due to thermal agitation, there is a significant attraction between the molecules which allows the existence of a denser phase. Molecules exhibiting a nonuniformity of their electric charge are polar molecules. The separation between regions which are positively and negatively charged then induces the existence of permanent dipoles which mutually attract through a dipole–dipole interaction. Even though the liquid is formed from nonpolar molecules, such as hydrocarbons for instance, there is still an attraction between the molecules due to the so-called dispersive van der Waals forces. The purpose of this section is to give a fairly basic explanation of the origin of these intermolecular attractive forces.[1]

Consider the simplest neutral molecule sketched out in Figure 6.2, consisting of a one-electron atom, whose electron of negative charge $-e$ orbits at a distance a around a proton with positive charge $+e$. Under the influence of an external electric field of magnitude E, the electron orbit is shifted by a distance ℓ from the nucleus. Owing to the electric charge $+e$ of the proton, this results in the formation of an induced dipole whose moment is the product $u_{\text{induced}} = \ell e$. The neutral atom has been polarized. The force at the origin of the shift has

---

[1] The following presentation is inspired by Israelachvili (1991) referred to in the Further Reading Section.

the intensity $F = eE$. In a vacuum the proton exerts a force in the opposite direction, whose intensity is $(e^2/4\pi\varepsilon_0 a^2) \times \ell/a$ where ε_0 is the vacuum permittivity. Mechanical equilibrium requires that the intensity of both forces is equal. This gives the distance ℓ. We finally obtain

$$u_{\text{induced}} = \ell e = \alpha E, \tag{6.1}$$

where

$$\alpha = 4\pi\varepsilon_0 a^3 \tag{6.2}$$

is the electric polarizability of the medium. The energy $\omega_{0\to\ell}$ required for the formation of the induced dipole is the work done by the force F:

$$\omega_{0\to\ell} = \int_0^\ell F\,dx = \int_0^\ell eE\,d(\alpha E/e) = \frac{1}{2}\alpha E^2. \tag{6.3}$$

The potential energy $\omega(u = e\ell)$ of a dipole u made up of the opposite charges $-e$ and $+e$ located at the infinitesimal distance ℓ from each other can be expressed in the form

$$\omega(u = e\ell) = -e\psi(r - \ell/2) + e\psi(r + \ell/2) \simeq e\frac{\partial\psi}{\partial r}\ell = -Eu, \tag{6.4}$$

where $\psi(r)$ is the electric potential producing the electric field of intensity E in the direction of the dipole u. From Equations (6.1)–(6.4) the total energy of interaction, $\omega_{\text{interaction}}$, of the induced dipole with the electric field E is given by

$$\omega_{\text{interaction}} = \omega_{0\to\ell} + \omega(u_{\text{induced}} = e\ell) = -\frac{1}{2}\alpha E^2. \tag{6.5}$$

For a dipole u oriented at an angle θ to the line joining it to the previous molecule which is a distance r away from the dipole (see Figure 6.2), it may be shown that the magnitude of the electric field of the dipole acting on the molecule is

$$E = u\left(1 + 3\cos^2\theta\right)^{1/2} / 4\pi\varepsilon_0 r^3. \tag{6.6}$$

From Equations (6.5) and (6.6) the dipole-induced dipole interaction energy is therefore

$$\omega_{\text{interaction}}(r, \theta) = -\frac{1}{2}\alpha E^2 = -u^2\alpha\left(1 + 3\cos^2\theta\right)/2(4\pi\varepsilon_0)^2 r^6. \tag{6.7}$$

Because the liquid is made of disordered molecules, the effective interaction energy is the angle-averaged energy. Since the angle average of $\cos^2\theta$ is $1/3$, from Equation (6.7) we derive

$$\omega_{\text{interaction}}(r) = -u^2\alpha/(4\pi\varepsilon_0)^2 r^6. \tag{6.8}$$

In the Bohr atom pictured as an electron orbiting around a proton, the smallest distance a between the electron and the proton is the radius at which the Coulomb electric potential $e^2/4\pi\varepsilon_0 a$ is equal to $2h\nu$, that is,

$$a = e^2/2(4\pi\varepsilon_0)h\nu, \tag{6.9}$$

where $h\nu$ is the smallest energy quantum and is equal to the energy needed to ionize the atom. The Bohr atom has no permanent dipole moment. However, at any instant there exists an instantaneous dipole of moment

$$u = ae. \tag{6.10}$$

Equations (6.2) and (6.9) defining the electric polarizability and the Bohr radius, respectively, combine with Equation (6.10) to give

$$u^2 = 2\alpha h\nu. \tag{6.11}$$

Substituting Equation (6.11) in Equation (6.8) we obtain

$$\omega_{\text{interaction}}(r) = -2h\nu\alpha^2/(4\pi\varepsilon_0)^2 r^6. \tag{6.12}$$

Remarkably, except for the factor 2 which has to be replaced by 3/4, Equation (6.12) agrees with the expression of the attractive interaction energy between two identical nonpolar atoms which London derived in 1930 using quantum mechanical perturbation theory. More generally the attractive interaction energy ω_{12} between two dissimilar atoms 1 and 2 can be written in the form

$$\omega_{12} = -\frac{k\alpha_1\alpha_2}{r^6}. \tag{6.13}$$

The attractive forces deriving from the interaction potential ω_{12} are the so-called dispersive van der Waals forces. For more complex molecules than the one-electron atom represented in Figure 6.2, with permanent dipoles arising from the nonuniformity of the electric charge within the molecule, other polarization interactions will contribute significantly to the net van der Waals force, namely, the dipole–dipole interaction – Keesom forces – and the dipole–dipole induced interaction – Debye forces. A similar approach to the one we carried out for the dispersive forces shows that the $1/r^6$ distance dependence still holds for the overall attractive interaction energy, and the general Equation (6.13) remains relevant.

6.1.3 Surface and interface energy

The attractive interaction energy ω_{12} defined by Equation (6.13) is relative to a pair of molecules of substances 1 and 2. The next step is to relate the pair interaction to the interaction between the molecules of substance 1 to a half-space formed by the molecules of substance 2. Consider the infinitesimal layer of substance 1 in Figure 6.3, having a unit extent in the direction parallel to the half-space of substance 2 and an infinitesimal thickness dZ in the normal direction. The layer surface, which faces the surface delimiting the half-space, is dA. Letting c_1 be the volumetric molecular concentration related to substance 1, this layer contains $c_1 dZ dA$ molecules of substance 1. Similarly, there are $c_2 \times 2\pi \varrho d z d\varrho$ molecules of substance 2 forming the circular ring – radius ϱ and cross-sectional area $dz d\varrho$ – that is located at a distance z away from the previous layer. The distance between the molecules forming the infinitesimal layer and the molecules forming the circular ring is $(z^2 + \varrho^2)^{1/2}$. Since the energy is additive, taking $r = (z^2 + \varrho^2)^{1/2}$ in Equation (6.13) for ω_{12}, the attractive interaction energy between these molecules is $c_1 dZ dA \times 2\pi c_2 \varrho d\varrho dz \times \omega_{12}$, and the integration of this energy with regard to z provides the net interaction $w_{12}(Z) dZ dA$ for the molecules forming the layer at a distance

Z away from the surface of the half-space. We obtain

$$w_{12}(Z)dZdA = -2\pi c_1 c_2 k\alpha_1\alpha_2 dZdA \int_{z=Z}^{z=\infty}\int_{\varrho=0}^{\varrho=\infty} \frac{\varrho d\varrho dz}{(z^2+\varrho^2)^3} = -\frac{\pi k\alpha_1\alpha_2}{6Z^3}c_1c_2 dZdA. \tag{6.14}$$

The net attractive interaction energy $W_{12}\,dA$ experienced by the molecules of substance 1, which constitute the half-space at a distance d away from the surface of the half-space made up of molecules of substance 2, is obtained by integrating $w_{12}(Z)\,dZdA$ from $Z = d$ to infinity. We derive

$$W_{12}dA = \int_{Z=d}^{Z=\infty} w_{12}(Z)dZdA = -\frac{\pi k\alpha_1\alpha_2}{12d^2}c_1c_2 dA. \tag{6.15}$$

The intensity of the interaction energy W_{12} between two media 1 and 2 can be quantified with the help of an intrinsic property, namely, the Hamaker constant:

$$A_H = \pi^2 k\alpha_1\alpha_2 c_1 c_2. \tag{6.16}$$

The interaction energy W_{12} is then written in the form

$$W_{12} = -\frac{A_H}{12\pi d^2}. \tag{6.17}$$

As shown in Figure 6.3, the volumetric interaction energy density w_{12}, because of the factor $1/Z^3$ involved in Equation (6.14), drastically decreases away from the surface as the distance Z becomes large compared with the separation length d. As a result, the net interaction energy

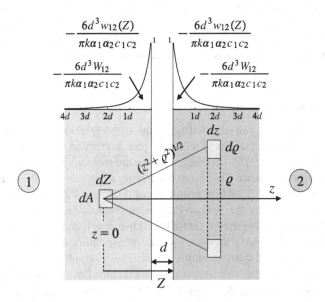

Figure 6.3 Integration of the attractive interaction energy between molecules to obtain the net interaction between two half-spaces of two distinct substances; the net interaction energy $W_{12}dA$ is shown to be mainly concentrated within a thin layer of molecules close to the surface

$W_{12}dA$, that is obtained by summing $w_{12}(Z)dZdA$, is mainly concentrated within a thin layer of molecules close to the surface.

If the two half-spaces are actually formed of the same substance J and if, in addition, the spacing length d is taken as being equal to the intermolecular distance as assumed from now on, the negative energy W_{JJ} is identified as having half the attractive cohesion energy of substance J per unit of surface. As a result, when the substance is separated into two parts in a vacuum, the (positive) opposite energy $-W_{JJ}dA$ is the energy gained by the molecules belonging to the thin layer forming the separation surface dA with regard to the bulk molecules some distance from the surface. The surface energy γ_J of substance J is then defined as

$$\gamma_J = -W_{JJ}. \tag{6.18}$$

According to the above analysis the interface energy γ_{JK} between two distinct substances J and K is given by

$$\gamma_{JK} = \gamma_J + \gamma_K + 2W_{JK}. \tag{6.19}$$

The first two positive terms γ_J and γ_K account for the energy gained by the molecules forming the surface layers of the two substances with respect to their companion bulk molecules, irrespective of the further formation of the interface. The negative term $2W_{JK}$ accounts for the negative attractive energy that these layers exert on each other when the two surfaces are brought together to form the interface. The successive definitions (Equations (6.13)–(6.15) and (6.18)) of ω_{JK}, w_{JK}, W_{JK} and γ_J, respectively, allow us to write

$$W_{JK} = -A_J A_K; \quad \gamma_J = A_J^2, \tag{6.20}$$

where A_J is a positive property of the substance J proportional to its electric polarizability. The net interaction between any pair of substances J and K is then always attractive and the interface energy is positive since the substitution of Equation (6.20) in Equation (6.19) gives

$$\gamma_{JK} = (A_J - A_K)^2 > 0. \tag{6.21}$$

6.1.4 Surface energy and cohesion

According to Section 6.1.3, the surface energy γ_L relates to an attractive property of any liquid surface: the two surfaces generated by the separation of a liquid into two parts tend to join together again, in order to lose their surface energy and recover their original state of lower energy. The surface energy γ_L represents half of the cohesion of the liquid due to intermolecular forces. The order of magnitude of the surface energy related to usual liquids is about $0.030\,\mathrm{J\,m^{-2}}$. The surface energy of liquid mercury, which is quite a cohesive liquid, is much higher, namely, $\gamma_L = 0.5\,\mathrm{J\,m^{-2}}$. The surface energy of water, which is a polar molecule, has a value $\gamma_L = 0.072\,\mathrm{J\,m^{-2}}$ higher than that of usual liquids because of the hydrogen bonds existing between the water molecules. The hydrogen bond, whose intensity is approximately a tenth of that of an ordinary chemical bond, forms because the oxygen atom O in the fragment $-\mathrm{O-H}\cdots$ attracts the electron pair of the chemical bond O–H so strongly that it leaves quasi unprotected the positive charge of the hydrogen nucleus. This positive charge is then attracted by neighboring electrons and, in the specific case of water, by those of an oxygen atom belonging to the neighboring molecules resulting in the formation of the hydrogen bond H···O.

In contrast to a liquid, cutting a solid body into pieces disrupts its bonds. If the cutting is carried out reversibly, according to the cleaved-bond model the energy that the cutting requires will be equal to the energy gained by the two newly created surfaces. The surface energy γ_S of a solid thus quantifies its surface excess energy compared with the bulk, and would be as for a liquid half of its energy of cohesion. In practice, the cleaved bond model is too simplistic, and most solid surfaces change their form, rearrange or react, so that energy is dissipated during the cutting process and the resulting energy excess of the surface created is less than the surface energy. Nevertheless, the surface energy remains a good assessment for half of the cohesion of a solid. Since W_{JJ} has been identified as half the attractive cohesion energy of any substance J per unit of surface, $2\gamma_S = -2W_{SS}$ can be identified with the cohesion of a brittle material. For such a material this remark gives the mean of an assessment of γ_J through the measurement of its toughness K_{IC}. According to Irwin's formula we have

$$2\gamma_S = \frac{1-\nu^2}{E} K_{IC}^2, \quad (6.22)$$

where E and ν are respectively the Young's modulus and Poisson's ratio. The right-hand side of Irwin's formula then represents the critical elastic energy release rate required per unit surface in the brittle fracture of a solid in two pieces. This critical energy release rate has therefore to be equated to the energy $2\gamma_S$ lost per unit of surface newly created. For instance for ice crystals C, taking $\nu = 0.33$, $E = 7.98 \times 10^3$ MPa, $K_{IC} = 0.12$ MPa m$^{\frac{1}{2}}$, we obtain $\gamma_S = 0.8$ J m^{-2}, that is, 10 times the value for liquid water. Basically, in most solids the intermolecular forces are stronger than they are in common liquids, so that the surface energy of solids is higher, often 10 or 20 times higher, than the surface energy of ordinary liquids. Indeed, the surface energy of structural solids is around 1.0 J m^{-2}, while that of diamond is equal to 5.14 J m^{-2}.

6.1.5 Surface energy and surface stress

According to Section 6.1.3 $\gamma_J dA$ is the surface energy associated with the material surface dA of a substance J in a vacuum. As a result, $\gamma_J dA$ is also the work which is needed, in a vacuum, to increase the external surface of substance J from A to $A+dA$.[2] For a fluid F, in most usual situations the infinitesimal area dA originating the extra surface energy $\gamma_F dA$ is provided by new fluid molecules which increase the current area A of the fluid surface. For a solid S, in addition to the creation of a new material solid surface by splitting or by solid–liquid phase transition, the extra area dA can also result from the deformation of the solid surface. Starting from some reference undeformed area A_0 of the solid surface, let A_{new} be the area which, in the undeformed configuration, consists of solid atoms newly belonging to the solid surface with regard to A_0. Furthermore, let A_{def} be the extra area generated by the deformation of the

---

[2] In contrast to the infinitesimal area dA where the symbol d stands for a spatial differentiation, dA is the variation with respect to time of the finite area A, d standing for a time differentiation. When the finite surface A undergoes nonuniform evolutions, the results of this section relative to A, and to its time variation dA, extend to an infinitesimal surface dA forming any part of A, and to its time variation $d(dA)$.

current undeformed solid surface $A_0 + A_{\text{new}}$. These definitions allow us to write

$$A = A_0 + A_{\text{new}} + A_{\text{def}}. \qquad (6.23)$$

The increment $d\epsilon_A$ of the current surface strain ϵ_A can be defined according to

$$d\epsilon_A \big|_{A_{\text{new}}} = \frac{dA_{\text{def}}}{A_0 + A_{\text{new}} + A_{\text{def}}}. \qquad (6.24)$$

Integration of Equation (6.24) with respect to A_{def} from zero to the current value gives

$$\epsilon_A = \ln \frac{A}{A_0 + A_{\text{new}}}. \qquad (6.25)$$

Because of Equation (6.25), ϵ_A is termed the logarithmic surface strain. A total differentiation of Equation (6.25) gives us the relation

$$\frac{A}{A_0 + A_{\text{new}}} d\epsilon_A = d\left(\frac{A}{A_0 + A_{\text{new}}}\right). \qquad (6.26)$$

For a deformable solid surface, because of the very definition of the surface energy γ_S^A, the overall surface energy $\mathcal{U}$ associated with the current deformed solid surface A is expressed in the form

$$\mathcal{U} = \gamma_S^A(\epsilon_A) \times A. \qquad (6.27)$$

The surface energy γ_S^A now depends on the surface strain ϵ_A[3] and is termed the Euler surface energy density, because it relates the surface energy to the current deformed solid surface A. Instead of Equation (6.27) we can also write

$$\mathcal{U} = \gamma_S^A(\epsilon_A) \times \frac{A}{A_0 + A_{\text{new}}} (A_0 + A_{\text{new}}). \qquad (6.28)$$

The surface energy density $\gamma_S^{A_0+A_{\text{new}}} = \gamma_S^A(\epsilon_A) \times A/(A_0 + A_{\text{new}})$ is termed the Lagrange surface energy density because it relates the surface energy to the current undeformed solid surface area $A_0 + A_{\text{new}}$. Differentiating Equation (6.28) and making use of Equation (6.26), we derive

$$d\mathcal{U} = \gamma_S^A \frac{A}{A_0 + A_{\text{new}}} dA_{\text{new}} + d\mathcal{W}, \qquad (6.29)$$

where $d\mathcal{W}$ is the infinitesimal surface strain work defined by

$$d\mathcal{W} = \sigma_S^A A \, d\epsilon_A. \qquad (6.30)$$

In Equation (6.30) σ_S^A is the surface stress, whose expression is given by the Shuttleworth equation:

$$\sigma_S^A = \gamma_S^A + \frac{d\gamma_S^A}{d\epsilon_A}. \qquad (6.31)$$

---

[3] A more general expression would be to make γ_S^A depend upon all the components of the deformation tensor of the solid surface and not only on ϵ_A.

The energy balance Equation (6.29) amounts to applying the first law of thermodynamics to the solid surface in isothermal evolutions. In Equation (6.29) $d\mathcal{U}$ represents the isothermal variation of the Helmholtz free energy of the solid surface. On the right-hand side, the first term accounts for the contribution to $d\mathcal{U}$ due to the formation of the new undeformed surface dA_{new}, whose area transforms into $[A/(A_0 + A_{\text{new}})] \, dA_{\text{new}}$ after deformation; the second term accounts for the strain work related to the solid surface under the action of the surface stress σ_S^A. In turn, Equation (6.31) can be recognized as the elastic constitutive equation of the surface.

It is worthwhile noting that the Euler surface energy γ_S^A and the surface stress σ_S^A are quite different properties. The surface energy γ_S^A is an energy property associated with the creation of a surface, whereas the surface stress σ_S^A is exerted along the existing surface A because of the deformation of the latter. Indeed, the surface strain work dW supplied to the material surface is similar to the volumetric strain work supplied to the material volume Ω in Equation (3.47). Therefore, the surface stress σ_S^A, like any stress, applies in the direction of the strain ϵ_A, whose σ_S^A is the energy conjugate in Equation (6.30) of the strain work dW.

A special case of interest is when the only effect of the surface deformation on the surface energy is to spread out the energy on the current surface A. This results in no new surface energy due to the deformation. The change $d\mathcal{U}$ in surface energy can then only be caused by the creation of a new material surface dA_{new}. As a result, the infinitesimal surface strain work dW must vanish in Equation (6.29) and the surface stress σ_S^A must therefore be zero. Equating σ_S^A to zero in Equation (6.31), we derive the relation $\gamma_S^A(\epsilon_A) = \gamma_S \exp(-\epsilon_A)$, where γ_S is the solid surface energy when the surface strain ϵ_A is zero. In the general case, where the deformation does affect the overall surface energy A, with no loss of generality we can therefore write

$$\gamma_S^A(\epsilon_A) = \gamma_S \, g(\epsilon_A) \exp(-\epsilon_A); \quad g(0) = 1, \tag{6.32}$$

where the function $g(\epsilon_A)$ accounts for the extra change in $\mathcal{U}$ due to the deformation, in addition to the spreading effect. The solid surface can then be viewed as an elastic membrane whose properties differ from those of the bulk. In infinitesimal transformations where $\epsilon_A \ll 1$ and $A_{\text{def}} \ll A_0 + A_{\text{new}}$, when keeping only the terms of the main order of magnitude with respect to ϵ_A, from Equations (6.27), (6.31) and (6.32), we obtain

$$\gamma_S^A \simeq \gamma_S; \quad \sigma_S^A \simeq \gamma_S \frac{dg}{d\epsilon_A}\bigg|_{\epsilon_A=0}; \quad \mathcal{U} \simeq (\gamma_S + \sigma_S^A \epsilon_A)(A_0 + A_{\text{new}}). \tag{6.33}$$

The approximations (6.33) show the distinct roles played by the surface energy, $\gamma_S^A \simeq \gamma_S$, and by the surface stress, σ_S^A, respectively. The surface energy γ_S accounts for the initial surface energy per unit of the undeformed surface $A_0 + A_{\text{new}}$, while $\sigma_S^A \epsilon_A$ is the strain work required to deform the latter. The surface stress σ_S^A acts as a prestress related to the solid surface $A_0 + A_{\text{new}}$.

According to Equation (6.19), Equation (6.27) for the surface energy $\mathcal{U}$ extends to a solid–fluid interface of current area $A_{\text{SF}} \equiv A$ in the form

$$\mathcal{U} = \gamma_{\text{SF}}^A(\epsilon_A) \times A = \left[\gamma_S^A(\epsilon_A) + \gamma_F + 2W_{\text{SF}}^A(\epsilon_A)\right] \times A. \tag{6.34}$$

When the solid remains undeformed, as for instance in Figures 6.5, 6.6 or 6.8, the area change dA belongs to the dA_{new}-type for both the solid and the fluid. The change in the solid–fluid

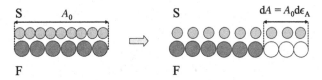

Figure 6.4 Sketch of the solid–fluid interface where, on the solid side, the interface is made of the same solid particles so that the area change is due to the deformation, whereas on the fluid side the area change is due to new fluid particles coming to the interface

interface energy is then $d\mathcal{U} = \gamma_{SF} dA$, where γ_{SF} is the constant interface energy given by

$$\gamma_{SF} = \gamma_S + \gamma_F + 2W_{SF}. \tag{6.35}$$

When the solid phase (for instance ice) is in contact with the liquid phase of the same substance (for instance liquid water), Equation (6.35) still applies. The common area change dA then belongs to the dA_{new}-type since it is achieved by the melting or the solidification of the interface.

In contrast, let us now examine the situation where, in the absence of further splitting or phase transition, a deformable solid S made of the same solid particles always remains in contact with the same fluid F. The situation is sketched in Figure 6.4 and, for instance, is the case analyzed in Section 7.1 where a deformable porous solid remains saturated by the same fluid. On the solid side, the common value dA of the area change is only due to the deformation. Taking $A_{\text{new}} = 0$ in Equation (6.25) and again assuming infinitesimal transformations, that is, $\epsilon_A \ll 1$, so that $A \simeq (1 + \epsilon_A) A_0$, the area change dA reduces to

$$dA = A_0 \, d\epsilon_A. \tag{6.36}$$

Differentiating Equation (6.34) and taking into account Equation (6.36), we obtain

$$d\mathcal{U} = d\mathcal{W}, \tag{6.37}$$

where $d\mathcal{W}$ is the interface infinitesimal strain work,

$$d\mathcal{W} = \sigma_{SF}^A A_0 \, d\epsilon_A, \tag{6.38}$$

providing σ_{SF}^A is defined by extending the Shuttleworth Equation (6.31) in the form

$$\sigma_{SF}^A = \gamma_{SF}^A + \frac{d\gamma_{SF}^A}{d\epsilon_A}, \tag{6.39}$$

which is also

$$\sigma_{SF}^A = \sigma_S^A + \gamma_F + 2\left(W_{SF}^A + \frac{dW_{SF}^A}{d\epsilon_A}\right). \tag{6.40}$$

Comparing Equations (6.38) and (6.39) with Equations (6.30) and (6.31) respectively, σ_{SF}^A can be termed the interface stress. Its explicit expression given in Equation (6.40) shows its hybrid nature, since σ_{SF}^A is the sum of the surface stress σ_S^A, the fluid surface energy γ_F, and an extra surface stress coming from the solid–fluid interaction. In short, like the surface stress σ_S^A,

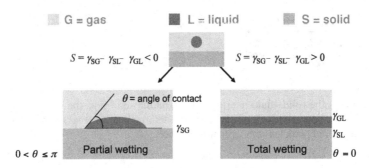

Figure 6.5 The wetting is total or partial according to the sign of the spreading coefficient S, which determines the most favourable state regarding the surface energy

the interface stress σ_{SF}^A acts as a prestress. Its effect upon porous solids will be examined in Section 7.1.

6.1.6 Wettability, angle of contact and the Young–Dupré equation

A drop of a liquid (subscript L) will completely spread on a solid flat surface (subscript S) if the complete spreading minimizes the interface energy and therefore corresponds to a stable state. The third fluid in the picture is a gas (subscript G). The complete spreading will then occur whenever the spreading coefficient S has a positive value,[4] that is,

$$S = \gamma_{SG} - \gamma_{SL} - \gamma_{GL} > 0. \tag{6.41}$$

In other words, when the condition of Equation (6.41) is met, the creation of an interface S–G having a given area has a cost in surface energy which is greater than the one associated with the creation of two separate interfaces, S–L and G–L, having the same area. As a result, the solid and the gas will therefore not come into contact. The wetting associated with complete spreading is termed total wetting. This situation is illustrated in the right-hand side of Figure 6.5.

When the spreading coefficient S is negative, the wetting becomes partial. When only a small amount of liquid is involved, a liquid droplet forms, which makes a contact angle θ with the solid substrate. This situation is illustrated in the left-hand side of Figure 6.5. The angle of contact θ ranges from zero – total wetting – to π – no wetting at all. When θ lies between 0 and $\pi/2$ the liquid is the wetting phase, whereas the gas is the nonwetting phase. It is the

---

[4] A rough approach to discover the sign of S may be the following. According to Equation (6.15) the cohesion of a gas, $\gamma_G = -W_{GG}$, is proportional to c_G^2, while the intermolecular interaction energies, W_{SG} and W_{GL}, are proportional to $c_S c_G$ and $c_G c_L$, respectively. Because of the low molecular concentration c_G of a gas, when compared with the concentration c_S or c_L of a solid or a liquid, the gas cohesion γ_G, as well as W_{SG} and W_{GL}, can be neglected with respect to the other interaction energies involved in Equation (6.19) for γ_{SG} or γ_{GL}. This allows us to make the approximations $\gamma_{SG} \sim \gamma_S$ and $\gamma_{GL} \sim \gamma_L$, whose substitution in Equation (6.41), together with the expression of γ_{SL} provided by Equation (6.19), leads to the approximation $S \sim -2\gamma_L - 2W_{SL} = 2(W_{LL} - W_{SL})$. Equation (6.15) for W_{JK} then shows that the sign of S is the same as $\alpha_S - \alpha_L$. When the order of magnitude of the concentrations $c_{J=1,2,3}$ is the same, from Equation (6.15) and Equation (6.19) it can be shown that $S = \gamma_{13} - \gamma_{12} - \gamma_{23}$ has the same sign as that of $(\alpha_1 - \alpha_2)(\alpha_2 - \alpha_3)$.

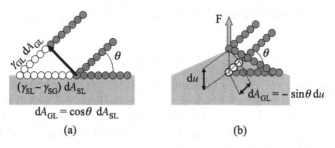

Figure 6.6 (a) Contact angle and the Young–Dupré equation – the Young–Dupré Equation (6.43) is derived by writing that the surface energy must be at a minimum when the equilibrium is achieved; (b) vertical equilibrium of the triple line – the equilibrium condition is derived by writing that the overall potential energy of the interface is at a minimum

opposite when θ lies between $\pi/2$ and π. For instance, when compared with liquid mercury, the mercury vapor is the wetting phase, with an angle of contact close to 130°. The relative wettability of two liquids may be similarly compared by means of their contact angle with a given solid substrate. In Figure 6.5 the gas is then replaced by one of the two liquids. For instance,[5] on a polystyrene flat surface a nonpolar oil like n-decane is the wetting phase when compared with a drop of liquid water, the contact angle of the water drop being close to 130°.

The actual value of θ corresponds to the most favorable state regarding the interface energy. The equilibrium is therefore achieved when the overall interface energy $\mathcal{U}$ is at a minimum, so that its variation, $d\mathcal{U}$, must be zero for any further motion of the triple line on the solid substrate. This condition allows us to write

$$d\mathcal{U} = (\gamma_{SL} - \gamma_{SG})\,dA_{SL} + \gamma_{GL}dA_{GL} = 0, \qquad (6.42)$$

where, as illustrated in Figure 6.6(a), $(\gamma_{SL} - \gamma_{SG})dA_{SL}$ accounts for the interface energy required to form the new infinitesimal area dA_{SL} of the interface between the solid substrate and the liquid to the detriment of the gas–solid interface. Similarly, $\gamma_{GL}dA_{GL}$ is the energy required to form the new infinitesimal area dA_{GL} of the gas–liquid interface. We have $dA_{GL} = \cos\theta\,dA_{SL}$. Substituting this in the energy balance Equation (6.42), we obtain the Young–Dupré equation

$$\gamma_{SG} = \gamma_{SL} + \gamma_{GL}\cos\theta, \qquad (6.43)$$

which accounts for the horizontal equilibrium of the triple line. The vertical equilibrium of the triple line involves the vertical linear force F originating from attractive intermolecular forces, which is transmitted to the solid substrate. In turn, the solid substrate exerts upon the triple line the opposite force $-F$. So let u be the vertical displacement of the triple line. When the equilibrium is achieved, the overall potential energy $\mathcal{U}+Fu$ of the interface must be at a minimum. In the displacement field represented in Figure 6.6(b), we have $d\mathcal{U} = \gamma_{GL}dA_{GL} = -\gamma_{GL}\sin\theta\,du < 0$. Indeed, when considering only the main order of magnitude, the only change in interface area is the one of the liquid–gas interface which is induced by the removal

---

[5] See Paunov, V. N. (2003) Novel Method for Determining the Three-Phase Contact Angle of Colloid Particles Adsorbed at Air–Water and Oil–Water Interfaces, *Langmuir*, **19**, 7970–7976.

of the molecules shown in dotted lines. As a result, the minimum condition $d\mathcal{U} + F du = 0$ provides us with the condition

$$F = \gamma_{GL} \sin \theta. \qquad (6.44)$$

The determination of the vertical displacement u involves both the constitutive equation and the geometry of the solid substrate, which govern how the latter deforms in response to the vertical force F per unit length located at the triple line. Because they separately account for the equilibrium of the triple line, the Young–Dupré Equation (6.43) and Equation (6.44) hold irrespective of the form of the droplet.

For larger liquid amounts, if the spreading parameter S is still negative, the wetting phase eventually forms a puddle. The main part of the puddle consists of a liquid layer of constant thickness e and surface area A, whose edges join the solid substrate according to the wetting angle. The actual value of the puddle thickness e corresponds to the most favorable state regarding energy. Here, the energy to consider is the overall potential energy $\mathcal{E}_p$, which encompasses the potential energy $\mathcal{V}_p$ related to the gravity forces and the overall Helmholtz free energy F. Providing the liquid compressibility is neglected, the Helmholtz free energy of the system reduces to the interface energy. When neglecting the edge effects, apart from a constant the overall potential energy $\mathcal{E}_p$ is then given by

$$\mathcal{E}_p = \mathcal{V}_p + F = \frac{1}{2}\rho_L g e^2 A - SA. \qquad (6.45)$$

The stable equilibrium of the puddle is achieved when $\mathcal{E}_p$ is at a minimum.[6] The variation $d\mathcal{E}_p$ must be zero for any further motion of the edges of the puddle on the solid substrate. Since the liquid compressibility is neglected, the volume eA of the puddle remains constant. Using the relation $d(eA) = 0$ when expressing the minimum condition $d\mathcal{E}_p = 0$, from Equation (6.45) we obtain

$$e = \sqrt{-2S/\rho_L g} = \sqrt{2\gamma_{GL}(1 - \cos\theta)/\rho_L g}. \qquad (6.46)$$

If R is the radius of the spherical droplet consisting of the same liquid amount, the puddle actually forms when R is much larger than the capillary length ℓ_{cap} defined by

$$\ell_{cap} = \sqrt{\gamma_{GL}/\rho_L g}. \qquad (6.47)$$

The equilibrium height h of a liquid rising in a capillary tube S of radius r above a reservoir can be determined in a similar way. Apart from a constant the overall potential energy $\mathcal{E}_p$ is given by

$$\mathcal{E}_p = \frac{1}{2}\pi r^2 \rho_L g h^2 + 2\pi r h (\gamma_{SL} - \gamma_{SG}). \qquad (6.48)$$

---

[6] The first and second laws applied to isothermal evolutions ($dT = 0$) of the system state $dE = dF + TdS = \delta W + \delta Q$ and $TdS \geq \delta Q$, where E and S are the system internal energy and entropy, while δW and δQ are respectively the infinitesimal mechanical work and heat supplied to the system. Besides, the infinitesimal work δW of external forces deriving from the potential $\mathcal{V}_p$ can be expressed as $\delta W = -d\mathcal{V}_p$. Therefore, from the first and second laws $d\mathcal{E}_p = d(\mathcal{V}_p + F) \leq 0$: the overall potential energy $\mathcal{E}_p$ can only spontaneously decrease, so that the stable equilibrium is achieved when $\mathcal{E}_p$ is at a minimum.

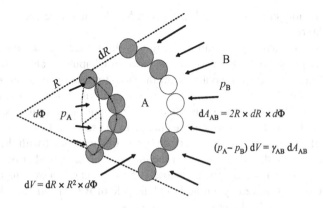

Figure 6.7 Derivation of the Laplace equation

The first term is positive and accounts for the energy cost related to the gravity forces which oppose the rise of the liquid in the capillary tube. In contrast, the second term is negative and is favorable to the rise of the liquid. It accounts for the interface energy loss related to the replacement of the gas G contained in the tube by the rising liquid L. The equilibrium is achieved when $\mathcal{E}_p$ is at a minimum with respect to the height h. The condition $\partial \mathcal{E}_p / \partial h = 0$ and the Young–Dupré Equation (6.43) provide the Jurin rule or Jurin height h of the capillary rise:

$$h = \frac{2\gamma_{GL} \cos \theta}{\rho_L g r}. \qquad (6.49)$$

6.1.7 The Laplace equation

Now we need to express the mechanical equilibrium of a curved fluid–liquid interface between a fluid — or liquid — A and a liquid — or fluid — B, under the combined action of its interface energy and the pressure difference. As illustrated in Figure 6.7, in a displacement field resulting in the infinitesimal increase dR of the curvature radius R, the work dW produced by the pressure difference $p_A - p_B$ applying to the infinitesimal spherical segment of the initial surface $R^2 d\Phi$, where $d\Phi$ is the infinitesimal solid angle, is

$$dW = (p_A - p_B)\, R^2 dR d\Phi. \qquad (6.50)$$

For the same displacement field the interface energy change $\gamma_{AB}\, dA_{AB}$ due to the change dA_{AB} in the contact area between fluids A and B is given by

$$\gamma_{AB}\, dA_{AB} = 2\gamma_{AB}\, R\, dR d\Phi. \qquad (6.51)$$

The equilibrium of the interface requires that its potential energy should be at a minimum or, equivalently, dW and $\gamma_{AB} dA_{AB}$ should be equal, resulting in the Laplace equation, namely,

$$p_A - p_B = \frac{2\gamma_{AB}}{R}. \qquad (6.52)$$

If the interface is no longer spherical, in Equation (6.52) the curvature $1/R$ must be replaced by the mean curvature.

Instead of the mechanical equilibrium of a liquid–fluid interface, we may also need to express that of a fluid–solid interface, as for instance in Chapters 8 and 9, where the solid phase of a substance, say ice, is in contact with the liquid phase of the same substance, say liquid water. The expression of γ_{AB} to use in Equation (6.52) is then the interface energy γ_{SF} given by Equation (6.35). However, if, as illustrated in Figure 6.4, no new solid molecules are added to the solid part of the interface, the change in the solid surface energy results from its deformation only. Providing the solid surface S remains in contact with the same fluid F, the energy associated with the change dA_{SF} of the interface area, instead of being $\gamma_{SF} dA_{SF}$, is $\sigma_{SF}^A dA_{AB}$, where σ_{SF}^A is the interface stress, whose expression is given by Equation (6.40). As a result, if the role of A is played by a fluid F and the role of B by a solid S in Figure 6.7, the Laplace equation related to this latter situation becomes

$$p_F - p_S = \frac{2\sigma_{SF}^A}{R}. \qquad (6.53)$$

Equation (6.53) also accounts for the mechanical equilibrium of an interface involving a liquid membrane with a nonzero surface stress. An instructive illustration is provided by the equilibrium of two air-filled soap bubbles connected by a pipe. Since the gas is free to move from one bubble to the other, the excess pressure $p_{bubble} - p_{atm}$ within the two bubbles, as well as their radius R, initially will be the same. However, if the equilibrium of the membrane delimiting the bubbles is governed by Equation (6.52), with a constant surface energy, the system of bubbles is unstable. Indeed, a spontaneous transfer of air from one bubble to the other through the pipe will cause the former bubble to shrink and its excess pressure $p_{bubble} - p_{atm}$ to increase. In contrast, the other bubble will swell and its excess pressure will decrease, so that the pressure difference between the two bubbles triggered in this way will carry on increasing. The system of bubbles is therefore unstable. The system can only be stable if the excess of pressure satisfies the condition $d(p_{bubble} - p_{atm})/dR > 0$. This condition can be fulfilled providing $p_{bubble} - p_{atm}$ is governed by Equation (6.53) instead of Equation (6.52). Noting that $d\epsilon_A = 2dR/R$, the derivation of the condition that the surface stress $\sigma_{SF}^A(\epsilon_A)$ must satisfy is left to the reader.

6.1.8 Pore invasion and interface energy change

In an unconfined environment the Laplace Equation (6.52) holds no matter what the fluids A and B are. Only the sign of the pressure difference $p_A - p_B$ imposes the direction of the A–B interface curvature, and it is possible to have both a bubble of fluid A embedded in fluid B and vice versa. In a confined environment where a solid is also present, such as the truncated conical pore represented in Figure 6.8, this is no longer true because of the contact angle θ resulting from the relative wettability of the fluids with regard to the solid. The equilibrium of the junction line between the three constituents (point P in Figure 6.8) is then governed by the Young–Dupré Equation (6.43), that is,

$$\gamma_{SnW} = \gamma_{SW} + \gamma_{nWW} \cos\theta, \qquad (6.54)$$

Surface Energy and Capillarity

where the subscripts W and nW refer to the wetting fluid and the nonwetting fluid, respectively. The interface is curved in the direction of the wetting fluid W so that the Laplace Equation (6.52) takes the more specific form

$$p_{n\text{W}} - p_\text{W} = \frac{2\gamma_{n\text{WW}}}{R}. \tag{6.55}$$

In the case of the truncated conical pore represented in Figure 6.8, noting

$$\beta = \pi/2 - \alpha - \theta, \tag{6.56}$$

where α stands for half the cone angle, and letting r be the radius of the cone section at the interface between the two fluids, we have

$$r = R \sin \beta. \tag{6.57}$$

For a cylindrical pore $\alpha = 0$, and Equations (6.55) and (6.57) then combine to give us the relation

$$p_{n\text{W}} - p_\text{W} = \frac{2\gamma_{n\text{WW}} \cos \theta}{r}, \tag{6.58}$$

which is the cornerstone of porosimetry as we will see in Section 6.2.3.

Let us now state the energy balance governing the motion of the interface between the nonwetting and wetting fluids along the symmetry axis of the conical pore of Figure 6.8. The infinitesimal displacement dL along the conical pore wall of the fluid interface requires the infinitesimal variation

$$d\mathcal{U} = (\gamma_{S n\text{W}} - \gamma_{S\text{W}}) dA_{S n\text{W}} + \gamma_{n\text{WW}} dA_{n\text{WW}} \tag{6.59}$$

of the overall interface energy $\mathcal{U}$. In Equation (6.59) $A_{S n\text{W}}$ and $A_{n\text{WW}}$ stand for the area between the solid and the nonwetting fluid, and the area of the interface between the two fluids, respectively. As shown in Figure 6.8, $A_{S n\text{W}}$ is the area of the truncated cone ranging in radius from r_0 to r and whose height is $H = L \cos \alpha$, whereas $A_{n\text{WW}}$ is the area of the spherical cap having R as radius and h as height:

$$A_{S n\text{W}} = \pi (r + r_0) L; \quad A_{n\text{WW}} = 2\pi R h. \tag{6.60}$$

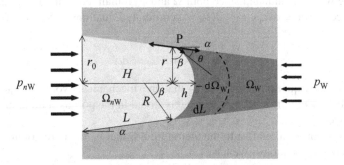

Figure 6.8 The Laplace equation and related energy balance in a confined environment

Because both angles α and θ are constant, there is only one independent length among r, L, H, h and R. Indeed, in addition to Equation (6.57), we have the relations

$$h = R(1 - \cos\beta); \quad H = L\cos\alpha = (r_0 - r)\cot\alpha. \tag{6.61}$$

Substituting Equation (6.60) in Equation (6.59), while making use of Equations (6.54), (6.57) and (6.61), we obtain

$$d\mathcal{U} = \left[-\frac{\sin^2\beta}{\sin\alpha}\cos\theta + 2(1 - \cos\beta)\right] \times 2\gamma_{n\text{WW}} \times \pi R \, dR. \tag{6.62}$$

Let $\Omega_{n\text{W}}$ be the current volume of the nonwetting fluid. It is the sum of the volume $\Omega_{n\text{W}}^{\text{cone}}$ of the previous truncated cone of height H and the volume $\Omega_{n\text{W}}^{\text{cap}}$ of the spherical cap of height h. We have

$$\Omega_{n\text{W}}^{\text{cone}} = \frac{\pi H}{3}\left(r_0^2 + r_0 r + r^2\right); \quad \Omega_{n\text{W}}^{\text{cap}} = \pi h^2(R - h/3) = \frac{4\pi R^3}{3}\phi(\beta), \tag{6.63}$$

where $4\phi(\beta)$ is the function

$$4\phi(\beta) = (1 - \cos\beta)^2(2 + \cos\beta). \tag{6.64}$$

Using Equations (6.57), (6.56) and (6.61), from Equation (6.63), we obtain

$$d\Omega_{n\text{W}} = d\Omega_{n\text{W}}^{\text{cone}} + d\Omega_{n\text{W}}^{\text{cap}} = \left(-\frac{\cos\alpha \sin^3\beta}{\sin\alpha} + 4\phi(\beta)\right) \times \pi R^2 dR. \tag{6.65}$$

Noting the identity

$$(1 - \cos\beta)^2(2 + \cos\beta) = 2(1 - \cos\beta) - \cos\beta\sin^2\beta, \tag{6.66}$$

and making use of Equations (6.56), Equations (6.62) and (6.65) combine to

$$\frac{d\mathcal{U}}{d\Omega_{n\text{W}}} = \frac{2\gamma_{n\text{WW}}}{R}. \tag{6.67}$$

Then let Ω be the overall volume of the schematic porous solid represented in Figure 6.8, and let ϕ_0 be its porosity. Since we have $\phi_0\Omega = \Omega_{n\text{W}} + \Omega_{\text{W}}$, where Ω_{W} stands for the volume occupied by the wetting fluid, and since both Ω and ϕ_0 remain constant during the motion of the fluid interface, we have $d\Omega_{n\text{W}} = -d\Omega_{\text{W}}$. Accordingly Equations (6.55) and (6.67) produce the relation

$$-(p_{n\text{W}} - p_{\text{W}})d\Omega_{\text{W}} = d\mathcal{U}. \tag{6.68}$$

The energy balance Equation (6.68) states that the mechanical work $-(p_{n\text{W}} - p_{\text{W}})d\Omega_{\text{W}}$ produced by the pressure difference $p_{n\text{W}} - p_{\text{W}}$ is stored in the porous solid in the form of the interface energy $d\mathcal{U}$.

Let S_{W} be the saturation relative to the wetting fluid, that is, the pore volume fraction the wetting fluid currently occupies. We have

$$S_{\text{W}} = \Omega_{\text{W}}/\phi_0\Omega. \tag{6.69}$$

Substitution of Equation (6.69) in Equation (6.68) gives

$$p_{n\text{W}} - p_{\text{W}} = -\frac{dU}{dS_{\text{W}}}, \qquad (6.70)$$

where $U = \mathcal{U}/\phi_0 \Omega$ stands for the overall interface energy per unit of porous volume $\phi_0 \Omega$. The interface energy balance Equation (6.70) controls the pore invasion.

The derivation of Equation (6.70) amounts to finding the minimum of the potential energy $\mathcal{E}_p = U - (p_\text{W} - p_{n\text{W}}) S_\text{W}$ with regard to S_W. The pressure difference $p_\text{W} - p_{n\text{W}}$ is the monitored variable, whilst the energy conjugate variable S_W is the driven variable. An alternative way to derive Equation (6.70) is to consider the opposite case, where the volume Ω_W of the injected wetting fluid is the monitored variable. The minimization of the interface energy U must then be expressed in the form

$$dU + \mathcal{L} dS_\text{W} = 0, \qquad (6.71)$$

where $\mathcal{L}$ is the Lagrange multiplier associated with Equation (6.69) imposed upon S_W relative to the monitored volume Ω_W of the injected wetting fluid. Comparing Equation (6.71) with Equation (6.70), the Lagrange multiplier $\mathcal{L}$ is then identified with the pressure difference $p_{n\text{W}} - p_\text{W}$.

As we will see in Section 6.2, the Laplace Equation (6.58) and the interface energy balance Equation (6.70) constitute the key relations for the understanding of the origin of the macroscopic capillary pressure curve related to complex porous solids.

6.1.9 Interface energy and adsorption

Up to now, when addressing surface energy effects, we have always considered pure substances. When only considering a solid–fluid mixture interface, molecules of the components forming the fluid mixture can be adsorbed or bound on the solid substrate. Since the binding of molecules modifies the composition of the molecules forming the interface, adsorption impinges on the fluid mixture–solid interface energy γ_SF. In detergents, for instance, the efficiency of surfactants on the dirt particles is governed by their capability to change the interface energy through adsorption. However, according to the Laplace Equation (6.53) a change in the interface properties affects the interface stress σ_SF^A and the pressure that is transmitted to the solid. Adsorption can therefore make a porous solid deform in the absence of any applied external stress. This will be addressed in Section 7.1.2. The purpose of this section is to revisit how the interface energy change due to adsorption can be assessed through the Gibbs adorption isotherm. Consider the volume V of Figure 6.9 filled with a fluid mixture. In the absence of any solid substrate the volumetric concentration c_J of constituent J is homogeneous within the volume V. The total number $\mathcal{N}_J$ of molecules of constituent J present in volume V is therefore

$$\mathcal{N}_J = c_J V. \qquad (6.72)$$

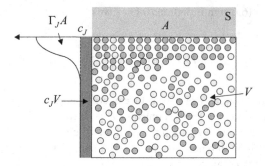

Figure 6.9 Definition of surface excess amount Γ_J of constituent J involved in the Gibbs adsorption isotherm

Assuming isothermal processes for the sake of simplicity,[7] the energy balance related to the volume V assumed to remain constant is given by taking $dV = 0$ in Equation (2.4). We obtain

$$d\mathcal{F} = \sum_J \mu_J d\mathcal{N}_J, \tag{6.73}$$

where $\mathcal{F}$ is the free Helmholtz energy related to the volume V. Because of adsorption the presence of a solid surface A on the border of V affects the concentration of constituent J close to the surface. As illustrated in Figure 6.9, there is a transition layer of molecular dimensions where the concentration increases from the concentration c_J prevailing a long way from the surface A in the bulk volume V. As a result, the total number of molecules of constituent J is no longer $\mathcal{N}_J$ defined by Equation (6.72), but $N_J > \mathcal{N}_J$. Assuming that no molecules of the fluid mixture penetrate the solid, the surface amount excess N_J^A of constituent J is then defined by

$$N_J^A = N_J - \mathcal{N}_J. \tag{6.74}$$

Instead of a tricky measurement of the number of adsorbed molecules in the transition layer, the surface amount excess N_J^A can be assessed experimentally by determining both the total number N_J of molecules present in the volume V, and the concentration c_J a long way from the solid surface A, the latter providing the value of $\mathcal{N}_J$ defined by Equation (6.72). In the presence of the solid surface A assumed to be undeformable, the isothermal energy balance related to the constant volume V is now derived in the form

$$dF = \sum_J \mu_J dN_J + \gamma_{SF} dA, \tag{6.75}$$

where $\gamma_{SF} dA$ accounts for the infinitesimal work associated with the area change dA of the surface A, while F is the Helmholtz free energy of the volume V in the presence of the solid surface A. Since the solid surface A is assumed undeformable, γ_{SF} is the interface energy whose expression is given by Equation (6.35). The energy balance only related to the surface A is obtained by assessing the energy variation $dF^A = d(F - \mathcal{F})$ introduced by the presence

---

[7] Basically, all the surface properties considered in this book are isothermal properties. A more refined approach would have to consider the entropy surface excess too.

of the surface. From energy balance Equations (6.73) and (6.75), and the definition given in Equation (6.74) for the surface amount excess, we obtain

$$dF^A = \sum_J \mu_J dN_J^A + \gamma_{SF} dA. \tag{6.76}$$

Since the changes in the area A are due to the addition of new molecules, A is an extensive quantity. Similarly to Equation (2.11), this remark allows us to write

$$F^A\left(\lambda N_J^A, \lambda A\right) = \lambda F^A\left(N_J^A, A\right). \tag{6.77}$$

Using Euler's theorem, in the same way as we derived Equation (2.13) from Equation (2.11), from Equation (6.77) we derive

$$F^A = \sum_J \mu_J N_J^A + \gamma_{SF} A. \tag{6.78}$$

Substitution of Equation (6.78) in Equation (6.76) gives the Gibbs adsorption isotherm governing the interface energy change due to adsorption, namely,

$$d\gamma_{SF}|_A = -\sum_J \Gamma_J d\mu_J, \tag{6.79}$$

where

$$\Gamma_J = N_J^A/A \tag{6.80}$$

is the number of moles in excess of component J, which are adsorbed per surface unit of the interface.

Considering a fluid made of only a single component and noting that γ_S does not depend on the fluid state, substitution of Equation (6.35) in the Gibbs adsorption isotherm of Equation (6.79) gives the relation

$$d\left(\gamma_F + 2W_{SF}\right)|_A = -\Gamma d\mu. \tag{6.81}$$

When the fluid phase reduces to a vacuum, the chemical potential μ goes to $-\infty$ and $\gamma_F + 2W_{SF}$ vanishes too. As a result, integration of Equation (6.81) leads to

$$\gamma_F + 2W_{SF} = -\int_{-\infty}^{\mu} \Gamma d\overline{\mu}. \tag{6.82}$$

When the fluid adjoins a vacuum, the interaction energy W_{SF} is zero. Since $\gamma_F = -W_{FF}$ is positive, Equation (6.82) shows that the surface density Γ of moles in excess must then be negative. Because of the surface energy the molecules close to the surface facing the vacuum gain an intermolecular energy greater than that of the molecules in the bulk fluid. This larger intermolecular energy has the effect of reducing the molecular density in the region of the surface, resulting in $\Gamma < 0$. Indeed, the molecular density must reduce in order that the chemical potential in the region of the surface, thanks to a smaller thermal contribution, remains equal to the chemical potential μ prevailing in the bulk fluid. When the fluid adjoins a solid, the interaction attractive energy $2W_{SF}$, which is negative, opposes the effect of the fluid surface energy γ_F and tends to increase Γ. The current density is then governed by the energy balance Equation (6.82). In turn, if the determination of the density Γ of moles in excess can be carried out independently of the Gibbs adsorption isotherm of Equation (6.82), the current interaction energy $\gamma_F + 2W_{SF}$ is then known. This is the object of the next paragraph.

At early stages of adsorption the adsorbed layer reduces to a monolayer.[8] Adsorption then corresponds to the trapping of molecules in energy wells existing on the solid wall in well-identified sites, whose intensity depends on both the gas and the solid substrate. When all the energy wells have the same depth $-E_0$, and do not interact with each other, the expression of Γ related to an ideal gas is provided by Langmuir's adsorption isotherm in the form

$$\Gamma = \frac{\Gamma_\infty p}{p + \varpi_0}, \qquad (6.83)$$

where p and ϖ_0 are respectively the gas pressure and a reference pressure depending only on the temperature and proportional to the Boltzmann factor $\exp(-E_0/kT)$; Γ_∞ is the number of energy traps per unit area of the solid surface so that $\Gamma = \Gamma_\infty$ as p tends to infinity. For a gas G = F, the cohesion γ_G is zero. Substituting both Equation (6.83) and the differentiated state equation of ideal gases, that is, $d\mu = RT dp/p$, in Equation (6.81), we can integrate Equation (6.81) in the form

$$2W_{SG} = -RT\Gamma_\infty \ln(1 + p/\varpi_0), \qquad (6.84)$$

which satisfies $W_{SG} = 0$ for $p = 0$, that is, for a solid matrix in the presence of a vacuum.

To derive the Gibbs adsorption isotherm of Equation (6.82) governing the variations $d\gamma_{SF}$ of the interface energy γ_{SF}, the solid surface is assumed to be undeformable so that the area change dA is made of both new solid and new fluid particles. In contrast, from the solid side, when the surface change is only due to deformation (as illustrated in Figure 6.4), instead of the energy balance Equation (6.76) we now write

$$dF^A = \sum_J \mu_J dN_J^A + dW = \sum_J \mu_J dN_J^A + A_0 \sigma_{SF}^A d\epsilon_A, \qquad (6.85)$$

where, in agreement with Equation (6.38), the term $dW = A_0 \sigma_{SF}^A d\epsilon_A$ accounts for the interface strain work produced by the interface stress σ_{SF}^A. In addition, invoking the additivity of energy, in a way consistent with Equation (6.34), Equation (6.78) for interface free energy F^A is replaced by[9]

$$F^A = \sum_J \mu_J N_J^A + \gamma_{SF}^A (\epsilon_A, \mu_J) A. \qquad (6.86)$$

When the solid surface is rigid Equation (6.86) reduces to Equation (6.78); when there is no adsorption Equation (6.86) reduces to $F^A = \mathcal{U}$, where $\mathcal{U}$ is the interface energy whose expression is given in Equation (6.34). Noting that ϵ_A and μ_J are independent state variables

---

[8] At later stages the formation of an adsorbed layer of a gas can be viewed as the early formation of a precondensed liquid film with a nonzero thickness e. In Section 8.5.2 of Chapter 8 devoted to unconfined phase transitions, this approach will allow us to derive the expression of $\Gamma = e/\overline{V}_L - e/\overline{V}_G$. The related Equation (8.103) for the effective interfacial energy γ_{SV}^{eff} may then be used for γ_{SF}.

[9] It is worthwhile noting that the derivation of Equation (6.86) cannot be worked out in the same way as the derivation of Equation (6.78) for F^A from Euler's theorem applied to Equation (6.77). This is because the area change $dA = Ad\epsilon^A$ of the solid surface is not due to the addition of new solid particles, but to the deformation of A. As a result, Equation (6.77) can no longer be claimed to hold, so that Euler's theorem cannot be invoked to obtain Equation (6.86) for F^A.

and using Equation (6.36), substitution of Equation (6.86) in Equation (6.85) allows us to recover Equation (6.39) already obtained for the interface stress σ_{SF}^A, providing $d/d\epsilon_A$ is replaced by $\partial/\partial\epsilon_A$. In addition, the Gibbs adsorption isotherm related to the deformable interface becomes

$$d\gamma_{SF}^A|_{\epsilon_A} = -\sum_J \Gamma_J d\mu_J. \qquad (6.87)$$

In the case of a deformable solid surface, use of Equation (6.34) in Equation (6.87) then provides a similar relation to Equation (6.82), namely,

$$d\left(\gamma_F + 2W_{SF}^A\right)|_{\epsilon_A} = -\Gamma d\mu. \qquad (6.88)$$

In the case of a deformable solid surface, we can approximately assume that the deformation does not significantly impinge on the density of adsorption sites and, consequently, on the density Γ of moles in excess.[10] As a result, Equation (6.83) for Γ still applies and Equation (6.84) found for W_{SG} extends to W_{SG}^A:

$$2W_{SG}^A = -RT\Gamma_\infty \ln(1 + p/\varpi_0). \qquad (6.89)$$

6.1.10 The disjoining pressure

Up to now, we have restricted ourselves to the situation depicted in Figure 6.3, where only two substances are involved so that there is no interaction with a third substance. This is not the case when a thin layer of liquid is inserted between a solid substrate and a third substance. This is, for instance, the case that we will encounter in Chapters 8 and 9, when accounting for the thin liquid layer remaining unfrozen at the interface between a solid crystal and a solid wall. Consider then a unit surface of the sandwich system of Figure 6.10, where a film of a wetting liquid (subscript L) of thickness e lies between a solid (subscript S) and a third nonwetting phase (subscript J), so that the related spreading coefficient is positive, reading

$$S = \gamma_{SJ} - \gamma_{SL} - \gamma_{JL} > 0. \qquad (6.90)$$

When the film thickness e vanishes the interface energy $\mathcal{U}$ simply reduces to the interface energy γ_{SJ}. For thick films, with large e, the profiles of attraction energy density, w_{SL} and w_{JL}, as defined by Equation (6.14), do not overlap, so that the interface energy $\mathcal{U}$ reduces to the sum $\gamma_{SL} + \gamma_{JL}$, as illustrated in Figure 6.10(a). In contrast, for thin films, with intermediary values of the thickness e, the former density profiles, w_{SL} and w_{JL}, do overlap, as illustrated in Figure 6.10(b). As a result, a corrective energy term W must be introduced, which depends on the thickness e and ensures the transition between the extreme cases. We write the interface energy $\mathcal{U}$ of the sandwich system in the form

$$\mathcal{U} = \gamma_{SL} + \gamma_{JL} + W(e); \quad W(0) = S; \quad W(e \to \infty) = 0. \qquad (6.91)$$

---

[10] In order to account for the effect of the deformation, in a similar way to Equation (6.32) we can write $W_{SF}^A = W_1(\mu) + W_2(\mu)\gamma(\epsilon_A)\exp(-\epsilon_A)$. From Equation (6.88) we then derive

$$\gamma_F + 2W_1(\mu) = -\int_{-\infty}^{\mu} \Gamma_1(\mu)d\mu; \quad 2W_2(\mu) = -\int_{-\infty}^{\mu} \Gamma_2(\mu)d\mu,$$

where $\Gamma_1(\mu)$ and $\Gamma_2(\mu)$ are functions such as Γ can be expressed in the form $\Gamma = \Gamma_1(\mu) + \Gamma_2(\mu)\gamma(\epsilon_A)\exp(-\epsilon_A)$.

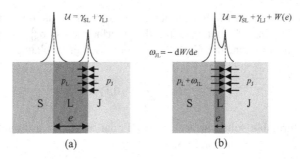

Figure 6.10 Sketch of the origin of the disjoining pressure ω_{JL}. (a) when the liquid film is thick enough, the overall interfacial energy is the sum of the two interface energies γ_{SL} and γ_{LJ} considered separately; (b) when the liquid film becomes thinner and thinner this is no longer the case since the interaction energy profiles $w_{SL}(Z)$ and $w_{LJ}(Z)$ defined by Equation (6.14) overlap. The disjoining pressure ω_{LJ} then derives from the corrective energy term $W(e)$

Let us now consider isothermal and isovolume processes with infinitesimal changes ($dT = dV = 0$), where the unit surface of the sandwich open system exchanges dN_L moles of the wetting liquid phase L and dN_J of the nonwetting phase J with two large reservoirs made of the same respective phases. The Helmholtz free energy balance relative to these infinitesimal evolutions of the unit surface of the sandwich system is provided by taking $dT = dV = 0$ in Equation (2.4). We obtain

$$dF = \mu_J dN_J + \mu_L dN_L. \tag{6.92}$$

Since energy is additive, the Helmholtz free energy of the sandwich system is the sum of the bulk Helmholtz free energy of each of the phases and of the interface energy. With the help of Equation (2.20) we write

$$F = a_J N_J + a_L N_L + \mathcal{U} = \left(\mu_J - p_J \overline{V}_J\right) N_J + \left(\mu_L - p_L \overline{V}_L\right) N_L + \mathcal{U}. \tag{6.93}$$

In Equation (6.93) p_L and $\overline{V}_L$ stand for the pressure and the molar volume of the liquid phase in the large reservoir, respectively. However, when the equilibrium is achieved, the chemical potential μ_L of the liquid in the large reservoir and that of the liquid in the thin layer are the same. In this way, the pressure p_L can also be viewed as the 'equivalent' thermodynamic pressure of the thin layer of liquid.[11] Assuming that the molar volume of the liquid phase is not significantly affected by the confinement, because there is no overall volume change of the sandwich system we also write

$$d\left(\overline{V}_L N_L\right) = -d\left(\overline{V}_J N_J\right) = de. \tag{6.94}$$

Substituting Equation (6.91) in Equation (6.93), and the resulting equation in Equation (6.92), while taking into account both Equation (6.94) and the Gibbs–Duhem Equation (2.18) applied to each phase, the energy balance Equation (6.92) reduces to

$$dW = -(p_J - p_L) de. \tag{6.95}$$

---

[11] As in Section 2.3.7, the term 'equivalent' recalls that the large reservoir acting as a thermodynamic pressuremeter may be fictitious.

This energy balance can be rewritten in the form

$$p_J = p_L + \varpi_L, \tag{6.96}$$

where ϖ_L is the extra pressure defined by

$$\varpi_L = -dW/de. \tag{6.97}$$

Owing to the presence of intermolecular forces, Equation (6.96) which accounts for the mechanical equilibrium of the J–L interface does not reduce to the equality of the thermodynamic pressures p_J and p_L. The mechanical equilibrium involves an extra pressure ϖ_L, the so-called disjoining pressure. In other words, even though the equilibrium chemical potential $\mu_L(p_L)$ and, thus, the equivalent thermodynamic pressure p_L, will be the same within the thin liquid film and a large reservoir freely exchanging molecules with the film, the film and the reservoir will not yet be subjected to the same mechanical pressure. The mechanical pressure of the liquid in the large reservoir will also be p_L, differing from the mechanical pressure $p_L + \varpi_L$ applying on the thin liquid film. It is worthwhile noting that the situation is similar to that encountered in Section 2.3.7, when analyzing the origin of the excess of osmotic pressure induced by electrostatics effects. Indeed, a comparison of Equation (6.96) with Equation (2.85) shows that the disjoining pressure, ϖ_L, and the excess of osmotic pressure, ϖ_{EL}, play an analogous role with respect to the thermodynamic liquid pressure and the solution thermodynamic pressure, respectively, both noted p_L.

Since $W(e)$ is a decreasing function of e, resulting in $dW/de < 0$, the disjoining pressure ϖ_L is positive. According to Equation (6.97) the assessment of the intensity ϖ_L of the disjoining pressure requires the knowledge of the energy corrective term W. The previous analysis of the dispersive van der Waals forces allows us to suggest

$$W = S(d/e)^2, \tag{6.98}$$

where the order of magnitude of the characteristic distance $d \leq e$ is that of the intermolecular distance. However, Equation (6.98) for W must be used with caution, since it only holds if the intermolecular forces at the origin of the interface energy have the same nature as those which lead to Equation (6.13) for the interaction energy.

In Section 6.1.7 we ignored the possible effect of the disjoining pressure. This effect can be accounted for by replacing γ_{SnW} in Equations (6.54)–(6.59) by the modified interface energy $\gamma_{SW} + \gamma_{nWW} + W(e)$, whereas the balance energy Equation (6.70) is formally preserved. Besides, in Chapters 8 and 9 the disjoining pressure ϖ_L will play a major role in the understanding of phase transitions regarding the solid walls enclosing the space where they take place. In turn, it will be shown that the experimental determination of the energy function $W(e)$ and, therefore, the intensity ϖ_L of the disjoining pressure, can be carried out through the effects it induces on the phase transitions — see in particular Equation (8.100) in Section 8.5.2. Interestingly, considering gas adsorption as a phase transition, by means of a quite distinct approach which has lead us to Equation (6.79), in Section 8.5.2 we will retrieve the Gibbs adsorption isotherm related to a single component, by involving the energy function $W(e)$ and the associated disjoining pressure ϖ_L.

6.2 Capillarity in Porous Solids

6.2.1 Capillary pressure curve and interface energy

Let us now consider a porous solid, whose porous space is occupied by two immiscible fluids. We would like to determine the law governing the invasion process of the porous space by the nonwetting phase nW, to the detriment of the wetting phase W. So let ϕ_J be the partial porosity related to the fluid J ($= n$W or W), defined in such a way that $\phi_J d\Omega_0$ is the volume which the fluid J occupies. Accordingly, we have

$$\phi_{n\text{W}} + \phi_\text{W} = \phi_0, \qquad (6.99)$$

where ϕ_0 is the overall porosity, which still remains constant since the porous solid is assumed to be undeformable throughout this chapter. It is then convenient to introduce the saturation S_J related to the fluid J, which represents the fraction of the porous space occupied by the fluid J. Accordingly, we write

$$\phi_{n\text{W}} = S_{n\text{W}}\phi_0; \quad \phi_\text{W} = S_\text{W}\phi_0; \quad S_{n\text{W}} + S_\text{W} = 1. \qquad (6.100)$$

For isothermal evolutions, that is, for $dT = 0$, in the absence of any deformation, that is, for $\varepsilon_{ij} = \varepsilon_{kk} = 0$, but now with two fluids present, the Helmholtz free energy balance Equation (3.72) extends in the form

$$da_S = p_{n\text{W}} d\phi_{n\text{W}} + p_\text{W} d\phi_\text{W}. \qquad (6.101)$$

Substituting Equation (6.100) in the free energy balance Equation (6.101), we derive

$$da_S = -\phi_0 (p_{n\text{W}} - p_\text{W}) dS_\text{W}. \qquad (6.102)$$

The free energy a_S is recalled to be the free energy of the porous solid which is left when the bulk fluids nW and W have been removed. Since the porous solid is undeformable and the evolutions considered are isothermal, the free energy a_S reduces to the free energy of the interfaces between the solid matrix and the two fluids nW and W. So let U be the free energy of the interfaces per unit of the infinitesimal porous volume $\phi_0 d\Omega_0$. We therefore write

$$a_S = \phi_0 U. \qquad (6.103)$$

Substitution of Equation (6.103) in Equation (6.102) gives

$$U = U(S_\text{W}); \quad p_\text{cap} = -\frac{dU}{dS_\text{W}}, \qquad (6.104)$$

where p_cap is the capillary pressure defined by

$$p_\text{cap} := p_{n\text{W}} - p_\text{W}. \qquad (6.105)$$

Equations (6.104) and (6.105) extend to porous solids the interface energy balance of Equation (6.70) which we found for an isolated conical pore. It shows that the capillary pressure p_cap is a state function of the wetting phase saturation S_W related to the interface energy associated with the fluids saturating the porous space. This state function is called the capillary pressure curve. The capillary pressure $p_\text{cap}(S_\text{W})$ decreases from a maximum value corresponding to the absence of the wetting fluid ($S_\text{W} = 0$) to the 'entry pressure' p_entry. The entry pressure is the threshold value required for the saturation degree S_W to start decreasing from the complete saturation value $S_W = 1$. As detailed later on in Section 6.2.3 devoted to porosimetry, the entry pressure is the minimum pressure required for the nonwetting fluid nW to start entering the material within the pores having the largest access radius according to the Laplace Equation (6.58). In order to account for the entry pressure p_{entry} the capillary pressure curve can be expressed in the form

$$\langle p_{cap} - p_{entry} \rangle = \pi (S_W); \quad \pi (1) = 0, \tag{6.106}$$

where $\langle p_{cap} - p_{entry} \rangle$ stands for the positive part of $p_{cap} - p_{entry}$ and where π is a property depending on the porous solid and the saturating fluids. Various experimental expressions have been proposed for the function $\pi (S_W)$. Originally determined from experiments performed on soil samples saturated by air and liquid water, acting as respectively the nonwetting fluid nW and the wetting fluid W, a useful expression often used is

$$\pi (S_W) = \pi_0 \left(S_W^{-1/m} - 1 \right)^{1-m}; \quad 0 < m < 1, \tag{6.107}$$

where π_0 is a pressure reference.[12] It is worthwhile noting that the interface energy change with regard to the saturated state can be assessed from the area beneath the capillary pressure curve since Equation (6.104) gives

$$U(S_W) = U(1) + \int_{S_W}^{1} p_{cap}(S) \, dS. \tag{6.108}$$

Figure 6.11 illustrates these various aspects for an experimental capillary pressure curve relative to a Berea sandstone.

6.2.2 Capillary rise

For sufficiently permeable materials, such as rocks and sands, the capillary pressure curve can be determined directly using a drainage experiment where, starting from complete saturation, the capillary pressure is increased to make the nonwetting fluid progressively invade the porous space. Alternatively, the capillary pressure curve can be obtained from the capillary rise resulting from a natural imbibition experiment. In natural imbibition a dry sample lying in the upper half-space, $z > 0$, comes into contact with the wetting fluid along the plane $z = 0$, the latter remaining at atmospheric pressure. The capillary pressure acts upwards against the vertical gravity forces, resulting in a capillary rise and an eventual equilibrium imbibition profile which matches the capillary pressure curve. More precisely the vertical equilibrium of the wetting fluid in the z direction requires

$$-\frac{\partial p_W}{\partial z} - \rho_W g = 0, \tag{6.109}$$

where ρ_W is the mass density of the wetting fluid.[13] At equilibrium the pressure of the nonwetting fluid (assumed to be the air) is equal to the atmospheric pressure p_{atm} everywhere.

---

[12] See van Genuchten, M. Th. (1980) A closed-form equation for predicting the hydraulic conductivity of unsaturated soils, *Soil Sci. Soc. Am. J.*, **44**, 892–898.

[13] Equilibrium Equation (6.109) can be recovered by taking $q_W = 0$ in Darcy's law (Equation (6.129)). It is worthwhile noting that the equilibrium condition is the same in the saturated and the unsaturated situations.

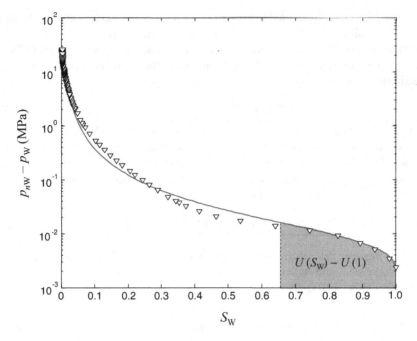

Figure 6.11 Experimental data relative to the capillary pressure curve of a Berea sandstone sample; the nonwetting fluid nW is air, while the wetting fluid W is liquid water. The solid line is obtained by choosing $p_{\text{entry}} = 0.0023$ MPa, $\pi_0 = 0.0084$ MPa and $m = 0.38$ in Equation (6.107); the hatched area beneath the capillary pressure curve represents the interfacial energy change with regard to the saturated state

The height h of the saturated zone is then determined by the integration of Equation (6.109), with boundary conditions $p_W = p_{\text{atm}}$ at $z = 0$ and $p_W = -p_{\text{entry}} + p_{\text{atm}}$ at $z = h$, where p_{entry} is the entry pressure involved in Equation (6.106). For $z > h$ the remaining capillary fringe of imbibition is determined by the integration of Equation (6.109), where we take $p_W = -p_{\text{cap}}(S_W) + p_{\text{atm}}$, and by requiring the solution to match the condition $p_W = -p_{\text{entry}} + p_{\text{atm}}$ at $z = h$. The procedure provides us with the saturation profiles of natural imbibition:

$$0 < z < h = p_{\text{entry}}/\rho_W g : \quad S_W = 1;$$
$$h = p_{\text{entry}}/\rho_W g < z : \quad p_{\text{cap}}(S_W) - p_{\text{entry}} = \rho_W g (z - h). \qquad (6.110)$$

According to Equation (6.110), the saturation profile of natural imbibition, $z = z(S_W)$, identifies with the profile of the capillary pressure $p_{\text{cap}} = p_{nW} - p_W = p_{\text{cap}}(S_W)$ (represented, for instance, in Figure 6.11) providing the capillary pressure values are divided by $\rho_W g$. For $z > h$ the capillary fringe of imbibition is scaled by the length $\pi_0/\rho_W g$ that is an intrinsic property of the porous solid, π_0 being the capillary reference pressure involved in Equation (6.107). Similar to the usual capillary length ℓ_{cap} defined by Equation (6.47), the characteristic length $\pi_0/\rho_W g$ scales the intensity of the capillary forces exerting within the porous solid with respect to the intensity of the gravity ones. The saturation profile of natural imbibition, $z = z(S_W)$, is the extension of relation Equation (6.49), which gives the height of the

Surface Energy and Capillarity 135

capillary rise in a unique capillary tube of radius R, to a great number of capillary tubes with distinct radii R. This remark constitutes the basis for the porosimetry methods presented in Section 6.2.3.

6.2.3 Porosimetry

The capillary pressure curve, such as the one of Figure 6.11, is a macroscopic state function related to the porous solid as a whole. This macroscopic state function can receive a microscopic interpretation, at the basis of porosimetry, by invoking the Laplace Equation (6.58) holding at the pore scale.

Consider a drainage process where a porous solid sample initially saturated by the wetting fluid W is subjected to an increasing capillary pressure $p_{nW} - p_W$. As the latter increases the nonwetting fluid nW progressively invades the porous solid, whereas the wetting fluid W is expelled from the sample and the wetting fluid saturation S_W decreases according to

$$S_W = \pi^{-1}\left(\langle p_{nW} - p_W - p_{\text{entry}}\rangle\right), \tag{6.111}$$

where π^{-1} is the inverse function of the function π involved in the capillary pressure curve of Equation (6.106). According to the Laplace Equation (6.58), for a given value $p_{nW} - p_W$ of the capillary pressure the pore space that can be invaded by the nonwetting fluid nW is the pore space whose pore-entry radius is greater than

$$r = \frac{2\gamma_{nWW}\cos\theta}{p_{nW} - p_W}. \tag{6.112}$$

This is sketched in Figure 6.12(b). The last two relations combine to provide

$$S_W = \pi^{-1}\left(\langle 2\gamma_{nWW}\cos\theta/r - p_{\text{entry}}\rangle\right) = S(r). \tag{6.113}$$

The function $1 - S(r)$ represents the pore volume fraction having a pore-entry radius greater than r, and it can therefore be determined from the knowledge of the capillary pressure curve related to any pair of nonwetting fluid nW and wetting fluid W, whose interface energy γ_{nWW} and contact angle θ are both known. For instance, using Equation (6.107), from Equation (6.113) we derive

$$S(r) = \left(1 + \left\langle\frac{r_*}{r} - \frac{p_{\text{entry}}}{\pi_0}\right\rangle^{1/(1-m)}\right)^{-m} \quad ; \quad r_* = 2\gamma_{nWW}\cos\theta/\pi_0, \tag{6.114}$$

where r_* is a characteristic pore-entry radius whose experimental assessment is provided by the expression given above. When taking $\gamma_{nWW} = 73\,\text{mJ}\,\text{m}^{-2}$ and $\theta = 0$, and the value of π_0 related to the Berea sandstone of Figure 6.11, we obtain $r_* = 17.4\,\mu\text{m}$. For this Berea sandstone Figure 6.12(a) shows the pore volume fraction $1 - S(r)$ derived from the air–liquid water capillary pressure curve shown in Figure 6.11. In contrast to the ordinary cement paste whose data is reported in Figure 1.2, for this sandstone and any rock of the same kind, the pore-entry radius does not go down significantly below $0.1\,\mu\text{m}$. Once determined the pore volume fraction $1 - S(r)$, which is intrinsic to the porous solid, the capillary pressure curve, which is related to any pair of a nonwetting fluid nW and a wetting fluid W, can be derived from Equation (6.113). For instance the air–liquid water capillary pressure curve of

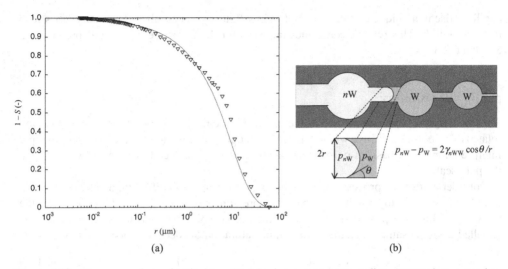

Figure 6.12 The pore volume fraction $1 - S(r)$ having a pore entry radius greater than r can be determined from the knowledge of the capillary pressure curve related to any pair of nonwetting fluid nW and wetting fluid W, whose interfacial energy γ and contact angle θ are known. This is illustrated for the Berea sandstone whose air–liquid water capillary pressure curve is given in Figure 6.11. The solid line represents the fitting curve provided by Equation (6.114) with the data reported in Figure 6.9

Figure 6.11 was actually derived from mercury porosimetry, where the nonwetting fluid is liquid mercury and the wetting fluid mercury vapor ($\gamma_{n{\rm ww}} = 485\,{\rm mJ\,m^{-2}}, \theta = 130°$).

6.2.4 Capillary hysteresis

When a sample of porous material is subjected to a drainage–imbibition cycle, a loop of hysteresis is usually observed in the $(S_{\rm W}, p_{\rm cap})$ plane. As a result, the link between the capillary pressure and the wetting fluid saturation $S_{\rm W}$ cannot actually reduce to a one-to-one relationship between $p_{\rm cap}$ and $S_{\rm W}$ as assumed in Equation (6.106). When starting from complete saturation and increasing the capillary pressure, the saturation degree $S_{\rm W}$ decreases and the corresponding point $(S_{\rm W}, p_{\rm cap})$ follows the drainage curve defined by

$$p_{\rm cap} = p_{\rm cap}^{\rm DRA}(S_{\rm W}). \tag{6.115}$$

At the end of a drainage process an imbition process can be carried out, by letting the capillary pressure retrieve its previous lower values. The point $(S_{\rm W}, p_{\rm cap})$ then follows the imbition curve defined by

$$p_{\rm cap} = p_{\rm cap}^{\rm IMB}(S_{\rm W}). \tag{6.116}$$

The imbition curve differs from the drainage curve: the saturation $S_{\rm W}$ related to the same capillary pressure $p_{\rm cap}$ is lower in the imbition process than in the drainage process. Both curves conjointly form a loop of hysteresis: when the direction of variation of the capillary pressure is inverted during a drainage process, the point $(S_{\rm W}, p_{\rm cap})$ leaves the drainage curve to meet the imbition curve at nearly the same saturation $S_{\rm W}$; conversely, when the direction of

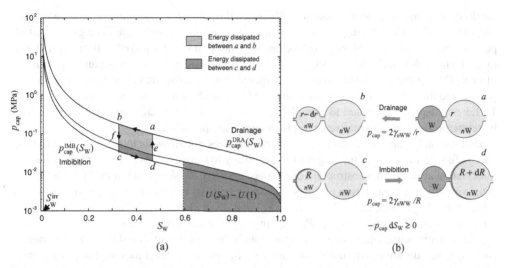

Figure 6.13 Capillary hysteresis and the ink bottle effect: the imbibition is governed by the radius of the large pores (bottle reservoir), whereas the drainage is governed by the radius of the connecting capillary tubes (bottle neck)

variation of the capillary pressure is inverted during an imbibition process, the point (S_W, p_{cap}) leaves the imbibition curve to meet the drainage curve at nearly the same saturation S_W. This rough description of capillary hysteresis is sketched out in Figure 6.13(a).

The existence of a hysteresis implies that the drainage or the imbibition of porous solid is a dissipative process. According to the second law of thermodynamics, instead of the equality of Equation (6.104) we now write the inequality

$$-p_{cap} dS_W - dU \geq 0. \tag{6.117}$$

The energy balance Equation (6.117) states that during the invasion of the porous solid by the nonwetting fluid nW, only the part dU of the work input $-p_{cap} dS_W$ is stored in the form of interface energy per unit of porous volume, the remaining part being dissipated in the form of heat. More precisely the definitions of the drainage and imbibition curves, corresponding to $dS_W > 0$ and $dS_W < 0$, respectively, and Equation (6.117) combine into the inequality

$$p_{cap}^{IMB} \leq -\frac{dU}{dS_W} \leq p_{cap}^{DRA}, \tag{6.118}$$

so that the current capillary pressure p_{cap} always differs from $-dU/dS_W$. As a result, in the (S_W, p_{cap}) plane the point $(S_W, -dU/dS_W)$ follows a curve lying between the drainage and the imbibition curves. During a capillary pressure cycle such as $abfcdea$ in Figure 6.13(a), the amount of energy dissipated through capillary hysteresis into heat can now be identified. Along the drainage path ab the dissipated energy per unit of porous volume is the area between ab and the path ef followed by the point $(S_W, -dU/dS_W)$ during the same drainage process. This dissipated energy per unit of porous volume is the part of the mechanical work $-p_{cap} dS_W$, where $dS_W < 0$, that is not stored in the interface energy form dU, but is dissipated into heat form, for instance, through microscopic viscous flow. Conversely, along the imbibition path cd

the dissipated energy is the area comprised between the path cd and the path fe followed by the point $(S_W, -dU/dS_W)$ during the same imbibition process. This dissipated energy per unit of porous volume is the part of the interface energy $-dU$ that is not released to the surroundings in the form of the mechanical work $-p_{cap}dS_W$, where $dS_W > 0$, but is dissipated in the form of heat. During the cycle $abfcdea$ the dissipated energy is the area comprised between the part ab and the part cd of the drainage curve and the imbibition curve, respectively. In Section 10.1.3 of Chapter 10 devoted to poroplasticity, the capillary hysteresis described above will be reinterpreted as an instructive example of hardening plasticity.

Unfortunately, the above macroscopic approach does not allow us to identify the interface energy function $U(S_W)$ or the origin of the capillary hysteresis. This requires more pieces of information about the microstructure of the porous space. As sketched out in Figure 6.13(b) a usual simplified representation of the porous space consists of sets of spherical pores with the same increasing radius, $... < R_i < R_{i+1} < ...$, and connected to each other by thin capillary tubes of much smaller radius $r_i \ll R_i$ and ranged in the same order, $... < r_i < r_{i+1} < ...$ The thin tubes have negligible volumes compared with those of the spherical pores they connect. Although they do not contribute significantly to the saturation level they will play a key role in the hysteresis analysis given below.

According to this simplified representation of the porous space the imbibition process is achieved by the filling of the spherical pores. Consider for instance the current imbibition state c of Figure 6.13, and let R be the current upper bound of the radius of pores that are already filled and contribute to the current saturation degree S_W. The value of radius R is provided by the Laplace equation ruling the mechanical equilibrium of the interface between the nonwetting fluid and the wetting fluid. Indeed, assuming a zero contact angle $\theta = 0$, R is equal to the curvature radius of the interface between the wetting fluid and the nonwetting fluid (see Figure 6.13(b)). Accordingly we write

$$p_{cap}^{IMB} = \frac{2\gamma_{nW W}}{R}. \qquad (6.119)$$

This results in a one-to-one relationship between the given capillary pressure p_{cap}^{IMB} and the current radius R of the largest pores contributing to the current saturation degree S_W since, according to Equation (6.119), there is no possible equilibrium between the wetting fluid and the nonwetting fluid within pores having a radius less than R or more than R. Consider now an infinitesimal decrease in capillary pressure so that the representing point on the imbibition curve in Figure 6.13(a) moves from point c to point d. The updated largest value $R + dR$ of the radius of the pores which are filled consecutively to the drop in capillary pressure has to fit the updated capillary pressure according to Equation (6.119) where R is replaced by $R + dR$. This is achieved by the imbibition by the wetting fluid W of the part $\phi_0 dS_W \, d\Omega_0$ of the overall porous volume $\phi_0 \, d\Omega_0$, resulting in the saturation increase $dS_W > 0$. The ratio of the surface area of a sphere of radius R to its volume is $3/R$. As a result, if we neglect the volume of the capillary tubes connecting the spherical pores, the area dA_{SW} of the newly wetted solid walls relates to the volume $\phi_0 dS_W \, d\Omega_0$ newly invaded by the wetting fluid through the relation

$$dA_{SW} = \frac{3}{R} \phi_0 dS_W \, d\Omega_0. \qquad (6.120)$$

In the infinitesimal filling process the area variation of the interface between the wetting and the nonwetting fluid is a second-order term with regard to dA_{SW}. Prior to the invasion of the volume $\phi_0 dS_W d\Omega_0$ by the wetting fluid W the surface dA_{SW} was in contact with the nonwetting fluid nW. The change $\phi_0 dU d\Omega_0$ in interface energy associated with the saturation increase dS_W is therefore

$$\phi_0 dU\, d\Omega_0 = (\gamma_{SW} - \gamma_{S_nW})\, dA_{SW}. \quad (6.121)$$

Substituting the Young–Dupré Equation (6.54) in Equation (6.121), and taking $\theta = 0$, we obtain

$$\phi_0 dU\, d\Omega_0 = -\gamma_{nWW} dA_{SW}. \quad (6.122)$$

Equations (6.116), (6.119), (6.120) and (6.122), combine to give

$$-\frac{dU}{dS_W} = \frac{3}{2} p_{cap}^{IMB}(S_W), \quad (6.123)$$

so that U can be obtained by a simple integration of the imbibition curve in the form

$$U(S_W) - U(1) = \frac{3}{2} \int_{S_W}^{1} p_{cap}^{IMB}(S)\, dS. \quad (6.124)$$

However, it must be pointed out that such an attractive quantitative identification of the interface energy relies upon the initial assumption of spherical microscopic pores, which lead in particular to the factor 3/2 in Equations (6.123) and (6.124). Any quantitative identification of the interface energy curve $U(S_W)$ will always be specific to the morphology of the porous space.

The representation of the porous space by large spherical pores connected to each other by thin capillary tubes of much smaller radii provides a standard interpretation of the capillary hysteresis, the so-called ink bottle effect pictured in Figure 6.13(b). In this interpretation, as just depicted the imbibition process is sketched out by filling up sets of spherical pores by decreasing the capillary pressure and leading for instance to the state represented by the point d on the imbibition curve of Figure 6.13(a). If now the capillary pressure is again increased from point d to point a, the drainage of the set of the largest pores of radius R filled with the wetting fluid can start only for a capillary pressure fitting the radius r of the thin capillary tubes allowing the access to the pore as sketched out in Figure 6.13(b). Consequently, as actually observed the saturation S_W remains almost constant along the path da, while the Laplace equation provides the value of the capillary pressure at which the drainage actually starts, that is

$$p_{cap}^{DRA}(S_W) = \frac{2\gamma_{nWW}}{r}. \quad (6.125)$$

The further increase of the capillary pressure makes the representing point move from point a to point b on the drainage curve shown in Figure 6.13(a). The updated greatest value $r-dr$ of the thin capillary tubes has then to fit the new capillary pressure according to Equation (6.125), where r is replaced by $r-dr$. This is achieved by the drainage of the wetting fluid W from the part $\phi_0 dS_W d\Omega_0$ of the overall porous volume $\phi_0 d\Omega_0$, resulting in the saturation decrease $dS_W < 0$. A similar analysis to the one that lead to Equation (6.123) gives the relation

$$-\frac{dU}{dS_W} = \frac{3r}{2R} p_{cap}^{DRA}(S_W). \quad (6.126)$$

Figure 6.13(a) also illustrates the generally observed existence of an irreducible saturation S_W^{irr} to the wetting fluid. For this saturation the wetting fluid is then trapped within dead-end pores, from which it cannot be expelled even if the capillary pressure indefinitely increases.

6.3 Transport in Unsaturated Porous Solids

6.3.1 Relative permeability

Extending Equation (5.23), the positiveness of the dissipation associated with the fluid transport of two immiscible fluids within a porous medium is written in the form

$$-\nabla p_{nW} \cdot \underline{q}_{nW} - \nabla p_W \cdot \underline{q}_W \geq 0, \tag{6.127}$$

where $\underline{q}_J$ is the filtration vector associated with the fluid J. Ignoring a possible coupling between the flows of the wetting and nonwetting fluids, Darcy's law given in Equation (5.30) extends to the unsaturated case in the form

$$\underline{q}_J = -\frac{\varkappa k_{rJ}(S_J)}{\eta_J} \nabla p_J. \tag{6.128}$$

When considering gravity forces, in the same way that Equation (5.33) related to the saturated case, the previous Darcy's law extends in the form

$$\underline{q}_J = -\frac{\varkappa k_{rJ}(S_J)}{\eta_J} \nabla (p_J + \rho_J gz). \tag{6.129}$$

In the two previous equations $k_{rJ}(S_J)$ stands for the relative permeability related to the nonwetting fluid (subscript J = nW) and the wetting fluid (subscript J = W), respectively, and satisfying

$$0 = k_{rJ}(0) \leq k_{rJ}(S_J) \leq k_{rJ}(1) = 1. \tag{6.130}$$

Initially also determined from experiments performed on soil samples saturated by air and liquid water, acting as the nonwetting fluid nW and the wetting fluid W, respectively, expressions commonly used in association with Equation (6.107) for the capillary pressure curve, that is, referring to the same value of coefficient m, are[14]

$$k_{rnW}(S_{nW} = 1 - S_W) = \sqrt{1 - S_W}\left(1 - S_W^{1/m}\right)^{2m}; \tag{6.131a}$$

$$k_{rW}(S_W) = \sqrt{S_W}\left[1 - \left(1 - S_W^{1/m}\right)^m\right]^2. \tag{6.131b}$$

In Figure 6.14 we have plotted k_{rnW} and k_{rW} against S_W for various values of m. For values of m close to one the relative permeability k_{rW} related to the wetting fluid decreases rapidly down to zero as S_W decreases from one. This can constitute a convenient way to account for the progressive disconnection of the wetting phase. Indeed, nonconnected wetted pores result in a zero associated permeability.

---

[14] See Luckner, L., van Genutchen, M. Th. and Nielsen, D. R. (1989) A consistent set of parametric models for the two-phase flow of immiscible fluids in the subsurface, *Water Res. Res.*, **25**, 2187–2193.

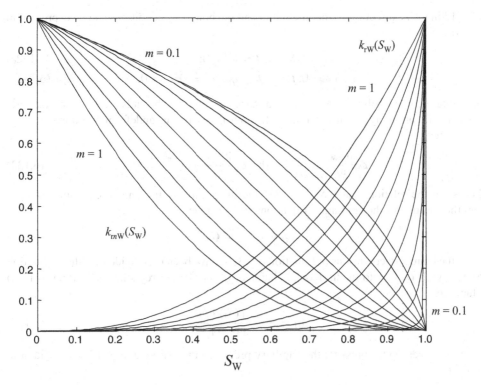

Figure 6.14 Relative permeability for the wetting and nonwetting fluids

6.3.2 Injection

In order to illustrate how advection and diffusion may compete in the transport of fluids within an unsaturated porous medium, let us analyze the injection problem. In an injection problem a wetting fluid is injected into a porous medium and displaces a nonwetting fluid which initially saturated the porous medium. An example of significant importance is the recovery of a rock reservoir of oil by means of water flooding. With regard to durability issues, another example of concern is provided by the imbibition of cement-based materials, initially saturated by air and invaded by seawater supplying aggressive agents such as chloride ions. In monitored injection the volume injection rate Q of the wetting phase is imposed and leads to an advection–diffusion process, with a vanishing diffusion front propagating at finite speed. This section aims to analyze the formation and the propagation of this diffusion front.

Formation of an injection front

Let us consider a semi-infinite layer, $x \geq 0$, formed by an undeformable porous solid, which is initially saturated by a nonwetting fluid (subscript nW). At time $t = 0$ a volume of wetting fluid (subscript W) is suddenly injected at a constant rate Q on the border $x = 0$. Incompressible flow is assumed for both fluids. Then let q_J be the component of the filtration vector $\underline{q}_J$ relative to fluid J in the $\underline{e}_x$ direction so that the initial and the boundary conditions can be expressed in the form

$$S_W(x, t = 0) = 0; \tag{6.132a}$$

$$q_W(x = 0, t) = Q; \quad q_{nW}(x = 0, t) = 0. \tag{6.132b}$$

Assuming incompressible flows so that the mass density of both fluids are constant, the one-dimensional continuity equations related to both fluids for an undeformable homogeneous medium read

$$\phi_0 \frac{\partial S_W}{\partial t} + \frac{\partial q_W}{\partial x} = 0; \quad \phi_0 \frac{\partial S_{nW}}{\partial t} + \frac{\partial q_{nW}}{\partial x} = 0. \tag{6.133}$$

The use of the relation $S_{nW} + S_W = 1$ and the integration of the unique equation resulting from the addition of the continuity Equations (6.133) give

$$q_{nW} + q_W = Q, \tag{6.134}$$

where the boundary conditions in Equation (6.132b) have been used to identify the integration constant with Q. Due to the horizontal character of the flow Darcy's law (Equation (6.128)) reduces to

$$q_J = -\frac{\varkappa k_{rJ}(S_J)}{\eta_J} \frac{\partial p_J}{\partial x}. \tag{6.135}$$

Assuming a zero entry pressure the capillary pressure curve (Equation (6.106)) is written in the form

$$p_{nW} - p_W = \pi_0 \Pi(S_W), \tag{6.136}$$

where, as in Equation (6.107), π_0 stands for a reference capillary pressure so that $\Pi(S_W)$ is a dimensionless capillary pressure. Elimination of the different unknown fields from Equations (6.133)–(6.135) and (6.136) to the unique benefit of S_W allows us to write the continuity equation relative to the wetting fluid in the form

$$\phi_0 \frac{\partial S_W}{\partial t} - D \frac{\partial}{\partial x} \left(\delta(S_W) \frac{\partial S_W}{\partial x} - \frac{\zeta(S_W)}{\ell} \right) = 0, \tag{6.137}$$

where $\delta(S_W)$ and $\zeta(S_W)$ are the following diffusion and advection dimensionless functions

$$\delta(S_W) = -\frac{d\Pi(S_W)}{dS_W} \frac{\eta_W k_{rW}(S_W) k_{rnW}(S_{nW} = 1 - S_W)}{\eta_{nW} k_{rW}(S_W) + \eta_W k_{rnW}(S_{nW} = 1 - S_W)}; \tag{6.138a}$$

$$\zeta(S_W) = \frac{\eta_{nW} k_{rW}(S_W)}{\eta_{nW} k_{rW}(S_W) + \eta_W k_{rnW}(S_{nW} = 1 - S_W)}. \tag{6.138b}$$

In addition, in Equation (6.137) D and ℓ are respectively the diffusion coefficient and the advection–diffusion length defined by

$$D = \varkappa \pi_0 / \eta_W; \quad \ell = D/Q. \tag{6.139}$$

According to Equation (6.137) the injection process is governed by an advection–diffusion equation. The injection profile is shaped by diffusion through the competition existing between the capillary pressure gradient which drives the flow, and the viscous forces which oppose it.

While the diffusion function δ accounts for the gradient of the capillary pressure through the factor $d\Pi/dS_W$, according to its definition in Equation (6.139), the diffusion coefficient D scales its order of magnitude (numerator of D proportional to π_0) compared with the resistant viscous forces (numerator of D proportional to $\eta_W/\varkappa$). The driving force of the injection that makes the injection profile move ahead is advection. The advection–diffusion length $\ell = D/Q$ assesses the distance over which the diffusion is active when compared with advection scaled by the injection rate Q: the smaller the length ℓ, the steeper the profile. Equations (6.138) of the diffusion and advection functions, $\delta(S_W)$ and $\zeta(S_W)$, show that they both vanish with the saturation S_W of the injected wetting fluid. Since S_W is zero in the porous medium before the injection starts, it can therefore be inferred that an actual front of injection whose head is defined by $S_W = 0$ may form and propagate at a finite speed within the porous medium. Let us then determine the speed of this front.

Speed of the injection front

Since the length ℓ assesses the distance over which the diffusion is active when compared with advection, at large times $t \gg \ell^2/D$ diffusion will only be significant at the front so that the injection becomes mainly governed by advection. In the limit of the advection approximation $t \gg \ell^2/D$, the transport Equation (6.137) reduces to the celebrated Buckley–Leverett equation

$$\frac{\partial S_W}{\partial t} + \frac{D}{\phi_0 \ell} \frac{\partial \zeta(S_W)}{\partial x} = 0. \tag{6.140}$$

The advection approximation amounts to ignoring the capillary effects. Accordingly, taking $p_{nW} \simeq p_W$ and, therefore, $\partial p_{nW}/\partial x = \partial p_W/\partial x$, Equations (6.134), (6.135) and (6.138b) give us the following expressions for the filtration vectors q_W and q_{nW}:

$$q_W = \zeta(S_W) Q; \quad q_{nW} = \zeta(S_W) \frac{\eta_W k_{rnW}(S_{nW} = 1 - S_W)}{\eta_{nW} k_{rW}(S_W)} Q. \tag{6.141}$$

Because of Equations (6.130) and (6.141), the boundary conditions given in Equation (6.132b) can conveniently be replaced by

$$S_W(x=0, t) = 1. \tag{6.142}$$

A possible solution of Equation (6.140) satisfying the initial condition of Equation (6.132a) is

$$S_W = S_W^{up} H(x - ct); \quad c = c_0 \zeta'\left(S_W^{up}\right); \quad c_0 = D/\phi_0 \ell = Q/\phi_0, \tag{6.143}$$

where H is the Heaviside step function defined by Equation (4.90), and where ζ' stands for the derivative of ζ. The solution given in Equation (6.143) represents a step function moving at speed c and admitting S_W^{up} as upstream saturation.

Now let $[[\cdot]]$ denote the jump that the quantity $\cdot$ between the brackets undergoes when passing the injection front. During time dt the infinitesimal surface dA normal to the $\underline{e}_x$ direction and moving with the injection front at the same speed c sweeps out the porous volume $\phi_0 c\, dA dt$. The volume variation of the wetting fluid this volume contains is therefore $[[\phi_0 S_W c]] dA dt$. During the same time dt the volume of the wetting fluid flowing through this volume is $[[q_W]] dA dt$. Since the flow is incompressible, mass conservation requires that both volumes are equal, resulting in the Rankine–Hugoniot jump condition

$$[[q_W - \phi_0 S_W c]] = 0. \tag{6.144}$$

Due to the initial condition of Equation (6.132a), the height $[[S_W]]$ of the injection front reduces to the upstream saturation S_W^{up} so that Equations (6.141), (6.143) and (6.144) combine to provide the following expression for the speed c:

$$c = c_0 \zeta \left(S_W^{up} \right) / S_W^{up}. \tag{6.145}$$

Equating Equations (6.143) and (6.145) we separately found for the speed c of the discontinuity front, we conclude that the solution given in Equation (6.143) can actually develop if the upstream saturation satisfies

$$\zeta' \left(S_W^{up} \right) = \zeta \left(S_W^{up} \right) / S_W^{up}. \tag{6.146}$$

However, even though such a steep injection front can actually propagate, diffusion effects will inexorably spread out the front head, where a transition layer will form. Within this thin layer, whose thickness is scaled by ℓ,[15] the saturation S_W increases abruptly from zero at the moving front to the upstream value S_W^{up}. The stability of this moving transition layer therefore depends on the speed $c_0 \zeta (S_W) / S_W$ that Equation (6.145) associates with each saturation of the layer as a possible upstream value. The stability condition is

$$0 \leq S_W < S_W^{up}: \quad \zeta (S_W) / S_W < \zeta \left(S_W^{up} \right) / S_W^{up}. \tag{6.147}$$

The condition in Equation (6.147) turns out to be the 'entropy criterion' for admissible shocks in the mathematical sense, while the step function in Equation (6.143), where c is given by Equation (6.145), is the standard 'entropy weak solution' towards which the solution of Equation (6.137) converges when the diffusion term vanishes. If no such interval $\left[0, S_W^{up} \right]$ exists, no autonomous injection front can form: whatever the upstream saturation S_W^{up}, the speed $c_0 \zeta (S_W) / S_W$ of some saturation heads of the layer will always go faster than the upstream flow so that no stable layer can exist. For instance this is the case when the viscosity ratio η_W / η_{nW} tends to zero, so that function $\zeta (S_W)$ reduces to the step function $H (S_W)$. Conversely, when such an interval $\left[0, S_W^{up} \right]$ exists, although the upstream flow goes faster than the front transition layer, the localized diffusion occurring ahead slows down the upstream flow so that a stable layer can develop.[16] An extreme case where the thickness of the layer becomes zero corresponds to a viscosity ratio η_W / η_{nW} going to infinity; function $\zeta (S_W)$ then reduces to the step function $H (S_W - 1)$, so that the couple $\left(S_W^{up} = 1, c = c_0 \right)$ becomes suitable with regard

---

[15] Taking $k = 1\,\mu m^2$ and $Q = 1\,m\,s^{-1}$, and considering that the wetting fluid is water, that is, $\eta_W \simeq 10^{-3}\,kg\,m^{-1}\,s^{-1}$, the order of magnitude of length ℓ (m) is $10^{-3}\pi_0$ where the reference pressure π_0 is expressed in MPa.

[16] As shown in Coussy (2004) referred to in the Further Reading Section a refined analysis shows that the saturation profile of the stabilized transition layer is given by

$$\int_0^{S_W} \frac{\delta (S) / S}{\zeta (S) / S - \zeta \left(S_W^{up} \right) / S_W^{up}} \, dS = \frac{x - ct}{\ell}.$$

The integral converges as soon as $\delta (S_W \to 0) \equiv a S_W^\gamma$ with $a, \gamma > 0$. This condition is met whatever the value of m, when adopting for δ the expression given by Equations (6.107) and (6.131). When the integral diverges, the intensity of the capillary pressure gradient is so intense that diffusion spreads out the penetration front over the whole x axis. The profile of the penetration front is then provided by replacing the lower bound 0 by $S_W^{up}/2$ in the previous integral.

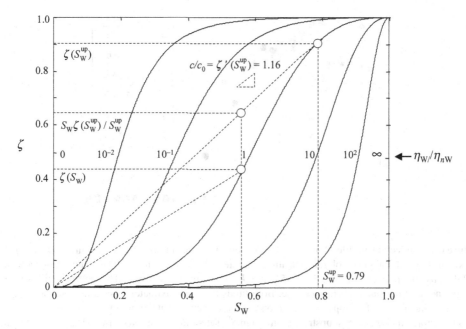

Figure 6.15 Function $\zeta(S_W)$ for various values of the viscosity ratio η_W/η_{nW} resulting from Equation (6.131) for the relative permeability k_{rnW} and k_{rW} with $m = 0.85$. The determination of the largest upstream saturation S_W^{up} consistent with Equation (6.147) and satifying Equation (6.146) is illustrated for $\eta_W/\eta_{nW} = 1$. The value of S_W^{up} and the associated normalized speed $\zeta'(S_W^{up})$ respectively increase and decrease for an increasing viscosity ratio η_W/η_{nW}

to the condition of Equation (6.147). As a result, the autonomous injection profile of Equation (6.143) can develop fully since it meets the upstream boundary condition of Equation (6.142). For intermediary values of the viscosity ratio η_W/η_{nW} ranging from zero to infinity, the largest value S_W^{up} that is consistent with Equation (6.147) is provided by solving Equation (6.146). This is illustrated in Figure 6.15 for various values of the viscosity ratio η_W/η_{nW} and for the function $\zeta(S_W)$ provided by taking $m = 0.85$ in Equations (6.107) and (6.131) relative to $\Pi(S_W)$, $k_{rnW}(S_{nW} = 1 - S_W)$ and $k_{rW}(S_W)$. The value of the largest suitable upstream saturation S_W^{up} and the associated speed $c_0\zeta'(S_W^{up})$ respectively increase and decrease as the viscosity ratio η_W/η_{nW} increases: the lower the viscosity ratio, the more difficult the penetration of the nonwetting fluid and, consequently, the lower the possible maximum upstream height of penetration S_W^{up}.

Without having exactly solved the transport Equation (6.137), we are now able to summarize the different stages of an injection process. At the start and in a region close to $x = 0$ – that is, for both $t \ll \ell^2/D$ and $x < \ell$ – diffusion starts spreading out the profile of the saturation degree of the wetting phase within a layer of finite extent, from complete saturation ($S_W = 1$) down to zero saturation. Subsequently, advection makes the profile penetrate within the porous medium. Because of their greater speed, the higher saturation heads of the profile catch up with the lower saturation heads. As a result, the profile progressively steepens, and an autonomous traveling transition layer forms at the front as soon as Equation (6.146) is satisfied. Nonetheless,

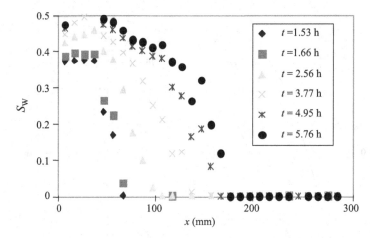

Figure 6.16 Injection profiles for a porous medium made of quartzitic grains with radius 100 μm, packed in a cylindrical glass column. The injection rate Q is close to $4\,\mathrm{m\,s^{-1}}$. The values of the porosity and the absolute permeability are $\phi_0 = 35\,\%$ and $k = 10\,\mu\mathrm{m}^2$, respectively. The wetting fluid is water and the nonwetting fluid is pure n-decane. Both fluids have approximately the same viscosity, $\eta_{nw} = \eta_w \simeq 10^{-3}\,\mathrm{kg\,m^{-1}\,s^{-1}}$. The apparent saturation threshold occurring at the inlet $x = 0$ is equal to $1/2$ and can be interpreted as the largest upstream saturation S_w^{up} satisfying Equation (6.147) (Data from Melean, Y., Broseta, D., Hasmy A. and Blossey, R. (2003) Dispersion of imbibition fronts, *European Physics Letters*, **62**, 505–511.)

a transient regime is always needed to match the autonomous traveling transition layer to the upstream the boundary condition of Equation (6.142). An alternative conclusion is that the boundary condition of Equation (6.142) is in practice tough to achieve, the saturation degree S_w^{up} representing a threshold which is difficult to overcome. These remarks agree well with the experimental observations reported in Figure 6.16, where the viscosity ratio η_w/η_{nw} is equal to unity while the experimental value S_w^{up} is found to be close to $1/2$.

Starting at the molecular scale and ending up at the macroscopic scale of fluid flow in the porous medium, this chapter has explored how surface energy governs the occupation of the pore volume of a porous solid by the nonwetting and wetting phases. Since these phases do not exert the same pressure upon the internal solid walls that delimit them, the capillary pressure curve determining the phase saturations will be an essential material property in Chapter 7, when addressing the mechanics of unsaturated deformable porous solids. In Chapter 9 interface energy and capillarity will also be the two keys for understanding how phase transitions are affected when they occur within a porous solid. They will also be central for assessing the shrinkage of porous materials induced by drying, or the deformation associated with their freezing.

Further Reading

Adamson, A. W. and Gast, A. P. (1997) *Physical Chemistry of Surfaces*, John Wiley & Sons, Ltd.
Bear, J. (1988) *Dynamics of Fluids in Porous Media*, Dover Publications, New York (reprint of 1972 edition published by Elsevier, New York).

Bear, J. and Cheng, A. H. -D. (2010) *Modeling Groundwater Flow and Contaminant Transport*, Series: Theory and Applications of Transport in Porous Media, Springer.

Butt, H. J., Graf, K. and Kappl, M. (2006). *Physics and Chemistry of Interfaces*, 2nd edn, Wiley-VCH Verlag GmbH.

Coussy, O. (2004) *Poromechanics*, John Wiley & Sons, Ltd.

de Gennes, P.-G., Brochard-Wyard, F. and Quéré, D. (2004) *Capillarity and Wetting Phenomena: Drops, Bubbles, Pearls and Waves*, Springer, New York.

Dullien, F. A. L. (1979) *Porous Media: Fluid Transport and Pore Structure*, Academic Press, New York.

Everett, D. H. (1988) *Basic Principles of Colloid Science*, Royal Society of Chemistry.

Hahn, R. (2005) *Pierre Simon Laplace, 1749-1827: A Determined Scientist*, Harvard University Press.

Hunter, J. (2001) *Foundations of Colloid Science*, 2nd edn, Oxford University Press.

Isenberg, C. (1992) *The Science of Soap Films and Soap Bubbles*, Unabridged Dover republication of the work published by Tieto Ltd., Clevedon, Avon, England.

Israelachvili, J. (1991) *Intermolecular and Surface Forces*, 2nd edn, Academic Press.

Maugis, D. (2000) *Contact, Adhesion and Rupture of Elastic Solids*, Springer Series in Solid-State Sciences, Volume 130.

Zinszner, B. and Pellerin, F.-M. (2007) *A Geoscientist's Guide to Petrophysics*, Editions Technip.

7

The Unsaturated Poroelastic Solid

In Chapter 6 we explored how surface energy governs the invasion–recession of a fluid phase partially occupying the porous space of a porous solid. However, we ignored the deformation of the porous solid accompanying the invasion–recession process. In Chapter 9 we will investigate the deformation of porous solids whose inner fluids are subjected to phase transitions occurring during their drying or freezing. This requires us to consider both the invasion–recession process and the simultaneous deformation of the porous solid. This is the main topic of this chapter devoted to unsaturated poroelasticity.

Unsaturated poroelasticity consists of extending the concepts of saturated poroelasticity to unsaturated conditions. In contrast to saturated conditions, unsaturated conditions involve capillary effects and, therefore, surface energy effects within an undeformable porous solid, which we have examined in Chapter 6 devoted to capillarity. As a result, unsaturated poroelasticity needs to combine the energy balance associated with the deformation of the poroelastic solid and the energy balance associated with the change in surface energy during the invasion or the recession of the nonwetting phase.

The key concept required for the appropriate combination of both energy balances is the concept of Lagrangian saturation. The saturation with respect to a given fluid is generally defined with respect to the current overall porous volume. Such a saturation can be termed the Eulerian saturation since it refers to the current configuration. Unfortunately, since the current configuration undergoes a deformation, there is a contribution of the current pore deformation to the Eulerian saturation. As a consequence, an infinitesimal change in the Eulerian saturation does not only relate to the further invasion or recession of the porous volume by the corresponding fluid, but also to the further deformation of the porous volume already invaded by the same fluid caused by the further pore pressure change. This precludes the use of the Eulerian saturation from providing an appropriate energy balance accounting for the energy involved in the saturation or desaturation process, separately from that involved in the deformation process. In contrast to the Eulerian saturation, the Lagrangian saturation refers the current saturation related to a given fluid to the initial configuration and, therefore, is not affected by the pore deformation. As a result the Lagrangian saturation can conveniently be used to keep track of the surface energy associated with the invasion or the recession of the nonwetting fluid within the porous solid, and the energy required to deform the porous solid separately.

Mechanics and Physics of Porous Solids Olivier Coussy
© 2010 John Wiley & Sons, Ltd

In this chapter we first revisit poroelasticity in saturated conditions with regard to surface energy effects. Even in saturated conditions, because of the interface stress existing along the solid matrix–fluid interface, the nature of the solution which is in contact with the internal walls of the porous solid does affect the pore pressure finally transmitted to the surrounding solid matrix and, therefore, its deformation. We then introduce in detail the concept of Lagrangian saturation. In the remaining part of the chapter this concept will allow us to extend the results of Chapters 4 and 6 to unsaturated poroelastic solids.

7.1 Interface Stress as a Prestress

In Chapters 4 and 5 devoted to saturated poroelasticity we ignored any effect related to the interface energy between the fluid and the solid. The interface energy was introduced in Chapter 6. Before addressing unsaturated poroelasticity where the interface energy plays a key role, this section revisits saturated poroelasticity in order to include the possible effects related to the changes in the interfacial energy between the fluid mixture and the solid matrix.

7.1.1 Interface energy and saturated poroelasticity

Without considering shear (for the sake of simplicity) and restricting ourselves to isothermal conditions, the energy balance Equation (4.1) related to a saturated poroelastic solid, from which the bulk fluid has been removed, is recalled here for convenience

$$\sigma d\epsilon + p d\varphi - da_S = 0. \tag{7.1}$$

Since only the energy of the bulk saturating fluid has been removed, the Helmholtz free energy a_S still includes the energy of the interface between the saturating fluid and the solid matrix. However, in spite of the interface energy, from Equation (7.1) it can be inferred that da_S is still an exact differential and therefore a_S has the overall volumetric strain ϵ and the Lagrangian porosity change φ as its only arguments. Nonetheless, the free energy a_S has now to be split into the contribution of the elastic energy w_S of the solid matrix and the contribution u of the interface energy, both per unit of the initial volume $d\Omega_0$, so we can write

$$a_S = w_S(\epsilon, \varphi) + u(\epsilon, \varphi). \tag{7.2}$$

Let $A_0 = s d\Omega_0$ be the initial area of the solid surface forming the internal walls of the porous network prior to any deformation of the initial volume $d\Omega_0$; s is therefore the area of the internal walls per unit volume of porous solid. The expression for the infinitisimal change in the overall interface energy, $d\mathcal{U} = du \times d\Omega_0$, is provided by Equations (6.37) and (6.38):

$$du = \sigma_{SF}^A s d\epsilon_A, \tag{7.3}$$

where σ_{SF}^A is the interface stress related to the solid–fluid interface and defined in Section 6.1.5, while ϵ_A is the surface strain undergone by the solid internal walls. Because of the linear elasticity of the solid matrix, in infinitesimal transformation ϵ_A must be a linear function of ϵ and φ. As a result, we write

$$\epsilon_A = \alpha \epsilon + \beta \varphi, \tag{7.4}$$

where α and β are intrinsic properties of the porous solid, depending both on its elastic properties and the geometry of its porous network. Substitution of Equation (7.4) in Equation (7.3) provides

$$du = \sigma_{SF}^A s\,(\alpha d\epsilon + \beta d\varphi). \tag{7.5}$$

Relations (7.2) and (7.5), when combined with Equation (7.1), provide us with the energy balance

$$(\sigma - \sigma_0)\,d\epsilon + (p - p_0)\,d\varphi - dw_S = 0, \tag{7.6}$$

where

$$\sigma_0 = \sigma_{SF}^A s\alpha; \quad p_0 = \sigma_{SF}^A s\beta. \tag{7.7}$$

Assuming linear poroelasticity, so that w_S is a quadratic form of its arguments ϵ and φ, from Equation (7.6) we derive

$$\sigma - \sigma_0 = K\epsilon - b(p - p_0); \tag{7.8a}$$

$$\varphi = b\epsilon + \frac{p - p_0}{N}. \tag{7.8b}$$

With regard to the standard constitutive Equations (4.6a) and (4.6b) of isotropic poroelasticity, Equations (7.7) and (7.8) show that a prestress σ_0 and an initial pore pressure p_0 have to be applied against the effects induced by the interface stress σ_{SF}^A in order to prevent any deformation and any porosity change with respect to the reference configuration.[1] Conversely, if we let $\sigma = p = 0$, but $\sigma_{SF}^A \neq 0$, we obtain a nonzero strain ϵ_0 and a nonzero change of porosity φ_0, whose expressions are

$$\epsilon_0 = -(\sigma_0 + bp_0)/K = -\sigma_{SF}^A s\,(\alpha + b\beta)/K; \quad \varphi_0 = b\epsilon_0 - \sigma_{SF}^A s\beta/N. \tag{7.9}$$

The constitutive Equations (7.8) can then be rewritten in the form

$$\sigma = K(\epsilon - \epsilon_0) - bp; \tag{7.10a}$$

$$\varphi - \varphi_0 = b(\epsilon - \epsilon_0) + \frac{p}{N}. \tag{7.10b}$$

When considering only infinitesimal transformations, the initial volumetric strain ϵ_0 and the porosity change φ_0 are the only effects associated with the interface stress σ_{SF}^A, whose change, in turn, will impinge on them.

Explicit expressions for ϵ_0 and φ_0 cannot be determined without additional information concerning the microstructure of the porous network. According to Equation (7.9), these expressions depend upon the deformation of the surface of the internal walls. Therefore, they involve not only the geometry of these walls but also the elastic properties of the solid matrix. In the particular case shown in Figure 7.1(a) of a granular porous solid made of spherical

---

[1] According to Equation (7.5), the volumetric density u of the interface energy depends only on the volumetric deformation ϵ and the change φ of porosity. This conclusion is not general and, in turn, shows that the energy balance Equation (7.1) precludes the possibility of u depending on shear. In a refined approach, yet remaining linear, the expression of u will have to include a contribution $\sigma_{SF}^A s_0 \underline{\underline{\Gamma}} : \underline{\underline{e}}$ of the deviatoric strain tensor $\underline{\underline{e}}$ (see Equation (3.18)), generating the shear prestress $\sigma_{SF}^A s_0 \underline{\underline{\Gamma}}$.

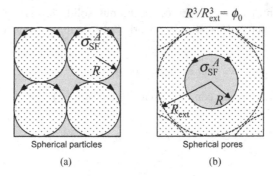

Figure 7.1 The volumetric strain due to the interface stress existing between the fluid and the solid depends strongly on the microstructure: (a) in the case of a solid matrix formed of spherical particles, the volumetric strain is proportional to the bulk modulus k_S of the solid matrix; (b) in the case of a porous network formed from spherical pores the volumetric strain is proportional to the the bulk modulus K of the porous solid

particles of radius R, the initial surface area of the internal walls of the porous network per unit of the initial volume $d\Omega_0$ is equal to $s = 4\pi R^2/8R^3 = 3(1-\phi_0)/R$, where ϕ_0 stands for the porosity. Besides, the interface stress σ_{SF}^A produces the volumetric deformation $\epsilon_S^0 = -2\sigma_{SF}^A/Rk_S$ of the spherical solid particles with k_S as bulk modulus. Since the deformation undergone by the particles is a homothetic transformation, ϵ_S^0 is also both the overall volumetric deformation ϵ_0 undergone by the porous solid and the deformation of its porous volume, that is φ_0/ϕ_0. We eventually write

$$\text{spherical solid particles: } \epsilon_0 = \epsilon_S^0 = -\frac{2}{3}\frac{\sigma_{SF}^A s}{(1-\phi_0)k_S}; \quad \varphi_0 = \phi_0\epsilon_S^0. \tag{7.11}$$

These particular values associated with spherical particles correspond to the case where the surface strain ϵ_A only depends on the solid matrix deformation according to

$$\epsilon_A = 2\epsilon_S^0/3. \tag{7.12}$$

From Equations (7.4), (7.11) and (7.12) we derive $\alpha = 2/3$ and $\beta = 0$. Substitution of these values in Equation (7.7) provides $\sigma_0 = 2\sigma_{SF}^A s/3$ and $p_0 = 0$.

Another case of particular interest is the case of the ideal isotropic porous solid made up of noninteracting hollow spheres of external radius R_{ext} and internal radius R, as shown in Figure 7.1(b). Since the resulting porosity of the porous solid must be ϕ_0, we require the radius R of the internal spheres delimiting the voids to meet the condition $R^3/R_{ext}^3 = \phi_0$. As for the granular porous solid shown in Figure 7.1(a), under the action of the interface stress the deformation of the porous solid in Figure 7.1(b) is still a homothetic transformation. The reasoning can then be restricted to a single hollow sphere. The initial volume of the external sphere is $d\Omega_0 = 4\pi R^3/3\phi_0$. In addition, since the initial area of the void surface is $4\pi R^2$, we have $s = 3\phi_0/R$. Under the action of the interface stress the radius R transforms into $R + \delta R$ and the volumetric surface area s into $s(1+\epsilon_A) = s + 8\pi R\delta R/d\Omega_0 = s(1+2\delta R/R)$. Noting that $\varphi_0 \simeq 4\pi R^2\delta R/d\Omega_0 = 3\phi_0\delta R/R$, the surface strain ϵ_A eventually depends on the void deformation φ_0/ϕ_0 only, according to

$$\epsilon_A = 2\varphi_0/3\phi_0. \tag{7.13}$$

From Equations (7.4) and (7.13) we derive $\alpha = 0$ and $\beta = 2/3\phi_0$. Substitution of these values in Equations (7.7) and (7.9) gives us $\sigma_0 = 0$, $p_0 = 2\sigma_{SF}^A s/3\phi_0$ and

$$\text{spherical pores:} \quad \epsilon_0 = -\frac{2b\sigma_{SF}^A s}{3\phi_0 K}; \quad \varphi_0 = -\frac{2\sigma_{SF}^A s}{3\phi_0}\left(\frac{b^2}{K} + \frac{1}{N}\right). \tag{7.14}$$

A comparison between Equations (7.11) and (7.14) shows that the order of magnitude of the contraction ϵ_0 resulting from the action of the interface stress can be quite different depending on the microstructure. For a granular porous solid formed of spherical particles the contraction is proportional to $1/k_S$ and, therefore, is usually negligible because of the poor compressibility of the solid particles. In contrast, for a porous solid with spherical pores the contraction is proportional to $1/K$ and can therefore be significant because of the softness of the porous solid as a whole as soon as its porosity is sufficiently large.

7.1.2 Adsorption-induced deformation

As shown by the analysis of Section 7.1.1, even though the pore pressure remains constant, a deformation will be induced by a change in the interface stress σ_{SF}^A. As analyzed in Chapter 6, adsorption does induce a change in the interface stress σ_{SF}^A. This was addressed through the Gibbs adsorption isotherm of Equation (6.87) in Section 6.1.9. When the fluid saturating the porous solid is an ideal gas (subscript G = F), substitution of $\gamma_G = 0$ and Equation (6.89) in Equation (6.40) gives us the following expression for the interface stress σ_{SG}^A:

$$\sigma_{SG}^A = \sigma_S^A - RT\Gamma_\infty \ln(1 + p/\varpi_0), \tag{7.15}$$

where σ_S^A is the surface stress given by Equation (6.33). Use of Equation (7.15) in Equations (7.9) and (7.10) shows that adsorption will introduce an extra nonlinear pore pressure contribution to the originally linear constitutive equations of a porous solid.

In addition, keeping the gas pressure constant, the replacement of the current saturating gas by another gas will paradoxically induce a deformation of the porous solid. Indeed, for the same solid substrate the pressure ϖ_0 involved in Langmuir's adsorption isotherm of Equation (6.83), through its dependency upon the energy depth $-E_0$, hinges on the gas, so that the gas replacement will affect the value of the interface stress σ_{SF}^A in Equation (7.9). For instance, the dipolar nature of carbon dioxide CO_2 induces larger values of E_0, when compared with the values related to methane CH_4. As a result, the adsorption of CO_2, enhanced by its penetration in the material, significantly lowers the surface energy and will induce the swelling of a coal sample where CH_4 was previously present. The injection of carbon dioxide in unmined deep coalbeds may help to mitigate the global warming due to the presence of carbon dioxide in the atmosphere. Nonetheless, the adsorption-induced swelling of the coal matrix causes the cleats of the coal seam to close and often hinders further injection.

7.2 Energy Balance for the Unsaturated Porous Solid

7.2.1 Lagrangian and Eulerian saturations

In Chapter 3 devoted to deformable saturated porous solids we introduced the Lagrangian porosity ϕ. According to Equation (3.49) the Lagrangian porosity ϕ relates the current porous volume $\phi d\Omega_0$ of a porous solid to the initial volume $d\Omega_0$ it occupied prior to its deformation. This turned out to be particularly useful when addressing various energy balances for deformable saturated porous solids. Also in Chapter 6 devoted to capillarity we explored the physical mechanics of a rigid unsaturated porous solid, whose porous space was filled by two distinct fluids. We will now consider the case where the porous solid is both deformable and unsaturated. Its initial porosity before deformation is ϕ_0, while its current Lagrangian porosity is ϕ in the deformed state. Referring the properties related to the two fluids by the subscript J (1 or 2), the current Lagrangian porosity ϕ can be split into the two partial Lagrangian porosities, ϕ_1 and ϕ_2:

$$\phi_1 + \phi_2 = \phi. \qquad (7.16)$$

Equation (7.16) extends Equation (6.99) related to rigid unsaturated porous solids to deformable unsaturated porous solids. The current partial porosity ϕ_J is such that $\phi_J d\Omega_0$ is the porous volume that the fluid J currently occupies. In the case of a rigid porous solid examined in Chapter 6, the volume $\phi_J d\Omega_0$ only resulted from the invasion of the porous volume, or the receding process from it, by the fluid J under the action of the pressure difference between the nonwetting and the wetting fluids that were in contact. In the case of a deformable porous solid, the volume $\phi_J d\Omega_0$ now results from a twofold process: the deformation of the porous volume that the fluid J currently occupies under the action of the pore pressure p_J, resulting in the partial porosity change φ_J; and the possible invasion of the porous volume or the receding process from it. Accordingly, the partial porosity ϕ_J can be split into two parts:

$$\phi_J = S_J \phi_0 + \varphi_J; \quad S_1 + S_2 = 1. \qquad (7.17)$$

The overall change $\varphi = \phi - \phi_0$ of the initial porosity ϕ_0 is eventually due to the deformation only. Substituting the decomposition of Equation (7.17) for ϕ_J in Equation (7.16), we find that $\varphi = \phi - \phi_0$ can be split into the contributions φ_1 and φ_2 associated with the deformation of the volume that each fluid respectively occupies:

$$\varphi = \phi - \phi_0 = \varphi_1 + \varphi_2. \qquad (7.18)$$

Equation (7.18) extends Equation (3.50) related to saturated deformable porous solids to unsaturated situations.

Equation (7.17) also extends Equation (6.100) related to a rigid unsaturated porous solid to a deformable unsaturated porous solids. Indeed, since φ_J in Equation (7.17) represents the contribution to the partial porosity ϕ_J due only to the deformation of the porous volume which the fluid J currently occupies, the term $\phi_0 S_J$ is the current partial saturation related to the fluid J prior to the deformation of the same volume. When the porous networks that the fluids occupy are disconnected, as shown in Figure 7.2, the part of the overall porous volume which each fluid occupies is always delimited by the same internal solid walls. As a result, there is no invasion or receding process, and S_J remains constant so that $dS_J = 0$. The variation of the

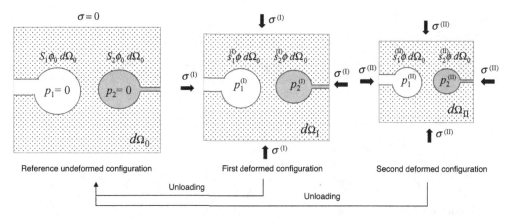

Figure 7.2 In the case of disconnected porous networks, when the material deforms only the Eulerian saturation s_J varies, whereas the Lagrangian saturation S_J remains constant whatever the applied stress and pore pressures

partial Lagrangian porosity ϕ_J is then only due to the deformation of the porous network J, resulting in $d\phi_J = d\varphi_J$.

This is no longer true in the case of connected porous networks. Each fluid, now either wetting (subscript J = W), or nonwetting (subscript J = nW), can then invade the porous solid, or recede from it. As a result, the fluids do not remain in contact with the same internal solid walls throughout the evolutions of the porous solid. As an effect, the partial saturation S_J does not remain constant and dS_J departs from zero. As shown in Figure 7.3 for a porous solid formed from connected pores embedded in a solid matrix, the partial saturation S_J is the fraction of the initial porous volume $\phi_0 d\Omega_0$ which the internal solid walls – delimiting the volume $\phi_J d\Omega_0$ in the current deformed configuration – would delimit in the undeformed configuration, after an unloading process restoring a zero pore pressure everywhere in the porous network. In other words, if the internal solid walls currently delimiting the volume $\phi_J d\Omega_0$ in the deformed state were painted with, say, a J color, $\phi_0 S_J d\Omega_0$ is the volume that the same internal solid walls painted with the J color would delimit after restoration of a zero pore pressure everywhere in the porous network. Note that, in contrast to the case of disconnected porous networks represented in Figure 7.2, in the case of connected porous networks represented in Figure 7.3 the wetting fluid and the nonwetting fluid can coexist within the porous solid only if their pressures differ. As a result, a zero pressure can be restored everywhere with only a single fluid present, the wetting fluid for instance as chosen in Figure 7.3. This latter condition appropriately defines the reference configuration regarding the saturation and the fluid pressures.

As illustrated in Figure 7.2 the current partial saturation s_J is more usually defined with regard to the deformed current configuration according to

$$\phi_J = \phi s_J; \quad s_1 + s_2 = 1. \tag{7.19}$$

In contrast to the saturation S_J defined by Equation (7.17), with which it coincides for undeformable porous solids only, the saturation s_J defined by Equation (7.19) will not remain constant, even in the case of disconnected porous networks, because of the deformation of the porous volume. Saturation s_J stands for the current volume fraction of fluid J relative to the

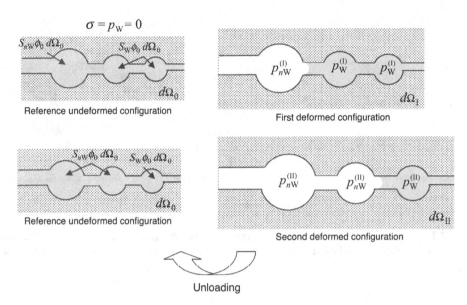

Figure 7.3 In contrast to the case of disconnected porous networks shown in Figure 7.2, in the case of a porous solid formed of connected pores embedded in a solid matrix the Lagrangian saturation S_J does not remain constant since each fluid, either wetting — subscript J = W — or nonwetting — subscript J = nW — can invade the porous solid or recede from it

current deformed porous volume $\phi d\Omega_0$ and, therefore, it can be termed the Eulerian saturation. In contrast, always referring to the undeformed initial volume $d\Omega_0$, the saturation S_J can be termed the Lagrangian saturation related to the fluid J.[2] The concept of Lagrangian saturation is not restricted to porous solids formed from connected pores embedded in a solid matrix. As illustrated in Figure 7.4, it also applies to a granular porous solid in the so-called funicular regime, where the wetting phase remains continuous. Similarly to the use of the porosity ϕ in saturated conditions, in the next section the use of the Lagrangian saturation S_J, instead of the Eulerian saturation s_J, will turn out to be more efficient to express free energy balances.

7.2.2 Lagrangian saturation and free energy balance

In Chapter 6, Equation (6.101) expressed the isothermal free energy balance for isothermal evolutions when a nonwetting fluid nW and a wetting fluid W were simultaneously present within an undeformable porous solid. Considering a deformable porous solid we have, in addition, to account for the strain work supplied to the porous solid. Accordingly, Equation (6.101) extends in the form

$$da_S = \sigma d\epsilon + p_{nW} d\phi_{nW} + p_W d\phi_W + s_{ij} de_{ij}. \tag{7.20}$$

---

[2] The concept of Lagrangian saturation has been introduced in Coussy, O. (2005) Poromechanics of freezing materials. *Journal of The Mechanics and Physics of Solids*, **53**, 1689–1718. See also Coussy, O., Pereira, J.-M. and Vaunat, J. (2007) Revisiting the thermodynamics of plasticity for unsaturated soils. *Computers and Geotechnics*, **37**, 207–215, from which Figure 7.4 is extracted.

The Unsaturated Poroelastic Solid

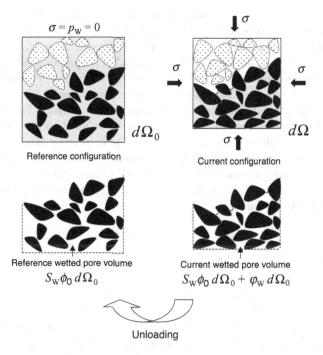

Figure 7.4 Illustration of the Lagrangian saturation S_W concept relative to the wetting fluid for a granular deformable porous solid

In turn, Equation (7.20) is the extension of the free energy balance Equation (3.72) related to a saturated porous solid. Substitution of Equation (7.17) into Equation (7.20) allows us to restate Equation (7.20) in the attractive form

$$da_S = \sigma d\epsilon + p_{nW} d\varphi_{nW} + p_W d\varphi_W + s_{ij} de_{ij} - \phi_0 (p_{nW} - p_W) dS_W. \qquad (7.21)$$

In Equation (7.21) the last term, that is $-\phi_0 (p_{nW} - p_W) dS_W$, identifies with the infinitesimal work made against the interfacial forces during the invasion of the porous space by the wetting phase or its receding process from it. This work is required to make the inner front between the nonwetting and wetting phases move within the porous solid. Because of the front propagation the part $\phi_0 dS_W$ of the porous volume, which was previously occupied by the nonwetting phase, is newly invaded by the wetting phase. From Equation (7.21) we derive

$$a_S = a_S \left(\epsilon, \varphi_{nW}, \varphi_W, e_{ij}, S_W \right); \qquad (7.22a)$$

$$\sigma = \frac{\partial a_S}{\partial \epsilon}; \quad p_{nW} = \frac{\partial a_S}{\partial \varphi_{nW}}; \quad p_W = \frac{\partial a_S}{\partial \varphi_W}; \quad s_{ij} = \frac{\partial a_S}{\partial e_{ij}}; \quad \phi_0 (p_{nW} - p_W) = -\frac{\partial a_S}{\partial S_W}. \qquad (7.22b)$$

It is worthwhile noting that the constitutive state Equations (7.22b) apply whatever the actual nature – gaseous or liquid – of the wetting and nonwetting constituents. These are the general constitutive equations of unsaturated poroelasticity. Without further specification they capitalize the general information provided by the first and second laws of thermodynamics about nondissipative changes occurring in unsaturated conditions.

Owing to the additive character of energy, the Helmholtz free energy a_S of the porous solid can be split into two parts: (i) the (elastic) free energy a_S^* of the porous solid considered alone, that is not including the interfaces; (ii) the surface energy U per unit of initial porous volume of the interfaces between the constituents forming the porous material. We will assume that U does not vary significantly when the porous solid deforms at saturation S_W held constant. As a result, the possible dependency of the surface energy U on the deformation variables – that is ϵ, φ_{nW}, φ_W and e_{ij} – will be ignored. This amounts to neglecting the effects analogous to those captured by Equation (7.5) in saturated conditions.[3] According to Equation (7.14), this assumption will be relevant as soon as the condition $\sigma_{SJ}^A s/\phi_0 K \ll 1$ is met. This condition is commonly satisfied because of the small values of $\sigma_{SJ}^A \gamma_{SJ}$ compared with the stiffness K of usual porous solids. We eventually write

$$a_S = a_S^* \left(\epsilon, \varphi_{nW}, \varphi_W, e_{ij}, S_W \right) + \phi_0 U (S_W). \tag{7.23}$$

Substituting Equation (7.23) into the last part of Equation (7.22b), we obtain

$$\phi_0 (p_{nW} - p_W) = -\frac{\partial a_S^*}{\partial S_W} - \phi_0 \frac{dU}{dS_W}. \tag{7.24}$$

In linear poroelasticity the energy a_S^* of the porous solid is a quadratic function of the strain arguments ϵ, φ_{nW}, φ_W and e_{ij}. The coefficients of this quadratic function are related to the current poroelastic properties and therefore they depend on the current saturation S_W, as explored later on in this chapter. As an effect, $\partial a_S^*/\partial S_W$ is the same quadratic form as a_S^* providing the coefficients of a_S^* are replaced by their derivative with respect to S_W. Accordingly, under the assumption of infinitesimal transformations, $\partial a_S^*/\partial S_W$ involves negligible terms. Indeed, as in Equation (4.5) for saturated isotropic porous solids, $\partial a_S^*/\partial S_W$ involves terms that are proportional to ϵ^2, $\epsilon \varphi_{nW}$, $\varphi_{nW} \varphi_W$ and $e_{ij} e_{kl}$. As a result, neglecting $\partial a_S^*/\partial S_W$ with respect to dU/dS_W in Equation (7.24), we derive

$$p_{nW} - p_W = -\frac{dU}{dS_W}, \tag{7.25}$$

that is, Equation (6.104). In short, for infinitesimal transformations and linear unsaturated poroelasticity, the effect of deformation on the capillary pressure curve is negligible.

In order to determine the remaining state equations, let us now proceed as we did at the beginning of Chapter 4. Similarly to Equation (4.2), we first write

$$\eta_S^* = a_S^* - p_{nW} \varphi_{nW} - p_W \varphi_W. \tag{7.26}$$

Substituting Equations (7.23) and (7.26) in Equation (7.21), and using Equations (7.22b) and (7.25), the energy balance Equation (7.21) reduces to

$$d\eta_S^* = \sigma d\epsilon - \varphi_{nW} dp_{nW} - \varphi_W dp_W + s_{ij} de_{ij}. \tag{7.27}$$

---

[3] Section 7.4.2 provides the means of extending Equation (7.5) and some results of Section 7.1.1 to any pore size distribution and to unsaturated conditions.

The Unsaturated Poroelastic Solid

Since the energy associated with the interfaces has been removed, the energy balance Equation (7.27) relates to the porous solid only. It shows that η_S^* depends only on ϵ, p_{nW}, p_W and e_{ij}, so that we obtain

$$\sigma = \frac{\partial \eta_S^*}{\partial \epsilon}; \quad \varphi_{nW} = -\frac{\partial \eta_S^*}{\partial p_{nW}}; \quad \varphi_W = -\frac{\partial \eta_S^*}{\partial p_W}; \quad s_{ij} = \frac{\partial \eta_S^*}{\partial e_{ij}}, \qquad (7.28)$$

which extend the state Equations (4.4) of saturated poroelasticity to unsaturated conditions.

7.3 The Linear Unsaturated Poroelastic Solid

7.3.1 Constitutive equations of unsaturated poroelasticity

The state Equations (7.28) of unsaturated poroelasticity are valid whether the porous networks are connected or not. The wetting or nonwetting character of the fluids matters only when considering capillary effects as, for instance, in Equation (7.25). Accordingly, for the sake of simplicity of notation, when considering general unsaturated poroelastic effects we will use the subscripts J = 1 and 2 rather than subscripts nW and W. For instance, instead of Equation (7.28) we now write

$$\sigma = \frac{\partial \eta_S^*}{\partial \epsilon}; \quad \varphi_1 = -\frac{\partial \eta_S^*}{\partial p_1}; \quad \varphi_2 = -\frac{\partial \eta_S^*}{\partial p_2}; \quad s_{ij} = \frac{\partial \eta_S^*}{\partial e_{ij}}. \qquad (7.29)$$

Considering only unsaturated isotropic linear poroelastic solids, the energy η_S will be a quadratic function of the pore pressures p_J and of the two first invariants of the strain tensor. As a result, we write

$$\eta_S^* = \frac{1}{2}K\epsilon^2 - b_1 \epsilon p_1 - b_2 \epsilon p_2 - \frac{1}{2}\frac{p_1^2}{N_{11}} - \frac{p_1 p_2}{N_{12}} - \frac{1}{2}\frac{p_2^2}{N_{22}} + G e_{ij} e_{ji}, \qquad (7.30)$$

which extends Equation (4.5) to unsaturated conditions. Substitution of Equation (7.30) in Equation (7.29) gives us the state constitutive equations of unsaturated linear poroelasticity in the explicit form:

$$\sigma = K\epsilon - b_1 p_1 - b_2 p_2; \qquad (7.31a)$$
$$\varphi_1 = b_1 \epsilon + p_1/N_{11} + p_2/N_{12}; \qquad (7.31b)$$
$$\varphi_2 = b_2 \epsilon + p_1/N_{12} + p_2/N_{22}; \qquad (7.31c)$$
$$s_{ij} = 2G e_{ij}. \qquad (7.31d)$$

7.3.2 Unsaturated microporoelasticity

Micro–macro relations between unsaturated poroelastic properties

When setting $p_1 = p_2 = p$ and $\varphi_1 + \varphi_2 = \varphi$ in the unsaturated constitutive Equations (7.31), the latter must reduce to the poroelastic constitutive Equations (4.6) related to saturated porous solids. From this remark stem the relations

$$b = b_1 + b_2; \quad 1/N_{11} + 2/N_{12} + 1/N_{22} = 1/N, \qquad (7.32)$$

where b and N are the thermoporoelastic properties related to the saturated porous solid and hence are still subjected to Equations (4.14) and (4.18).

However, at this stage b_J and N_{JK} are still not known separately. A further analysis is needed, similar to that we developed in Chapter 4 devoted to saturated poroelasticity, whose notations are again used. The macroscopic mean stress $\sigma = \frac{1}{3}\sigma_{ii}$ is the space average of the microscopic mean stress so that, similarly to Equation (3.66), we may write

$$\sigma = (1 - \phi_0)\sigma_S - \phi_0 S_1 p_1 - \phi_0 S_2 p_2. \tag{7.33}$$

In Equation (7.33) the interface stress associated with the surface energy of the solid–fluid interfaces is assumed to be negligible. As a result, p_J is the pressure actually transmitted to the solid matrix on the internal solid walls delimiting which the porous volume fluid J occupies. Consider then an experiment where

$$\sigma = -p; \quad p_1 = p_2 = p. \tag{7.34}$$

A combination of the last two equations gives

$$\sigma_S = -p, \tag{7.35}$$

so that the volumetric strain ϵ_S related to the solid matrix (assumed to be homogeneous) is

$$\epsilon_S = -p/k_S. \tag{7.36}$$

In the experiment related to the loading conditions of Equation (7.34), similarly to a spherical shell externally and internally loaded by the same pressure, the porous solid is loaded by a uniform pressure p on its whole boundary. We have already come across this situation in Section 4.1.2. According to Equation (4.17), it results in a homogeneous volumetric strain within the porous solid, causing the matrix and the porous space to deform in the same way:

$$\epsilon = \varphi_J/\phi_0 S_J = \epsilon_S. \tag{7.37}$$

Substituting Equation (7.37) in Equation (7.36), we obtain

$$\epsilon = \varphi_J/\phi_0 S_J = -p/k_S. \tag{7.38}$$

Use of Equations (7.34) and Equation (7.38) in Equation (7.31b) or (7.31c) provides us with the relation extending Equation (4.18) to unsaturated conditions, namely,

$$1/N_{JJ} + 1/N_{12} = (b_J - \phi_0 S_J)/k_S. \tag{7.39}$$

Pore volumetric isodeformation

Although the general relations of Equation (7.39) allow us to retrieve the second part of Equation (7.32), and give new information about unsaturated poroelastic properties, separate expressions for N_{JJ} and N_{12}, b_1 and b_2, are still lacking. While Equations (7.32) and (7.39) apply to any isotropic porous solid, the determination of extra equations requires some additional information specific to the porous solid under consideration. For instance, this may consist of the volumetric isodeformation of the two porous networks whenever they are both subjected to a zero pressure. This additional information can be expressed in the form

$$p_1 = p_2 = 0: \quad \varphi_1/\phi_0 S_1 = \varphi_2/\phi_0 S_2. \tag{7.40}$$

Substituting Equation (7.40) in Equations (7.31b) and (7.31c), we obtain

$$b_1/S_1 = b_2/S_2. \tag{7.41}$$

Substituting Equation (7.41) in the first part of Equation (7.32), we derive

$$b_J = b S_J. \tag{7.42}$$

Because of Equation (7.39), using Equation (7.42) it can eventually be shown that both volume fractions $\phi_0 S_1$ and $\phi_0 S_2$ keep deforming in the same way whenever they are both subjected to the same nonzero pressure p.

Even though separate assessments of b_1 and b_2 are provided by Equation (7.42), a further analysis is still needed to obtain separate expressions of N_{JJ} and N_{12}. Actually, there are no separate expressions for N_{JJ} and N_{12} valid for any porous solid. Nonetheless, separate assessments of N_{JJ} and N_{12} can be obtained when the porous solid, as in Section 4.1.2 and Figure 7.1, can be assimilated to a solid matrix embedding spherical voids as porous space. Since the overall volumetric strain ϵ is the average of the volumetric strain, letting ε_S be the volumetric strain of the solid matrix, we first write

$$\epsilon = (1 - \phi_0)\varepsilon_S + \varphi_1 + \varphi_2. \tag{7.43}$$

From the standard theory of elastic solids, a spherical void subjected to the pore pressure p_J, embedded in a linearly elastic solid matrix having k_S and g_S as bulk and shear moduli, respectively, and whose imposed volumetric strain some distance from the void is ε_S, undergoes a volumetric dilation $\varphi_J/\phi_0 S_J$ given by

$$\varphi_J/\phi_0 S_J = \left(1 + \frac{3k_S}{4g_S}\right)\varepsilon_S + \frac{3}{4g_S}p_J. \tag{7.44}$$

Substitution of Equation (7.44) in Equation (7.43) produces

$$\varepsilon_S = \frac{4g_S}{3\phi_0 k_S + 4g_S}\left[\epsilon - \frac{3\phi_0}{4g_S}(S_1 p_1 + S_2 p_2)\right]. \tag{7.45}$$

In turn, a substitution of Equation (7.45) in Equation (7.44) allows us to recover the constitutive Equations (7.31b) or (7.31c) of unsaturated poroelasticity, as well as Equations (7.39) and (7.42), but now with the benefit of the new separate relations

$$\frac{1}{N_{JJ}} = \frac{S_J^2}{N} + \frac{3S_1 S_2 \phi_0}{4g_S}; \quad \frac{1}{N_{12}} = \frac{S_1 S_2}{N} - \frac{3S_1 S_2 \phi_0}{4g_S}. \tag{7.46}$$

In order to interpret Equations (7.46), let us assume, for a while, that the porous volumes 1 and 2 still undergo the same volumetric deformation even though they are subjected to *distinct* pore pressures p_1 and p_2. Let us then substitute the pore volumetric isodeformation condition, that is $\varphi_1/\phi_0 S_1 = \varphi_2/\phi_0 S_2$, in Equations (7.31b) and (7.31c). If the resulting equations were to hold whatever the values of p_1 and p_2, the right-hand side of Equations (7.46) would have to reduce to the first term. Accordingly, in both right-hand sides of Equation (7.46) the extra term proportional to $S_1 S_2$ accounts for the difference of volumetric strains that the porous volumes 1 and 2 undergo whenever they are subjected to distinct pore pressures p_1 and p_2.

7.3.3 Double porosity approach to the brittle fracture of liquid-infiltrated materials

Experiments performed on brittle liquid-infiltrated materials, such as shales or concretes, have shown that the loading at failure depends on the loading rate:[4] the faster the loading rate, the larger the loading at failure. Brittle materials cannot store elastic or free energy beyond some critical threshold f_{cr}. When the threshold is reached, brittle fracture occurs through the abrupt and irreversible release of the whole stored energy a_S^* so that the failure criterion can be written

$$a_S^* = f_{cr}. \qquad (7.47)$$

Owing to the linearity of the stress–strain relationship, the free energy of an isotropic linearly elastic solid, subjected to the mean stress σ and undergoing the volumetric dilation ϵ, can be written

$$a_S^* = \frac{1}{2}\sigma\epsilon. \qquad (7.48)$$

For a brittle porous material the limit in energy storage applies to the solid matrix. For a dry linearly elastic porous material ($p = 0$), we have $\sigma = K\epsilon$, where K stands for the drained bulk modulus. Combining the stress–strain relationship $\sigma = K\epsilon$ with Equations (7.47) and (7.48), an intrinsic dry or porous solid compressive strength ϖ_{cr} can be defined as

$$\varpi_{cr} = \sqrt{2K f_{cr}}. \qquad (7.49)$$

Now let $-\sigma$ be the intensity of the mean compressive stress applied to the liquid-infiltrated material and increasing at constant rate $\overset{\circ}{\sigma}$. We write

$$-\sigma = \overset{\circ}{\sigma} t; \quad \overset{\circ}{\sigma} > 0. \qquad (7.50)$$

We would like to determine why the intensity ϖ_{fr} of the compressive stress causing the fracture departs from the compressive strength ϖ_{cr}, by depending on the loading rate $\overset{\circ}{\sigma}$.

Since the loading rate is involved, the explanation necessarily involves a rate-dependent viscous mechanism. For liquid-infiltrated porous materials a good candidate is the squirt viscous flow mechanism. According to this mechanism, the compression of the solid grains forming the matrix generates local gradients of the liquid pressure which, in turn, set the liquid in motion. In a first approach the squirt flow can be captured by considering a porous material exhibiting a double porosity network, the squirt flow occurring from one porous network to the other one. The former accounts for the porous space squeezed between the solid grains, while the latter represents larger pores which act as expansion reservoirs. Constitutive Equations (7.31) for unsaturated poroelasticity can then be applied to materials exhibiting a double porosity network by attributing to each network the subscript 1 or 2 as we did in the previous

---

[4] See for instance Rossi, P. et al. (1994) Effect of loading rate on the strength of concrete subjected to uniaxial tension. *Materials and Structures*, **27**, 260–264.

section. Owing to the presence of the internal pore pressures p_1 and p_2, instead of Equation (7.48), the elastic free energy that is stored within the solid matrix can be written in the form

$$a_S^* = \frac{1}{2}\sigma\epsilon + \frac{1}{2}p_1\varphi_1 + \frac{1}{2}p_2\varphi_2. \tag{7.51}$$

The changes considered here are externally undrained so that the liquid motion within the porous material results only from an exchange of liquid mass between the two porous networks. For the sake of simplicity assuming in addition the fluid incompressibility, the liquid mass conservation can be expressed in the form

$$\varphi_1 + \varphi_2 = 0. \tag{7.52}$$

The state Equations (7.31) and the continuity Equation (7.52) combine into

$$b\epsilon + p_1\left(\frac{1}{N_{11}} + \frac{1}{N_{12}}\right) + p_2\left(\frac{1}{N_{22}} + \frac{1}{N_{12}}\right) = 0. \tag{7.53}$$

Without having to specify the law governing the liquid flow between the porous networks 1 and 2, we can consider the two limit cases: the limit of infinitely slow loadings and the limit of infinitely fast loadings. In the limit of infinitely slow loadings, the liquid exchange between the two porous networks, which tends to decrease the pressure difference, occurs infinitely rapidly when compared with the loading rate $\overset{\circ}{\sigma}$. As an effect, in the limit $\overset{\circ}{\sigma} \to 0$, the two pressures remain constantly equal, that is, $p_1 = p_2 = p$. Inserting this equality in Equation (7.53) we obtain an expression for the common pressure value p as a linear function of the volumetric expansion ϵ. Substitution of this expression in Equation (7.31a) gives

$$\sigma = K_u \epsilon; \quad K_u = K + b^2 N. \tag{7.54}$$

The modulus K_u can be recognized as the undrained bulk modulus defined in Section 4.2.3, as if the porous networks were forming the same unique porous network. Since $p_1 = p_2 = p$, from the continuity Equation (7.52) it follows that Equation (7.51) for the free energy a_S^* reduces to Equation (7.48). Substitution of Equation (7.54) in Equation (7.48) and the use of the brittle fracture criterion of Equation (7.47) give us the intensity ϖ_{fr}^{slow} of the compressive stress $-\sigma$ causing the fracture in the limit of infinitely slow loadings:

$$\varpi_{fr}^{slow}/\varpi_{cr} = \sqrt{K_u/K}, \tag{7.55}$$

where ϖ_{cr} is the intrinsic dry compressive strength defined by Equation (7.49).

In the regime of fast loadings, the liquid exchange between the two porous networks occurs infinitely slowly when compared with the loading rate $\overset{\circ}{\sigma}$. As an effect, in the limit $\overset{\circ}{\sigma} \to \infty$, there is no liquid exchange between the two porous networks and the liquid mass content remains constant in each network, that is $\varphi_1 = \varphi_2 = 0$. Combining this with Equations (7.31b) and (7.31c), we derive separate expressions for p_1 and p_2 as linear functions of the volumetric expansion ϵ. With the help of Equations (7.32) and (7.54) of K_u, substitution of these expressions in Equation (7.31a) produces

$$\sigma = K_U \epsilon; \quad K_U = K_u + N\frac{[b_1(1/N_{22} + 1/N_{12}) - b_2(1/N_{11} + 1/N_{12})]^2}{1/N_{22}N_{11} - 1/N_{12}^2}. \tag{7.56}$$

The modulus K_U is the bulk modulus under undrained conditions with respect to both porous networks considered separately. Interestingly, when adopting the pore volumetric isodeformation assumption, which allows us to bring into play Equation (7.42), the equality $K_U = K_u$ then follows from Equations (7.39), (7.54) and (7.56). Indeed, under the assumption of the volumetric isodeformation of the two porous networks, we have $p_1 = p_2$ whatever the nature of the undrained conditions. In the double porosity approach to the squirt flow process, the pore volumetric isodeformation assumption is not relevant because the porous space squeezed between the solid grains and the larger pores acting as expansion reservoirs do not deform in the same way.

Since we have $\varphi_1 = \varphi_2 = 0$, the Equation (7.51) for the free energy a_S^* still reduces to Equation (7.48) in the limit of fast loadings. Substitution of Equation (7.56) in Equation (7.48) and use of the brittle fracture criterion of Equation (7.47) provide us with the intensity ϖ_{fr}^{fast} of the compressive stress $-\sigma$ causing the fracture in the limit of infinitely fast loadings:

$$\varpi_{fr}^{fast}/\varpi_{cr} = \sqrt{K_U/K}. \quad (7.57)$$

According to Equation (7.56), we have the inequality $K_U > K_u$ and therefore the successive inequalities $\varpi_{fr}^{fast} > \varpi_{fr}^{slow} > \varpi_{cr}$, as observed experimentally. Indeed, because of different internal conditions of drainage, the faster the loading regime, the greater the apparent stiffness of the porous material and, as a result, the greater the compressive stress intensity at failure supplying the same critical energy to the skeleton.[5]

7.4 Extending Linear Unsaturated Poroelasticity

7.4.1 The nonlinear poroelastic solid in unsaturated conditions

If we adopt the pore volumetric isodeformation assumption of Equation (7.42), the state Equations (7.31) and (7.39) combine to give

$$\sigma = K\epsilon - b(S_1 p_1 + S_2 p_2); \quad (7.58a)$$

$$\varphi = \varphi_1 + \varphi_2 = b\epsilon + (S_1 p_1 + S_2 p_2)/N, \quad (7.58b)$$

where b and N are the thermoporoelastic properties related to the saturated porous solid and, therefore, are still subjected to Equations (4.14) and (4.18). Interestingly, the state Equations (7.58) are exactly the same as the constitutive Equations (4.6) of a porous solid saturated by a single fluid at pressure $S_1 p_1 + S_2 p_2$. As a result, if the constitutive equations of the porous solid in saturated conditions are the nonlinear ones we have obtained in Section 4.1.3, they extend to unsaturated conditions by replacing the Terzaghi effective stress $\sigma' = \sigma + p$ by its unsaturated extension, namely, $\sigma' = \sigma + S_1 p_1 + S_2 p_2$ in Equations (4.34a) and (4.34b).

---

[5] This phenomenon can also be encountered at the scale of the structure as for instance in the analysis of a borehole breakdown in petroleum geophysics. See Garagash, D. and Detournay, E. (1997) An analysis of the influence of the pressurization rate on the borehole breakdown pressure. *Journal of Solids and Structures*, **34**, 3099–3118.

7.4.2 Accounting for interface stress effects upon deformation

In Section 7.3 the formulation of constitutive equations of unsaturated poroelasticity ignored the effects of the interface stress between the fluids and the internal solid walls delimiting the porous space. Because of these surface effects, the pressure that a fluid actually transmits to the solid matrix depends on the local curvature of the solid wall facing the fluid. Because of the difficulty of describing at each location the curvature of the solid surface which encloses each fluid and varies with their respective saturation, it is also difficult to account precisely for the deformation that the interface stress induces. However, when the pore volumetric isodeformation assumption can be adopted, a tractable situation is the one where the porous volume is made from spherical voids. According to the Laplace Equation (6.53), the pressure that a fluid J, with pressure p_J, actually transmits to the bulk solid matrix through the solid wall of a spherical pore of radius R is $p_J - 2\sigma_{SJ}^A/R$, where σ_{SJ}^A stands for the interface stress between the fluid and the solid matrix. As a result, in order to account for surface effects the unsaturated constitutive Equations (7.58) can be extended in the form

$$\sigma = K\epsilon - b\left[S_1\left(p_1 - p_{01}\right) + S_2\left(p_2 - p_{02}\right)\right]; \tag{7.59a}$$

$$\varphi = \varphi_1 + \varphi_2 = b\epsilon + \left[S_1\left(p_1 - p_{01}\right) + S_2\left(p_2 - p_{02}\right)\right]/N, \tag{7.59b}$$

where p_{0J} is the pressure which accounts for the averaged pressure shift induced by the interface stress, that is $2\sigma_{SJ}^A/R$, with respect to the possible values of the pore radius R :

$$p_{0J} = \frac{1}{S_J}\int_0^\infty \frac{2\sigma_{SJ}^A}{R}\frac{dS_J}{dR}dR. \tag{7.60}$$

In Equation (7.60) $dS_J(R)$ represents the infinitesimal fraction of the porous volume consisting of pores occupied by fluid J, which have their radius comprised between R and $R + dR$.

The constitutive Equations (7.59), where the initial configuration is free of interface stress, are the extension of the saturated constitutive Equations (7.8) for any pore radius distribution and for unsaturated conditions, albeit restricted to porous networks made of spherical pores. Indeed, in Section 7.1.1, for saturated conditions and spherical pores with the same radius R, we have shown that the prestress σ_0 and the initial pore pressure p_0 generated by the surface stress, as defined by Equation (7.14), could be specified in the form $\sigma_0 = 0$ and $p_0 = \sigma_{SF}^A s\beta = 2\sigma_{SF}^A s/3\phi_0$. Taking $\sigma_0 = 0$, the unsaturated state Equation (7.59a) extends the saturated state Equation (7.8a). The unsaturated state Equation (7.59b) extends the saturated state Equation (7.8b) and the relation $p_0 = \sigma_{SF}^A s\beta = 2\sigma_{SF}^A s/3\phi_0$, when recognizing that p_{0J} given by Equation (7.60) can be expressed in the alternative form

$$p_{0J} = \sigma_{SJ}^A \beta_J s_J; \quad \beta_J = \frac{2}{3\phi_0 s_J}; \quad ds_J = \frac{3\phi_0 dS_J}{R}, \tag{7.61}$$

where s_J is the volumetric area of the internal walls enclosing the fluid J.

7.4.3 The linear thermoporoelastic solid in unsaturated conditions

In order to account for thermal strains, the state Equations (4.73) of thermoporoelasticity extend to unsaturated conditions in the form

$$\sigma = K\epsilon - b_1 p_1 - b_2 p_2 - 3\alpha K (T - T_0); \qquad (7.62a)$$

$$\varphi_1 = b_1\epsilon + p_1/N_{11} + p_2/N_{12} - 3\alpha_{\varphi_1} (T - T_0); \qquad (7.62b)$$

$$\varphi_2 = b_1\epsilon + p_1/N_{12} + p_2/N_{22} - 3\alpha_{\varphi_2} (T - T_0). \qquad (7.62c)$$

where T_0 is an initial reference temperature. Quite similar developments to those that lead us to Equations (4.78) provide us with the relations

$$\alpha = \alpha_S; \quad \alpha_{\varphi_J} = \alpha_S (b_J - \phi_0 S_J). \qquad (7.63)$$

7.4.4 The linear unsaturated poroviscoelastic solid

Similarly to Equation (4.97), the unsaturated constitutive Equations (7.31) of poroelasticity extend to poroviscoelasticity in the form

$$\sigma = K \odot \epsilon - b_1 \odot p_1 - b_2 \odot p_2; \qquad (7.64a)$$

$$\varphi_1 = b_1 \odot \epsilon + N_{11}^{-1} \odot p_1 + N_{12}^{-1} \odot p_2; \qquad (7.64b)$$

$$\varphi_2 = b_2 \odot \epsilon + N_{12}^{-1} \odot p_1 + N_{22}^{-1} \odot p_2; \qquad (7.64c)$$

$$s_{ij} = 2G \odot e_{ij}, \qquad (7.64d)$$

where $\odot$ stands for the convolution product defined by Equation (4.89). Similarly to Equation (4.98), Equation (7.39) extends in the form

$$N_{JJ}^{-1} + N_{12}^{-1} = (b_J - \phi_0 S_J H) \odot k_S^{-1}, \qquad (7.65)$$

where H is the Heaviside step function of Equation (4.90).

In this chapter we have extended the constitutive equations of saturated poroelasticity to unsaturated poroelasticity. This extension was carried out by accounting for both the interfacial energy and the elastic energy stored in the solid matrix in the overall Helmholtz free energy of the porous solid. Introducing the key concept of the Lagrangian saturation, we were able to split the variation of the current saturation in the contribution due to the deformation of the internal solid walls and the contribution due to the invasion of the porous network by one fluid to the detriment of the other. In the same way as in Chapter 5, the constitutive equations of unsaturated poroelasticity can be now combined with the transport laws stated in Section 6.3 to address situations where the deformation of the porous solid and the flow of the fluids are coupled. For instance, these situations are encountered when dealing with the drying or the freezing of porous solids. Before addressing these complex situations in Chapter 9, we are going to revisit unconfined phase transitions in Chapter 8.

Further Reading

Coussy, O. (2004) *Poromechanics*, John Wiley & Sons Ltd.
Dormieux, L., Kondo, J. and Ulm, F.-J. (2006) *Microporomechanics*, John Wiley & Sons Ltd.

8

Unconfined Phase Transition

Because of the stable points they represent, phase transitions of water have been used as natural references in the early elaboration of scales of relative temperature. These purely empirical temperature scales originated from the phase transition points observed at atmospheric pressure and, therefore, ignored any effect on the phase transitions other than that of temperature. It was only in 1849 that James Thomson (1822–1892) predicted that the melting point of water would be lowered when a pressure was applied. Basically, his younger brother William Thomson, better known as Lord Kelvin (1824–1907), was putting forward the possibility that ice could be produced continuously, with no actual need of mechanical work, from two water reservoirs both held at $0\,°C$: the heat would be uninterruptedly extracted from one reservoir to the other. James Thomson opposed this possibility because this would contradict the second law of thermodynamics, still in its infancy, as formulated by Sadi Carnot (1796–1832). Indeed, mechanical work could then be produced indefinitely with no cost, by taking advantage of the pressure that can be exerted by the expansion of water during its solidification. This means of producing ice with no cost does fail if the pressure exerted on the solidifying water during its expansion lowers its melting point below $0\,°C$. This pressure melting effect was confirmed experimentally by Michael Faraday (1791–1867). Later on, in the paper entitled 'On the Equilibrium of Heterogeneous Substances', already mentioned at the beginning of Chapter 2, Josiah Willard Gibbs (1839–1903) explained the systematic effect of pressure upon phase transitions of matter, as well as the surface energy effects upon confined crystallization. It ultimately came to Johannes Diderik van der Waals (1837–1923) to open the way to the molecular interpretation of phase transitions, by showing that the transition from vapor to liquid results from the instability of the vapor phase, when the condensed state becomes more favorable with regard to energy because of the attractive energy related to intermolecular forces.

In preparation for exploring in Chapter 9 how phase transitions occur in a deformable porous solid, and what the related mechanical effects are, this chapter addresses the standard concepts related to unconfined phase transitions. We will start by revisiting the laws governing the usual phase equilibria. The equilibrium law between the liquid and solid phases of the same substance will require us to define the chemical potential of an elastic solid in a consistent way with the definition of the chemical potential of a fluid given in Section 2.1. Indeed, the chemical potential of an elastic solid has to account for the elastic energy associated with both volumetric and shear strains. In Chapter 9 this will be revealed as a key point to understanding how crystallization can occur in a pore. This chapter then recalls how phase transitions result from an instability that is triggered by the nucleation phenomenon. The chapter ends by looking into how surface energy effects induce the existence of precondensed and premelted liquid films at the surface of solid substrates. This formation involves the work produced by the disjoining pressure applied to thin liquid films and introduced in Section 6.1.10. In Chapter 9, the premelted liquid film will prove to play a major role in the mechanics of freezing porous materials.

8.1 Chemical Potential and Phase Transition

8.1.1 Phase equilibrium law

Since the celebrated works of Gibbs it is well known that the coexistence of two phases of the same substance requires for their chemical potential to be equal. Indeed, because of the very definition of the molar chemical potential μ_J related to a phase J, the free energy supplied to the phase throughout the formation of the new number of moles dn within the phase is $\mu_J dn$. Conversely, the loss in free energy undergone by the phase K throughout the extraction of the same number of moles dn is $-\mu_K dn$. If we assume that the exchange of the mole number dn between the phases J and K of the same substance occurs with no dissipation, the free energy cost must be zero, and we obtain the equality of the molar chemical potential related to each phase in the form

$$\mu_J = \mu_K. \tag{8.1}$$

This relation is analogous to Equation (2.10) governing the thermodynamic equilibrium of the same component belonging to two mixtures in contact, and has a quite similar meaning: whatever its nature, any transformation occurs spontaneously in the direction which makes the difference in the chemical potential of the involved components smaller; the transformation stops, and equilibrium is achieved, as soon as the chemical potential difference vanishes.

Phase transitions are usually classified according to whether they are first-order or second-order phase transitions. A first-order phase transition is a transition during which discontinuities affect the first derivatives of the chemical potential. Entropy and volume can be defined by appropriate first derivatives of the chemical potential, and they change discontinuously at a first-order phase transition. At sufficiently low temperatures the liquid–vapor and the liquid–solid transitions are first-order phase transitions, where the density and the entropy undergo a discontinuity. A second-order phase transition is characterized by continuous first derivatives and discontinuous second derivatives of the chemical potential. The approach to the critical point, where the first-order line of the liquid–vapor phase transition ends, is a second-order phase transition. In this chapter we will restrict ourselves to the above mentioned first-order phase transitions.

8.1.2 Chemical potential of a pure substance in any form

In Chapter 2 the chemical potential was introduced in the context of the thermodynamics of fluids and that of the mixtures which constitute the fluid. To investigate the consequences of the equilibrium Equation (8.1) whatever the phase transition under consideration, we need to define the chemical potential of a pure substance irrespective of its form, either fluid or solid. The free energy supply $\mu_J dn$ associated with the formation of new dn moles within any phase J can be split into two contributions:

$$\mu_J dn = a_J dn + dW_\rightarrow. \tag{8.2}$$

The first contribution, $a_J dn$, where a_J is the molar Helmholtz free energy, is the free energy supplied to the phase J by the extra dn moles that the phase will contain at the end of the formation process. The second contribution $dW_\rightarrow$ accounts for the additional work to be done against the already existing phase J to make room for the new dn moles in formation within the phase. When the substance considered is a fluid, if $\overline{V}_J$ is the molar volume, the volume that the new dn moles will finally occupy is $dV = \overline{V}_J dn$. The work $dW_\rightarrow$ is thus given by

$$dW_\rightarrow = p_J \overline{V}_J dn, \tag{8.3}$$

where p_J is the fluid pressure. Combining Equations (8.2) and (8.3), we obtain

$$\mu_J = a_J + p_J \overline{V}_J. \tag{8.4}$$

Therefore, the fluid molar chemical potential μ_J is identified with the fluid molar free enthalpy, often also referred to as its molar Gibbs free energy by comparison with Equation (2.5).

In order to determine $dW_\rightarrow$ whether the substance is fluid or solid, let the subscript 0 now refer to a reference state that is both undeformed and unstressed. In this reference state the dn moles occupy the volume $dV_0 = \overline{V}_J^0 dn$ while, as just stated, they will occupy the volume $dV = \overline{V}_J dn$ in the deformed current state. The distinction between the reference undeformed state and the current deformed state allows us to split the work $dW_\rightarrow$ into two contributions:

$$dW_\rightarrow = dW_{0 \rightarrow dV_0} + dW_{dV_0 \rightarrow dV}. \tag{8.5}$$

In Equation (8.5) $dW_{0 \rightarrow dV_0}$ accounts for the work produced against the already existing phase J to make room for the still undeformed volume $dV_0 = \overline{V}_J^0 dn$ which the dn moles would occupy prior to any deformation; $dW_{V_0 \rightarrow dV}$ accounts for the extra work, which is again produced against the phase J, to make the volume dV_0 deform into the final volume dV which the dn moles will ultimately occupy in the current strain state of phase J. The work $dW_{dV_0 \rightarrow dV}$ is therefore equal to the opposite of the strain work undergone by the moles whose initial volume dV_0 transforms into the final volume dV.

As defined by Equations (3.30) and (3.31), let σ_J and $\underline{\underline{s}}_J$ be the mean stress and the deviatoric stress tensor respectively. As defined by Equations (3.16) and (3.18), also let ϵ_J and $\underline{\underline{e}}_J$ be the current volumetric strain and the current deviatoric strain tensor related to phase J with respect to the reference state, respectively. In infinitesimal transformations both the volumetric strain

$$\epsilon_J = \left(\overline{V}_J - \overline{V}_J^0\right)/\overline{V}_J^0, \tag{8.6}$$

and the norm of $\underline{\underline{e}}_J$ are much smaller than one. Works $dW_{0 \rightarrow dV_0}$ and $dW_{dV_0 \rightarrow dV}$ are produced at stresses σ_J and $\underline{\underline{s}}_J$ held constant within the surrounding phase J where the extra dn moles are introduced. Owing to their definitions, these works can therefore be expressed in the form

$$dW_{0 \rightarrow dV_0} = -\sigma_J \overline{V}_J^0 dn; \quad dW_{dV_0 \rightarrow dV} = -\left(\sigma_J \epsilon_J + \underline{\underline{s}}_J : \underline{\underline{e}}_J\right) \overline{V}_J^0 dn. \tag{8.7}$$

Substituting Equation (8.7) in Equation (8.5), while using the definition of Equation (8.6) for ϵ_J, we derive the following expression for $dW_\rightarrow$:

$$dW_\rightarrow = -\left(\sigma_J \overline{V}_J + \underline{\underline{s}}_J : \underline{\underline{e}}_J \overline{V}_J^0\right) dn, \quad (8.8)$$

whose substitution in Equation (8.2) leads us to identify μ_J with

$$\mu_J = a_J - \sigma_J \overline{V}_J - \underline{\underline{s}}_J : \underline{\underline{e}}_J \overline{V}_J^0. \quad (8.9)$$

It is noteworthy that, when deriving Equation (8.9), we never require the substance in the solid form to be elastic. The derivation of Equation (8.9) only assumes infinitesimal transformations, that is, the validity of Equation (3.13), in order that the overall work $dW_\rightarrow$ defined by Equation (8.5) does not depend on the sequence of the loading steps, $0 \to dV_0$ and $dV_0 \to dV$.

When extended to any pure substance J, which now may be solid, but elastic, the Clausius–Duhem Equation (2.21) gives the energy balance

$$\sigma_J d\overline{V}_J + \overline{V}_J^0 \underline{\underline{s}}_J : d\underline{\underline{e}}_J - \overline{S}_J \, dT - da_J = 0. \quad (8.10)$$

Substituting Equation (8.9) in Equation (8.10) we derive the Gibbs–Duhem equation related to phase J:

$$\overline{V}_J d\sigma_J + \overline{V}_J^0 \underline{\underline{e}}_J : d\underline{\underline{s}}_J + \overline{S}_J dT + d\mu_J = 0, \quad (8.11)$$

from which we derive the state equations in the form

$$\overline{V}_J = -\frac{\partial \mu_J}{\partial \sigma_J}; \quad \overline{V}_J^0 \underline{\underline{e}}_J = -\frac{\partial \mu_J}{\partial \underline{\underline{s}}_J}; \quad \overline{S}_J = -\frac{\partial \mu_J}{\partial T}. \quad (8.12)$$

For a fluid, either in gaseous or liquid form, the stress state is spherical. We have $\sigma_J = -p$ and $\underline{\underline{s}}_J = 0$ so that Equation (8.11) reduces to the familiar Gibbs–Duhem Equation (2.18). As an effect, the definition of Equation (8.9) for the chemical potential and the Gibbs–Duhem Equation (8.11) constitute the extensions to a solid phase of Equation (8.4) and the Gibbs–Duhem Equation (2.18) related to a fluid phase, respectively.

An alternative way to derive Equation (8.11) consists of extending to any pure solid substance J – within which the stress sate remains uniform – the energy balance of Equation (2.6) in the form

$$dG = -V_J d\sigma_J - V_J^0 \underline{\underline{e}}_J : d\underline{\underline{s}}_J - S \, dT + \mu dN, \quad (8.13)$$

where V_J is the current solid volume, whereas V_J^0 is the related undeformed and unstressed reference volume which only changes when the substance amount changes. Using Euler's identity as we did to derive Equation (2.13), we similarly derive $G = N\mu$, irrespective of the shear contribution existing for a solid. Substitution of $G = N\mu$ in the energy balance Equation (8.13) produces again the Gibbs–Duhem Equation (8.11). However, because of the shear contribution, the stress state is generally not uniform within the components of a mixture of solids. As a result, neither Equation (2.13), that is, $G = \Sigma_J N_J \mu_J$, nor the Gibbs–Duhem Equation (8.11) simply extend to a mixture of solids.

8.1.3 Supersaturation

At temperature T, when both phases J and K are subjected to the same spherical stress state, the equilibrium pressure $p_{eq}(T)$ is obtained from Equation (8.1) by writing

$$\mu_J(p_{eq}, T) = \mu_K(p_{eq}, T). \tag{8.14}$$

However, the equilibrium Equation (8.1) makes possible the coexistence of two phases J and K of the same substance even though they are not subjected to the same pressure. In turn, a phase J can exist at a pressure p_J different from the equilibrium pressure $p_{eq}(T)$, while having a greater chemical potential than the phase K would have at the same pressure p_J: the phase J is in an equilibrium metastable state. If the phase J is the vapor phase and the phase K is the liquid phase, the vapor is then described as supersaturated. In the opposite case, the liquid phase is described as superheated. If the phase J is the liquid phase and the phase K is the solid phase, the liquid phase is described as supercooled.

The concept of supersaturation aims at quantifying the degree of metastability whatever the phase transition is. To define the supersaturation, from now on we define the mother phase as the phase having the largest entropy and we systematically refer to it by the subscript 1. Conversely, the daughter phase is defined as the phase having the lowest entropy and will systematically be referred to by the subscript 2. The supersaturation is the intensity of the force driving the transformation of the mother phase into the daughter phase. The supersaturation $\Delta\mu$ is defined as the difference between the chemical potential of the daughter phase and that of the mother phase, at the same temperature T and pressure p_1 of the mother phase:

$$\Delta\mu(p_1, T) = \mu_2(p_1, T) - \mu_1(p_1, T). \tag{8.15}$$

The supersaturation only depends on the current state of the mother phase through its pressure p_1. The definition of Equation (8.15) for the supersaturation and the phase equilibrium Equation (8.1), resulting here in $\mu_1(p_1, T) = \mu_2(p_2, T)$, combine together to give us the pressure p_2 of the daughter phase which can possibly form:

$$\Delta\mu = \mu_2(p_1, T) - \mu_2(p_2, T). \tag{8.16}$$

In the liquid–vapor transition the most ordered phase is the liquid phase, which is thereby identified with the daughter phase 2. In the liquid–solid phase transition the most ordered phase is the solid phase, which is therefore also identified with the daughter phase. Regarding the phase equilibrium both daughter phases can be considered as being poorly compressible, so that their molar volume $\overline{V}_2$ can be considered as nearly constant. In addition, assuming that a spherical stress state prevails in the daughter phase 2, we take J = 2, $\sigma_2 = -p_2$, $\underline{s}_2 = 0$ and $dT = 0$ in the Gibbs–Duhem equality (8.11). The integration of the resulting equation from p_1 to p_2 provides us with the following useful expression of the supersaturation defined by Equation (8.16):

$$\Delta\mu = -(p_2 - p_1)\overline{V}_2. \tag{8.17}$$

We can now make the remarks at the beginning of the section more quantitative. When not specifying the phase transition under consideration, the mother phase 1 is then described as supersaturated whenever its chemical potential is larger than the chemical potential which the daughter phase 2 would have at the same pressure p_1. According to the definition of

Equation (8.16), the supersaturation $\Delta\mu$ is then negative: the mother phase is either in a metastable state or in an unstable state, so that its transformation into the daughter phase becomes possible. Since $\Delta\mu < 0$, Equation (8.17) shows that the further possible coexistence of the mother and daughter phases requires the daughter phase to become more pressurized than the mother phase, resulting in $p_2 > p_1$. Indeed, with respect to the free energy $\mu_2(p_1, T)$ that the daughter phase would have at the lower pressure p_1 of the mother phase, a larger pressure p_2 allows the daughter phase to gain the required free energy difference $-\Delta\mu = \mu_2(p_2, T) - \mu_2(p_1, T) > 0$ in order to still meet the equilibrium condition of Equation (8.1). In turn, the energy amount $-\Delta\mu = (p_2 - p_1)\overline{V}_2 > 0$ is the work available for a mole of the daughter phase to form from the mother phase at pressure p_1 regarding the associated surface energy cost. As examined in detail in Section 8.5 for various situations, this balance between the available work and the surface energy cost will prove to govern the phase transition.

8.1.4 Phase transition as an instability

If Equation (8.1) provides us with the requirement for two phases of the same substance to coexist, it does not elucidate why and how a phase transition does occur. As the concept of supersaturation suggests, a phase transition occurs because the current phase becomes metastable or unstable, since a lower energy state related to the other phase turns out to be available, which also meets the constraints to which the substance is subjected. For instance, the transformation of a gas into a liquid results from the outcome of the competition between the disorder allowed by the thermal agitation, which is favorable to the gaseous state, and the negative intermolecular attraction energy, which is favorable to the liquid state. According to Equation (2.3), we have

$$p = -\left(\frac{\partial F}{\partial V}\right)_T = -\left(\frac{\partial E}{\partial V}\right)_T + T\left(\frac{\partial S}{\partial V}\right)_T. \tag{8.18}$$

For an ideal gas the internal energy E depends only upon temperature T, so that $\partial E/\partial V$ is zero. The state Equation (8.18) then shows that the elasticity of the gaseous state is entropic, that is it only depends on thermal agitation. For a liquid, in Equation (8.18) the quantity $T\partial S/\partial V$ becomes significantly lower than $\partial E/\partial V$, while for a solid it is often quite negligible. In both cases the elasticity is therefore mainly due to the intermolecular or interatomic energy. Accordingly, a phase transition is accompanied by a drastic change in the physical origin of the elasticity.

In order to be more quantitative let us now follow the historical approach of van der Waals of the liquid–vapor phase transition, and let us assume that the fluid, whatever its form, vapor or liquid, obeys his celebrated Equation (2.29), namely,

$$p = \frac{RT}{\overline{V} - B} - \frac{A}{\overline{V}^2}. \tag{8.19}$$

For large values of the molar volume $\overline{V}$ the molecules are some distance from each other. The thermal agitation results in a large entropy, and the related (negative) contribution to the chemical potential dominates, when compared with the contribution of the molecular interaction energy. The pressure is then mainly due to the first contribution on the right-hand side of Equation (8.19). Large values of the molar volume mean that the fluid is in a gaseous

Table 8.1 Properties at the critical point for various gases quantifying the departure from a van der Waals gas for which $RT_{cr}/p_{cr}\overline{V}_{cr} = 8/3 \simeq 2.7$

	H_2	He	N_2	CO_2	H_2O
$RT_{cr}/p_{cr}\overline{V}_{cr}$	3.0	3.1	3.4	3.5	4.5
T_{cr} (K)	33.2	5.2	126	304	647
p_{cr} (MPa)	1.3	0.23	3.4	7.4	22.1

or vapor state. In contrast, for small values of the molar volume $\overline{V}$ the (negative) contribution of the intermolecular interaction energy to the chemical potential dominates, when compared with the contribution due to thermal agitation. The liquid state, where the second contribution in the right-hand side of Equation (8.19) is no longer negligible, becomes more favorable than the previous gaseous state. A phase transition will occur when, for the appropriate values of temperature and pressure, the value of the chemical potential of the current phase is no longer lower than that associated with the other possible phase. The precise determination of the appropriate values of temperature and pressure can be looked into through the analysis of the isotherms associated with Equation (8.19).

The critical point is defined as the point where the first and second derivatives of p, with respect to $\overline{V}$, both vanish. For the van der Waals fluid, whose state equation is given by Equation (8.19), this is achieved for the following specific critical values p_{cr}, $\overline{V}_{cr}$ and T_{cr}:

$$p_{cr} = A/27B^2; \quad \overline{V}_{cr} = 3B; \quad T_{cr} = 8A/27BR; \quad RT_{cr}/p_{cr}\overline{V}_{cr} = 8/3. \tag{8.20}$$

Unfortunately the state Equation (8.19) does not hold whatever the gas. If it did, the value $RT_{cr}/p_{cr}\overline{V}_{cr} = 8/3 \simeq 2.7$ would have been universal. Table 8.1 shows the value of $RT_{cr}/p_{cr}\overline{V}_{cr}$ for various gases. For water molecules we have $RT_{cr}/p_{cr}\overline{V}_{cr} = 4.5$. This significant departure from 8/3 is explained by the existence of both permanent dipoles and hydrogen bonds.

In the $(p/p_{cr}, \overline{V}/\overline{V}_{cr})$ plane, Figure 8.1 shows the isotherms related to the van der Waals fluid and provides the meaning of the critical point. For $T > T_{cr}$ the pressure p, as a function of the molar volume $\overline{V}$, is a one-to-one relation; the fluid then changes continuously from a vapor to a liquid as $\overline{V}$ decreases from large values to small values. In contrast, for $T < T_{cr}$ and the same value of p, three values of the molar volume $\overline{V}$ are now allowed by Equation (8.19), and a first-order phase transition becomes possible. The equilibrium pressure p_{eq} at which the phase transition actually occurs can be determined by combining the equilibrium relation Equation (8.14), where we take $1 = $ V and $2 = $ L, and the Gibbs–Duhem Equation (2.18). Along an isotherm they give the relation

$$\mu_V\left(p_{eq}, T\right) - \mu_L\left(p_{eq}, T\right) = \int_L^V \overline{V}\,dp = 0. \tag{8.21}$$

Integrating the previous equation by parts, we obtain

$$p_{eq}\left(\overline{V}_V - \overline{V}_L\right) = \int_L^V p\,d\overline{V}. \tag{8.22}$$

According to Equation (8.22), the areas LL_1F and FV_1V, which are delimited by the isotherm and are located on each side of the pressure line p_{eq}, must be the same for the phase transition

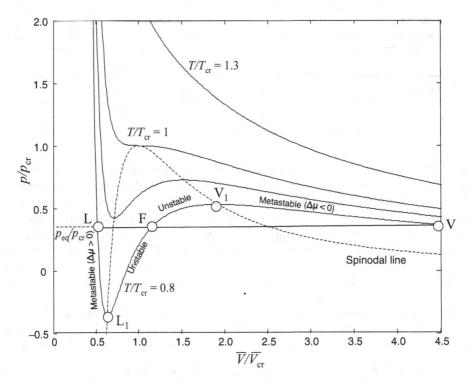

Figure 8.1 Van der Waals isotherms: Maxwell's rule states that the liquid at point L can transform into vapor at point V whenever areas LL_1F and FV_1V are equal; the fluid states lying below the so-called spinodal curve represented in dashed lines are unstable and cannot be observed

to occur at pressure p_{eq}. This property, which is known as Maxwell's rule, is shown in Figure 8.1. Starting from point L where its form is liquid, the fluid progressively transforms into its vapor. At point V, where the chemical potential has the same value, the phase transition is achieved. Let us now show that the fluid is either in a metastable state or in an unstable state between the points L and V belonging to the same isotherm.

- Along LL_1 the fluid is in liquid form. The value of the liquid chemical potential $\mu_L(p, T)$ is then larger than the value of the vapor chemical potential $\mu_V(p, T)$ along the same isotherm,[1] where the fluid is in vapor form at the same pressure p. Basically, starting from the same value $\mu_L = \mu_V$ related to points L and V, we have $d\mu_L > d\mu_V$ along LL_1. This inequality is obtained by noting that $\overline{V}_L = d\mu_L/dp < \overline{V}_V = d\mu_V/dp$, where dp is negative. Along LL_1 the liquid is therefore in a metastable state, where another possible (vapor) state of the fluid, associated with a lower chemical potential, exists simultaneously. The liquid is then superheated, and the associated supersaturation $\Delta\mu$ is found to be positive by setting $1 = V$ and $2 = L$ in Equation (8.15). It is worthwhile noting that the pressure becomes

---

[1] This part of the isotherm where the fluid is in vapor form is not represented in Figure 8.1 and is formed by the curve starting at point V and corresponding to pressure values lower than p_{eq}.

negative before the limit of metastability L_1 is reached. The liquid is then subjected to a tensile spherical stress. For instance, at ambient temperatures the maximum tensile stress associated with metastable states of liquid water is close to 150 MPa.[2]

- Along VV_1 the fluid is in vapor form. As can be shown in a similar way to the previous case, the value of the vapor chemical potential $\mu_V(p, T)$ is then larger than the value of the liquid chemical potential $\mu_L(p, T)$ along the same isotherm where the fluid is in liquid form at the same pressure p — corresponding to the higher part of the isotherm starting at L in Figure 8.1. Along VV_1 the vapor, in a metastable state, is supersaturated, and the related supersaturation $\Delta\mu$ is negative.
- Along L_1FV_1 we have $d\overline{V}/dp = d^2\mu/dp^2 > 0$, so that the molar volume increases with the pressure: the fluid is then unstable and the states corresponding to L_1FV_1 cannot be observed.

In a similar way to the stability analysis of regular solutions (see Figure 2.7), the curve delimiting the domain of unstable states is the spinodal curve. The limit of stability is governed here by the combination of the extremum condition, that is, $(\partial p/\partial \overline{V})_T = 0$, and Equation (8.19), providing the equation of the spinodal curve in the form

$$\frac{p}{p_{cr}} = \frac{3\overline{V}_{cr}^2}{\overline{V}^2} \times \left(1 - \frac{2\overline{V}_{cr}}{3\overline{V}}\right). \tag{8.23}$$

As generally defined by Equation (8.15) the supersaturation $\Delta\mu$ is the key to the precise analysis related to a phase transition. To go further we need to know the explicit expressions for the supersaturation related to each possible phase transition. This is done in the following sections for the main phase transitions of interest.

8.2 Liquid–Vapor Transition

8.2.1 The Kelvin equation

The phase having the lowest entropy is the most ordered phase. Therefore, for the liquid–vapor phase transition, the liquid phase is the daughter phase, while the vapor phase is the mother phase. Bearing in mind the previous notations, we set $1 = V$ and $2 = L$. The equilibrium Equation (8.1) is then written in the form

$$\mu_V(p_V, T) = \mu_L(p_L, T). \tag{8.24}$$

Assuming sufficiently slow changes, in order that thermodynamic equilibrium between the two phases is achieved at any time, the previous relation can be differentiated. Using the Gibbs–Duhem Equation (2.18) relative to each phase and differentiating Equation (8.24), we obtain[3]

$$\overline{V}_V dp_V - \overline{V}_L dp_L = \left(\overline{S}_V - \overline{S}_L\right) dT. \tag{8.25}$$

---

[2] See for instance Mercury, L. and Tardy, Y. (2001) Negative pressures of streched liquid water, Geochemistry of soil capillaries. *Geochimica et Cosmochimica Acta*, **65**, 3391–3408.
[3] When taking $p_V = p_L = p$, Equation (8.25) reduces to the Clausius–Clapeyron equation.

With regard to the usual variations of the liquid pressure p_L the liquid compressibility does not affect the equilibrium Equation (8.24) significantly, and the liquid molar volume $\overline{V}_L$ can be considered as constant. As far as the vapor is concerned, according to the ideal gas Equation (2.23) we write

$$p_V = RT/\overline{V}_V. \qquad (8.26)$$

Substitution of Equation (8.26) in the differential equilibrium Equation (8.25), followed by an integration at temperature T held constant from the atmospheric pressure p_{atm} taken as the reference pressure for the liquid pressure p_L, gives us the Kelvin equation:

$$p_L - p_{atm} = \frac{RT}{\overline{V}_L} \ln \frac{p_V}{p_{VS}(T)}. \qquad (8.27)$$

In the Kelvin Equation (8.27) $p_{VS}(T)$ stands for the saturated vapor pressure related to the temperature T, when the liquid pressure p_L is equal to the atmospheric pressure p_{atm}. The physical interpretation of the Kelvin Equation (8.27) can be given as follows. The saturated vapor pressure $p_{VS}(T)$ is the equilibrium vapor pressure satisfying Equation (8.24) when the liquid pressure p_L is equal to the atmospheric pressure p_{atm}. At temperature T held constant, when the vapor pressure decreases from $p_{VS}(T)$, the decrease in the vapor molar chemical potential μ_V results from the increase in the vapor entropy due to the larger molar volume and, therefore, to the greater number of states accessible for the vapor molecules. Indeed, from Equations (2.27) and (2.28) it can be shown that $\mu_V(p_V, T) - \mu_V(p_{VS}, T) = -T\left(\overline{S}(p_V, T) - \overline{S}(p_{VS}, T)\right) = RT \ln(p_V/p_{VS})$. Simultaneously, in order to still meet the equilibrium condition of Equation (8.24) the liquid chemical potential has to decrease. Since the liquid is poorly compressible, unlike the vapor there is no possibility of the increase in the liquid entropy due to a larger molar volume. The decrease of the liquid chemical potential μ_L can only result from the pressure contribution, with $\mu_L(p_L, T) - \mu_L(p_{atm}, T) = (p_L - p_{atm})\overline{V}_L$. Equating both chemical potential variations, we return to the Kelvin Equation (8.27).

In order to derive an expression for the saturated vapor pressure $p_{VS}(T)$ as a function of the temperature, taking $p_L = p_{atm}$ and $p_V = p_{VS}(T)$ in Equation (8.25) and using Equation (8.26), we first obtain

$$p_L = p_{atm}; \quad p_V = p_{VS}(T): \quad RT\frac{dp_{VS}}{p_{VS}(T)} = \left(\overline{S}_V - \overline{S}_L\right) dT. \qquad (8.28)$$

Making use of Equation (2.27), we express $\overline{S}_L$ and $\overline{S}_V$ in the form

$$\left(\overline{S}_L\right)_{p=p_{atm}} = \overline{S}_{L0} + \overline{c}_L \ln \frac{T}{T_0}; \quad \left(\overline{S}_V\right)_{p_V=p_{VS}(T)} = \overline{S}_{V0} + \overline{c}_V \ln \frac{T}{T_0} - R \ln \frac{p_{VS}(T)}{p_{VS}(T_0)}, \qquad (8.29)$$

where $\overline{c}_L$ and $\overline{c}_V$ stand for the molar heat capacity of the fluid in liquid and vapor form (at constant pressure for the vapor), respectively, while $\overline{S}_{L0}$ and $\overline{S}_{V0}$ stand for the respective molar entropies at the reference temperature T_0. Substituting Equation (8.29) in Equation (8.28), we derive a differential relation between $p_{VS}(T)$ and T. Integrating this, we derive the Dupré formula,[4] namely,

$$\ln \frac{p_{VS}(T)}{p_{VS}(T_0)} = \frac{\Delta h + T_0(\overline{c}_L - \overline{c}_V)}{R}\left(\frac{1}{T_0} - \frac{1}{T}\right) - \frac{(\overline{c}_L - \overline{c}_V)}{R}\ln\frac{T}{T_0}, \quad (8.30)$$

where

$$\Delta h = L_{vap} = T_0\left(\overline{s}_{V0} - \overline{s}_{L0}\right) \quad (8.31)$$

stands for the molar enthalpy of vaporization related to the reference temperature T_0. The molar enthalpy of vaporization, also called 'latent heat' of vaporization and denoted by L_{vap}, is the energy input supplied in heat form which is required per mole to separate the liquid molecules from one another. The possible approximation $T_0(\overline{c}_L - \overline{c}_G) \ll \Delta h$ leads to neglecting the last term on the right-hand side of Equation (8.30). The Dupré formula then reduces to the Rankine empirical formula. When the liquid considered is pure water, the boiling point $T_0 = 373$ K obtained for $p_L = p_{atm}$ corresponds to $p_{VS}(T_0) = p_{atm}$, while $L_{vap} = 40.63$ kJ mol^{-1} ($\equiv 2257$ kJ kg^{-1}).

When the liquid considered is water, the ratio $p_V/p_{VS}(T)$ is the relative humidity h_R:

$$h_R = \frac{p_V}{p_{VS}(T)}, \quad (8.32)$$

so that the Kelvin Equation (8.27) can be rewritten in the familiar form

$$p_L - p_{atm} = \frac{RT}{\overline{V}_W}\ln h_R, \quad (8.33)$$

where $\overline{V}_W$ stands for the molar volume of liquid water. As soon as the atmosphere is no longer water vapor saturated, that is, as soon as $h_R < 100\%$, the coexistence of water in both liquid and vapor forms requires the liquid pressure to be lower than the atmospheric pressure.

The common value p_{eq} of the equilibrium pressure involved in Equation (8.21), and represented in Figure 8.1, can be assessed with the help of the Kelvin equation, as the solution of Equation (8.27) when we take $p_L = p_V = p_{eq}$. However, since Equation (8.27) holds irrespective of the equality between the liquid and vapor pressure, it applies when the vapor is only one of the components of an ideal gaseous mixture at pressure p_G. The pressure p_V is then identified with the partial pressure related to the vapor, and is only equal to p_G as a special case. In order to address the general situation the Kelvin Equation (8.27) can be conveniently rewritten in the form

$$p_L - p_G = \frac{RT}{\overline{V}_L}\ln\frac{p_V}{p_{0V}}, \quad (8.34)$$

where p_{0V} is the saturated vapor pressure associated with the condition $p_L = p_G$, and whose expression is obtained by taking $p_L = p_G$ in Equation (8.27):

$$p_{0V} = p_{VS}(T)\exp\frac{(p_G - p_{atm})\overline{V}_L}{RT}. \quad (8.35)$$

---

[4] Equation (8.30) for $p_{VS}(T)$ was historically proposed on purely empirical grounds in the form

$$\ln p_{VS}(T) = \alpha - \beta/T - \gamma \ln T.$$

The phase 1 of largest entropy being the vapor phase V, when the vapor is a component of a gaseous mixture, the supersaturation $\Delta\mu$ can be suitably defined as

$$\Delta\mu = \mu_L(p_G, T) - \mu_V(p_V, T). \tag{8.36}$$

Comparing Equation (8.36) with the definition of Equation (8.15) for $\Delta\mu$ related to pure phases, the liquid chemical potential μ_L is assessed in Equation (8.36) at the overall gas pressure p_G, and not at the partial pressure p_V of the vapor. Indeed, when the vapor is only one of the components of a gaseous mixture, the liquid phase related to this vapor must form within a gaseous mixture which is subjected to the overall gas pressure p_G, and not to the partial pressure p_V. Substitution of Equation (8.24) in Equation (8.36) provides

$$\Delta\mu = \mu_L(p_G, T) - \mu_L(p_L, T). \tag{8.37}$$

Still assuming that the liquid is poorly compressible, so that its molar volume $\overline{V}_L$ remains constant, from an integration of the isothermal Gibbs–Duhem Equation (2.18) applied to the liquid and from Equation (8.37) we derive

$$\Delta\mu = -(p_L - p_G)\overline{V}_L. \tag{8.38}$$

Comparing Equation (8.17) with Equation (8.38) regarding the supersaturation, the role of mother phase can finally be devoted to the gaseous mixture. Substitution of Equation (8.34) in Equation (8.38) gives us the expression for the supersaturation related to the liquid–vapor phase transition in the specific form

$$\Delta\mu = -RT \ln \frac{p_V}{p_{0V}}. \tag{8.39}$$

8.2.2 Effect of a solute

Let us now consider an aqueous solution where the solvent is liquid water. If the aqueous mixture is ideal and contains a single solute, whose molar fraction is denoted by x, Equation (2.61) allows us to write the molar chemical potential of liquid water μ_W in the form

$$\mu_W(p_L, T, x) = \mu_W^*(p_L, T) + RT \ln(1-x), \tag{8.40}$$

where we bear in mind that $\mu_W^*(p_L, T)$ is the chemical potential of pure liquid water at pressure p_L of the solution. Because of the Gibbs–Duhem Equation (2.18) and Equation (8.26), we can write

$$T = T_0: \quad d\mu_W^* = \overline{V}_W dp_L; \quad d\mu_V = \overline{V}_V dp_V = RT dp_V/p_V. \tag{8.41}$$

Differentiation of Equation (8.24), where we take L = W, provides the law governing the phase equilibrium in the form $d\mu_W = d\mu_V$. This latter relation and the two previous equations lead to the explicit isothermal differential relation

$$T = T_0: \quad \overline{V}_W dp_L - RT dx/(1-x) = RT dp_V/p_V, \tag{8.42}$$

whose integration from a reference state, where $p_L = p_{atm}$ and $x = 0$, to the current state provides the equilibrium relation

$$p_L - p_{atm} = \frac{RT}{\overline{V}_W} \ln \frac{h_R}{1-x}. \tag{8.43}$$

A comparison of Equation (8.33) with Equation (8.43) shows that, because of the entropy of mixing (see Section 2.3.3), the solute tends to reduce the relative humidity prevailing at equilibrium for the same temperature and the same pressure difference $p_L - p_{atm}$. In practice, according to Equation (8.43), when the solution pressure p_L remains atmospheric, the relative humidity h_R can be imposed through the use of salt solute whose molar fraction x is known. In turn, when the relative humidity h_R is the same, the presence of a solute increases the intensity of the supersaturation $\Delta\mu$, whose Equation (8.39) transforms into

$$\Delta\mu = -RT \ln \frac{p_V}{(1-x)p_{0V}}, \tag{8.44}$$

where p_{0V} is the reference value of Equation (8.39) corresponding to the absence of solute.

8.3 Liquid–Solid Transition

8.3.1 The Thomson equation

In the case of the liquid–solid transition the phase having the lowest entropy is the solid phase. In the following we will use the subscript C for the solid crystal phase, to avoid confusion with the subscript S which we have adopted for the porous solid throughout the book. We therefore have here $1 = L$ and $2 = C$, and the equilibrium Equation (8.1) applies in the form

$$\mu_L = \mu_C. \tag{8.45}$$

Like any elastic solid, the crystal can sustain shear stresses and store elastic energy. According to the Gibbs–Duhem Equation (8.11) the chemical potential μ_C relative to an elastic solid phase is a function of σ_J, $\underline{\underline{s}}_J$ and T. In linear elasticity μ_C is the sum of a linear form and of a quadratic form of the arguments σ_J, $\underline{\underline{s}}_J$ and T. When addressing the liquid–solid phase equilibrium, it is convenient to adopt as reference state for both phases the melting temperature T_m associated with the solid–liquid phase equilibrium where both phases are at atmospheric pressure. For an isotropic solid, omitting the subscript J for the stress components σ or s_{ij}, this choice turns out to be

$$\mu_C = \mu_C^0 - \overline{V}_C^0 (\sigma + p_{atm}) - (T - T_m)\left[\overline{S}_C^0 + 3\alpha_C \overline{V}_C^0 (\sigma + p_{atm})\right]$$
$$- \overline{V}_C^0 (\sigma + p_{atm})^2 / 2K_C - \overline{V}_C^0 s_{ij} s_{ji} / 4G_C - \overline{c}_C (T - T_m)^2 / 2T_m. \tag{8.46}$$

Substitution of Equation (8.46) in the state Equations (8.12) allows us to recover the standard constitutive equations related to any linear elastic solid, namely,

$$\overline{V}_C = \overline{V}_C^0 [1 + (\sigma + p_{atm})/K_C + 3\alpha_C (T - T_m)]$$
$$e_{ij} = s_{ij}/2G_C \tag{8.47}$$
$$\overline{S}_C = \overline{S}_C^0 + 3\overline{V}_C^0 \alpha_C (\sigma + p_{atm}) + \overline{c}_C (T - T_m)/T_m.$$

Accordingly, K_C and G_C are identified with the bulk and the shear moduli of the elastic solid crystals, whereas $\bar{c}_C$ and $3\alpha_C$ are their molar heat capacity and their volumetric dilation coefficient (at atmospheric pressure), respectively. For a compressible liquid, taking $\sigma_L = -p_L$ and removing the shear contribution, we similarly obtain

$$\mu_L = \mu_L^0 + \overline{V}_L^0 (p_L - p_{atm}) - (T - T_m)\left[\overline{S}_L^0 - 3\overline{V}_L^0 \alpha_L (p - p_{atm})\right]$$
$$- \overline{V}_L^0 (p_L - p_{atm})^2 / 2K_L - \bar{c}_L (T - T_m)^2 / 2T_m. \qquad (8.48)$$

From Equations (8.12) and (8.48) we can recover the state equations related to a compressible liquid in the familiar form

$$\overline{V}_L = \overline{V}_L^0 \left[1 - (p_L - p_{atm})/K_L + 3\alpha_L (T - T_m)\right]; \qquad (8.49)$$

$$\overline{S}_L = \overline{S}_L^0 - 3\overline{V}_L^0 \alpha_L (p_L - p_{atm}) + \bar{c}_L (T - T_m)/T_m. \qquad (8.50)$$

According to Equation (8.45), equating the expressions of Equations (8.46) and (8.48) which we obtained for μ_C and μ_L respectively, we derive the solid–liquid phase equilibrium condition in the form

$$-(\sigma + p_{atm}) - (p_L - p_{atm})\overline{V}_L^0/\overline{V}_C^0 - (\sigma + p_{atm})^2/2K_C$$
$$+ \left(\overline{V}_L^0/\overline{V}_C^0\right)(p_L - p_{atm})^2/2K_L - s_{ij}s_{ji}/4G_C =$$
$$\left[\Delta S_m - 3\alpha_C (\sigma + p_{atm}) + (\bar{c}_L - \bar{c}_C)(T_m - T)/2T_m \overline{V}_C^0 - 3\alpha_L (p - p_{atm})\left(\overline{V}_L^0/\overline{V}_C^0\right)\right](T_m - T), \qquad (8.51)$$

where ΔS_m is the melting entropy per phase volume unit defined by

$$\Delta S_m = \left(\overline{S}_L^0 - \overline{S}_C^0\right)/\overline{V}_C^0. \qquad (8.52)$$

To derive Equation (8.51) we took into account the equality $\mu_C^0 = \mu_L^0$, which derives from the definition of the melting temperature T_m. In the following we will now make use of the approximations

$$\overline{V}_L^0/\overline{V}_C^0 - 1 \simeq 1 - \overline{V}_C^0/\overline{V}_L^0 (\simeq 0.09 \text{ for water}) \ll 1; \quad (T - T_m)^2 / T_m^2 \ll 1. \qquad (8.53)$$

Neglecting the elastic energy on the left-hand side of Equation (8.51), that is, neglecting the quadratic terms, as well as the entropic terms related to the thermal dilation and the heat capacity on the right-hand side of Equation (8.51), we obtain the approximate equilibrium condition

$$-\sigma - p_{atm} - (p_L - p_{atm})\overline{V}_L^0/\overline{V}_C^0 = (T_m - T)\Delta S_m. \qquad (8.54)$$

Whenever a spherical stress $\sigma = -p_C$ prevails in the solid crystal phase, the equilibrium Equation (8.54) reduces to

$$p_C - p_{atm} - (p_L - p_{atm})\overline{V}_L^0/\overline{V}_C^0 = (T_m - T)\Delta S_m. \qquad (8.55)$$

Since the solid crystal is more ordered, its entropy $\overline{S}_C^0$ is smaller than the liquid entropy $\overline{S}_L^0$, so that $\Delta S_m > 0$. As a result, when the temperature T decreases continuously below the melting point T_m, the entropy contribution $-(T - T_m)\overline{S}_J^0$ to the increase of the chemical potential μ_J is more significant for the liquid chemical potential μ_L than for the solid chemical potential μ_C, that is, $-(T - T_m)\overline{S}_L^0 > -(T - T_m)\overline{S}_C^0 > 0$. In order to still meet the equilibrium condition of Equation (8.55) and counterbalance the previous effect, the pressure contribution to the increase of the chemical potential μ_J, that is, $(p_J - p_{atm})\overline{V}_J^0$, must act in the opposite direction, that is, $(p_C - p_{atm})\overline{V}_C^0 > (p_L - p_{atm})\overline{V}_L^0$.

As was first noticed by J. Thomson, the equilibrium Equation (8.55) shows that a higher pressure lowers the melting point. For this reason Equation (8.55) is known as the Thomson equation. Taking $p_C = p_L$ in Equation (8.55), we derive

$$(p_L - p_{atm})\left(1 - \overline{V}_L^0/\overline{V}_C^0\right) = (T_m - T_{0m})\Delta S_m, \tag{8.56}$$

where $T_m - T_{0m}$ is the shift in the melting point due to pressurization.[5] The lowering of the melting point due to pressurization is often wrongly used to explain why ice skating is possible. As apparently supported by Equation (8.56), the pressure that the skater exerts on the ice would lower the melting point and, thus, allow him to slide on melted ice. If we adopt 70 kg for the mass of the skater, and 2 mm $\times$20 cm for the surface of the skate in contact with ice, the intensity of the pressure the skater transmits to ice is 1.72 MPa. Since $\Delta S_m = 1.2$ MPa K^{-1} for water, substituting 1.72 MPa to $p_C - p_{atm}$ and p_{atm} to p_L in Equation (8.55), we find that this pressure lowers the melting point by only $T_m - T_{0m} = 1.43$ K. This is too small a decrease of the melting point to explain why ice skating is still possible in polar regions where the temperature T is much lower than 271.57 K. In fact, as we will see in Section 8.5.3, ice skating is made possible by the formation of a premelted film originated by the presence of intermolecular forces. Finally, combining the Thomson Equation (8.55) and Equation (8.56), we can rewrite the liquid–solid equilibrium law in the useful condensed form

$$p_C - p_L = (T_{0m} - T)\Delta S_m. \tag{8.57}$$

The approximate thermodynamic equilibrium Equation (8.54) does apply providing the elastic energy related to both the solid crystal and the liquid is actually negligible with respect to the other terms appearing in the exact equilibrium Equation (8.51), that is with regard to the terms appearing in Equation (8.54). Because of the mechanical equilibrium which the mean stress and the deviatoric stress both have to satisfy, the order of magnitude of $s_{ij}s_{ji}$ is the same as that of $(\sigma + p_{atm})^2$. According to Equation (8.54) this order of magnitude, as well as that of $(p_L - p_{atm})^2$, cannot exceed that of $[(T_m - T)\Delta S_m]^2$. As a result, the approximate Equation (8.54) will hold under the conditions

$$(T_m - T)\Delta S_m/2K_C, \ (T_m - T)\Delta S_m/4G_C, \ (T_m - T)\Delta S_m/2K_L \ll 1. \tag{8.58}$$

Similarly, the entropic terms related to the thermal dilation and the heat capacity will be negligible provided that

$$3|\alpha_C|(T_m - T), \ 3|\alpha_L|(T_m - T) \ll 1, \ |\bar{c}_L - \bar{c}_C|(T_m - T)/2T_m\Delta S_m\overline{V}_L^0 \ll 1. \tag{8.59}$$

---

[5] The presence of a solute will also contribute to a shift in the melting point. Using an analysis similar to that performed in Section 8.2.2, the presence of a solute with molar fraction x can be taken into account by adding the term $-\left(RT_{0m}/\overline{V}_C^0\right)\log(1 - x)$ to the left-hand side of Equation (8.55).

For usual substances these conditions hold for the whole range of realistic values of the cooling $T_m - T$, so that the approximate thermodynamic phase equilibrium Equation (8.54) applies with no actual restrictions.[6]

Since the phase with the largest entropy is the liquid phase, the supersaturation as defined by Equation (8.15) and associated with the spherical stress state $\sigma = -p_C$ can be written

$$\Delta\mu = \mu_C(p_L, T) - \mu_L(p_L, T). \quad (8.60)$$

Since we have here L = 1 and C = 2, from Equation (8.17) we derive

$$\Delta\mu = -(p_C - p_L)\overline{V}_C. \quad (8.61)$$

Combining Equations (8.57) and (8.61), we derive the expression for the supersaturation related to the liquid–solid phase transition in the specific form

$$\Delta\mu = -(T_{0m} - T)\overline{V}_C \Delta S_m. \quad (8.62)$$

8.3.2 Salt crystallization – the Correns equation

The crystallization of salts in solution relates to a solute–crystal phase transition, where the crystal is the phase having the lowest entropy. The solid crystal is then adopted as the daughter phase, with $2 = C$, while the solute is the mother phase, with $1 = $ sol. The phase equilibrium relation between the solute and the solid crystal can be written

$$\mu_C(p_C, T) = \mu_{sol}(p_L, T, x). \quad (8.63)$$

Equation (8.63) shows that, at a given temperature, the crystal pressure p_C is a function of the pressure p_L of the solution and the molar fraction x relative to the solute. With the aim of determining this function,[7] let us first note that the Gibbs–Duhem equation applied to the solid crystals and a differentiation of Equation (8.63) with the temperature held constant give us the relation

$$d\mu_C(p_C, T) = \overline{V}_C dp_C = d\mu_{sol}(p_L, T, x). \quad (8.64)$$

---

[6] For water we have $T_m = 273\,K$ and $\Delta S_m = 1.2\,MPa\,K^{-1} = 1.2 \times 10^6\,J\,m^{-3}\,K^{-1}$. Besides, $K_C = 7.81 \times 10^3\,MPa$ and $3\alpha_C = 155 \times 10^{-6}\,K^{-1}$ at 263 K for ice crystal, while a standard value for the Poisson coefficient of ice is 0.33, so that $G_C = 3 \times 10^3\,MPa$. Standard values for the molar heat capacity are $\overline{c}_C = 38\,J\,mol^{-1}\,K^{-1}$ and $\overline{c}_L = 75.3\,J\,mol^{-1}\,K^{-1}$. In addition, adopting $K_L = 1.79 \times 10^3\,MPa$ and $3\alpha_L = -286.3 \times 10^{-6}\,K^{-1}$ at 263 K for supercooled water, for water ($\overline{V}_L^0 = 18\,cm^3\,mol^{-1}$) we eventually find that the cooling $T_m - T$ expressed in K has to meet the conditions: $7.68 \times 10^{-5}(T_m - T) \ll 1$; $10^{-4}(T_m - T) \ll 1$; $3.35 \times 10^{-4}(T_m - T) \ll 1$, which are associated with the conditions of Equation (8.58), and the conditions: $1.55 \times 10^{-4}(T_m - T) \ll 1$; $2.86 \times 10^{-4}(T_m - T) \ll 1$; $3.16 \times 10^{-3}(T_m - T) \ll 1$, that are associated with the conditions of Equation (8.59). These explicit conditions are met for the whole range of realistic coolings, so that the approximate equilibrium Equation (8.54) can be adopted with no actual restriction. Nevertheless, it is worthwhile noting that the above specific conditions for water show that the effects on the liquid–solid phase equilibrium induced by the compressibility properties and the thermal properties, respectively, have the same order of magnitude. The effect having the largest order of magnitude is the one related to the heat capacity, which affects the entropy of each phase as the temperature varies. These various effects have the same order of magnitude whatever the phase considered, either solid or liquid, although the analogous restriction we would find for the liquid compressibility property when deriving the Kelvin Equation (8.27) governing the vapor–liquid equilibrium is usually ignored.

[7] See Coussy, O. (2006) Deformation and stress from in-pore drying induced crystallization of salt. *Journal of the Mechanics and Physics of Solids*, **54**, 1517–1547.

Now applying the Gibbs–Duhem Equation (2.17) to the aqueous solution, we obtain

$$\overline{V}_L dp_L = (1-x) d\mu_W + x d\mu_{sol}, \qquad (8.65)$$

where $\overline{V}_L$ is the molar fraction of the aqueous solution. Assuming that the solution is ideal, the chemical potential of liquid water, μ_W, satisfies Equation (2.43) and we have

$$d\mu_W = d\mu_W^* + RT d\ln(1-x) = \overline{V}_W dp_L + RT d\ln(1-x). \qquad (8.66)$$

Eliminating $d\mu_{sol}$ and $d\mu_W$ between Equations (8.64)–(8.66), we derive

$$dp_C = \left[1 + \frac{\overline{V}_L - (1-x)\overline{V}_W - x\overline{V}_C}{x\overline{V}_C}\right] dp_L + \frac{RT}{\overline{V}_C} \frac{dx}{x}. \qquad (8.67)$$

Because of Equation (8.63), the right-hand side of Equation (8.67) is required to be an exact differential of p_C with respect to p_L and x. Since the factor in front of dx does not depend on p_L, the factor in front of dp_L is required to be independent of x. This allows us to write

$$\overline{V}_L - (1-x)\overline{V}_W - x\overline{V}_C = -\delta x \overline{V}_C, \qquad (8.68)$$

where δ is a dimensionless coefficient which possibly depends on p_L. The left-hand side of Equation (8.68) can be recognized as the molar volume difference between the aqueous solution and the pure components forming it. Accordingly, δ is a dilation coefficient accounting for the volume change during the crystallization process, resulting in molecules less ($\delta > 0$) or more ($\delta < 0$) tightly packed within the crystal than within the pure solute prior to its mixing with the solvent. Indeed, since the solution is assumed to be ideal, the chemical potential of the solute, μ_{sol}, satisfies Equation (2.43):

$$\mu_{sol}(p_L, T, x) = \mu_{sol}^*(p_L, T) + RT \ln x, \qquad (8.69)$$

and, in the same way as Equation (8.66), we have

$$d\mu_{sol} = \overline{V}_{sol} dp_L + RT d\ln x, \qquad (8.70)$$

where $\overline{V}_{sol}$ is the molar fraction of the pure solute prior to its mixing with the solvent. Substitution of Equation (8.70) in Equation (8.64) gives Equation (8.67), provided that $\overline{V}_{sol} = (1-\delta)\overline{V}_C$, in agreement with the very definition of δ.

As in the previous section, the variations of $\overline{V}_C$ under the variations of the crystal pressure p_C can be neglected. Similarly, the variations of the dilation coefficient δ with the solution pressure will also be neglected. Besides, the convenient reference state is the state where $p_L = p_C = p_{atm}$, so that the related molar fraction x is equal to the salt solubility $x_{sat}(T)$ depending only on temperature. Taking $p_L = p_C = p_{atm}$ in Equations (8.63) and (8.69), the solubility can be expressed in the form

$$x_{sat}(T) = \exp\left(-\Delta G^*/RT\right), \qquad (8.71)$$

where, similarly to Equation (2.103), $\Delta G^* = \mu_{sol}^*(p_{atm}, T) - \mu_C(p_{atm}, T) > 0$ is the Gibbs free energy of dissolution. Substituting Equation (8.68) in Equation (8.67) and integrating the latter from the reference state to the current state we obtain the modified Correns equation[8]

$$p_C - p_{atm} - (1 - \delta)(p_L - p_{atm}) = \frac{RT}{\overline{V}_C} \ln \frac{x}{x_{sat}(T)}. \quad (8.72)$$

When x becomes larger than the solubility, $x_{sat}(T)$, the solution is supersaturated. The contribution of the solute to the entropy of mixing then decreases, resulting in $RT\,d(\ln x) > 0$ in Equation (8.70) and in an increase of the solute chemical potential μ_{sol}. The coexistence of the salts both in solute form and in solid crystal form then requires the crystals to become more pressurized than the solution in order that the pressure contribution $(p_C - p_{atm})\overline{V}_C$ to the crystal chemical potential is more significant than the one, $(p_L - p_{atm})\overline{V}_{sol}$, to the solute chemical potential. Since, according to Equation (8.71), the solubility $x_{sat}(T)$ increases with temperature, note that x can be made larger than the solubility by decreasing the temperature rather than increasing the solute concentration.

The equilibrium Equation (8.72) shows that a higher pressure, but the same for the crystal and the solution, increases the solubility. Indeed, taking $p_C = p_L$ in Equation (8.72), we derive

$$\delta(p_L - p_{atm}) = \frac{RT}{\overline{V}_C} \ln \frac{x_{0sat}}{x_{sat}(T)}, \quad (8.73)$$

where x_{0sat} is the updated solubility due to the pressurization. Combining the equilibrium Equation (8.72) and Equation (8.73), we derive

$$p_C - p_L = \frac{RT}{\overline{V}_C} \ln \frac{x}{x_{0sat}}. \quad (8.74)$$

The pressure p_C defined by Equation (8.74) is termed the crystallization pressure. It is worthwhile noting that the crystallization pressure relates to the solute–solid thermodynamic equilibrium, irrespective of any mechanical considerations.

The solution plays a similar role to that played by the gaseous mixture G in Equation (8.36). Indeed, the solid phase forming from the crystallization of the solute will be subjected to the overall solution pressure. Since the solute is the mother phase, the supersaturation related to salt crystallization is therefore conveniently defined as

$$\Delta\mu = \mu_C(p_L, T) - \mu_{sol}(p_L, T, x). \quad (8.75)$$

Taking into account Equation (8.64), the differentiation of Equation (8.75) provides the relation $d(\Delta\mu) = -\overline{V}_C d(p_C - p_L)$. Integrating the latter from the reference state where $\Delta\mu = 0$, $p_L = p_C = p_{atm}$ to the current state we obtain

$$\Delta\mu = -\overline{V}_C(p_C - p_L). \quad (8.76)$$

Combining Equations (8.74) and (8.76), we derive the expression for the supersaturation related to the solute–solid transition in the specific form

$$\Delta\mu = -RT \ln \frac{x}{x_{0sat}}. \quad (8.77)$$

---

[8] Correns, C. W. (1949) Growth and dissolution of crystals under linear pressure. *Discussions of the Faraday Society*, **5**, 267–271. The usual Correns equation does not consider dilation effects and amounts to taking $\delta = 0$ in Equation (8.72).

8.4 Gas Bubble Formation

The formation of gas bubbles from a solution supersaturated with respect to the gaseous solute can be addressed in a very similar way to salt crystallization. Let x be the molar fraction of a gaseous solute in a solution at pressure p_L. The equation governing the thermodynamic equilibrium between the gaseous solute and the gas bubbles at pressure p_G is derived by setting $J = G$ and $\overline{N}_J = x$ in the Henry Equation (2.59). We obtain

$$p_G = K_H(T)x. \tag{8.78}$$

As in salt crystallization, the role of the mother phase can be devoted to the solution and, similarly to Equation (8.75), the supersaturation $\Delta\mu$ can be written in the form

$$\Delta\mu = \mu_G(p_L, T) - \mu_{sol}(p_L, T, x). \tag{8.79}$$

The gas–solute equilibrium requires $\mu_{sol}(p_L, T, x)$ to be equal to $\mu_G(p_G, T)$, so that Equation (8.79) can be rewritten in the form

$$\Delta\mu = \mu_G(p_L, T) - \mu_G(p_G, T). \tag{8.80}$$

Adopting Equation (2.28) related to ideal gases for the expression of $\mu_G(p, T)$, from Equations (8.78) and (8.80) we obtain

$$\Delta\mu = RT \ln \frac{p_L}{K_H(T)x}. \tag{8.81}$$

When a sudden drop of the solution pressure p_L occurs, the solution becomes supersaturated ($\Delta\mu < 0$) with respect to the solute. In order to restore the thermodynamic equilibrium, gas bubbles form producing, for instance, the effervescence of a bottle containing a gaseous drink after its opening. However, the formation of gas bubbles requires the creation of new interfaces, between the aqueous solution and the gas trapped in the bubble. The creation of these interfaces has a surface energy cost opposing the bubble nucleation. The conditions governing the formation of nuclei of the daughter phase, whatever this phase is, is the topic of the next section.

8.5 Surface Energy and Phase Transition

In the previous sections we have established the laws governing the thermodynamic equilibrium between two bulk phases of the same substance. Since these laws concern bulk phases, we did not pay attention to surface effects. However, during a phase transition new interfaces are created, either between the mother phase and the nuclei of the daughter phase, or between the daughter phase and the solid substrate upon which the former forms. Because of the surface energy effects which we examined in the previous chapter, the creation of a new interface has an energy cost that is unfavorable to the phase transition. This section analyzes how the surface energy associated with the creation of new interfaces can affect unconfined phase transitions.

8.5.1 Nucleation

Nucleus critical radius and homogeneous nucleation

A phase transition occurs when the current phase is no longer stable owing to the possible existence of another phase of the same substance at a lower free energy. This energy condition requires the mother phase to be supersaturated. The supersaturation $\Delta\mu$ is then negative and is the driving force for the phase transition. The phase transition is triggered by the formation of nuclei of the daughter phase 2 within a mother phase 1, which is initially homogeneous.[9] In order to minimize the energy cost needed for their formation, these nuclei are spheres of radius R. The irreducible energy cost for the formation of a spherical nucleus of the daughter phase can be split into a bulk contribution and a surface contribution.

The assembling of $N = 4\pi R^3/3\overline{V}_2$ moles in the form of a sphere of radius R of the daughter phase 2, from the same number of moles of the mother phase 1 at pressure p_1, requires the bulk free energy $N(\mu_2(p_1, T) - \mu_1(p_1, T))$. Owing to the definition of Equation (8.15) for the supersaturation, this bulk contribution can be equated to $N\Delta\mu$.

The surface contribution to the energy cost related to the creation of a nucleus is due to the formation of the interface of area $4\pi R^2$ between the nucleus and the mother phase. This contribution is $4\pi R^2 \gamma_{12}$. The overall free energy cost $\Delta G(R)$ is therefore

$$\Delta G(R) = \left(4\pi R^3/3\overline{V}_2\right)\Delta\mu + 4\pi R^2 \gamma_{12}. \tag{8.82}$$

When the supersaturation is positive, that is, $\Delta\mu = \mu_2 - \mu_1 > 0$, the mother phase has the lowest energy state and is therefore stable: the nucleation process is inhibited. When the supersaturation becomes negative, that is when $\Delta\mu = \mu_2 - \mu_1 < 0$, the bulk contribution provided by the supersaturation to the energy cost becomes negative and favorable to the nucleation. The energy cost $\Delta G(R)$ can then be rewritten in the form

$$\Delta G(R) = \Delta G_{cr}\left[3(R/R_{cr})^2 - 2(R/R_{cr})^3\right], \tag{8.83}$$

where R_{cr} and $\Delta G_{cr} = \Delta G(R_{cr})$ are the critical quantities

$$R_{cr} = -2\gamma_{12}\overline{V}_2/\Delta\mu; \quad \Delta G_{cr} = -\left(2\pi R_{cr}^3/3\overline{V}_2\right)\Delta\mu = 4\pi \gamma_{12} R_{cr}^2/3. \tag{8.84}$$

As shown in Figure 8.2, ΔG_{cr} is the maximum of the energy cost $\Delta G(R)$, which is reached when the nucleus radius R equals the critical radius R_{cr}. The mechanical equilibrium of the interface of the spherical nucleus with the mother phase requires the Laplace Equation (6.52) to be satisfied, resulting in

$$p_2 - p_1 = 2\gamma_{12}/R. \tag{8.85}$$

Equation (8.17) for the supersaturation ensures that the critical radius $R = R_{cr}$ given by Equation (8.84) satisfies the mechanical equilibrium Equation (8.85). However, the equilibrium of the critical nucleus is unstable. Indeed, since the system tends to recover its lowest energy state, any radius fluctuation ΔR will make the critical nucleus either disappear ($\Delta R = R - R_{cr} < 0$),

---

[9] However, as considered before, the mother phase can be either a gaseous mixture for the liquid–vapor transition, or an aqueous solution for the solute–solid transition. The results derived hereafter can then be extended by replacing the index 1, either by G in the former case or by L in the latter case; and the index 2, by either L in the former case or C in the latter case. Instead of Equation (8.17), the relevant expression to use for the supersaturation $\Delta\mu$ is Equation (8.38) in the former case and Equation (8.76) in the latter case.

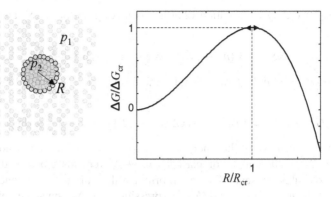

Figure 8.2 Gibbs free energy $\Delta G(r)$ required for the formation of a nucleus of radius r within a supersaturated mother phase. Resulting from the competition between a negative favorable bulk contribution and a positive unfavorable surface contribution, $\Delta G(r)$ exhibits a maximum $\Delta G(r_{cr})$ of the energy cost opposing the formation of the nucleus

or indefinitely spread ($\Delta R = R - R_{cr} > 0$). The equilibrium Equation (8.85) shows that the greater the intensity $|\Delta \mu|$ of the supersaturation, the smaller the critical radius R_{cr}. Therefore, when water droplets of various sizes form in the air from nonhomogeneous supersaturated vapor, the vapor tends to flow from the smaller droplets, where the vapor pressure is greater, toward the surface of the larger ones on whose surface the vapor finally condenses, resulting in the increase of their size. Feeding the larger droplets in this way, the smaller droplets gradually evaporate and disappear, whereas the droplets which are growing larger will rain down when they become too large to float in the air.

According to the previous analysis, whatever the phase transition involved, the maximum ΔG_{cr} accounts for the height of the energy barrier opposing the nucleation. The greater the intensity $|\Delta \mu|$ of the supersaturation, the smaller this energy barrier will become and, therefore, the more likely the nucleation and the phase transition will actually occur. According to Equation (8.84), it is noteworthy that the energy barrier height ΔG_{cr} is equal to a third of the unfavorable surface energy contribution $4\gamma_{12}\pi R_{cr}^2$, or to a half of the intensity of the favorable bulk energy contribution $-\left(4\pi R_{cr}^3/3\overline{V}_2\right)\Delta \mu$. It can be shown that this result is preserved for a polyhedral nucleus related to the crystallization process.

The analysis above has stated the conditions governing the formation of a single critical nucleus within a homogeneous mother phase whose supersaturation is $\Delta \mu$. We now have to determine the number n_{cr} of critical nuclei which can form within such a mother phase. Let $\mathcal{N}$ be the number of the remaining molecules of the mother phase. If we consider both the molecules and the nuclei as indivisible particles, the particle fraction x related to the critical nuclei is given by

$$x = \frac{n_{cr}}{n_{cr} + \mathcal{N}}. \tag{8.86}$$

The formation of n_{cr} critical nuclei is accompanied by an entropy increase ΔS due to the mixing of the critical nuclei of the daughter phase with the remaining $\mathcal{N}$ molecules of the mother phase. According to Equation (2.50), this mixing entropy ΔS can be expressed in the form[10]

$$\Delta S = -k \left(n_{cr} + \mathcal{N}\right) \left[x \ln x + (1-x) \ln(1-x)\right]. \tag{8.87}$$

The overall cost $\Delta \mathcal{G}$ in Gibbs free energy for the formation of n_{cr} critical nuclei is therefore given by

$$\Delta \mathcal{G} = n_{cr} \Delta \mathcal{G}_{cr} - T \Delta S = (n_{cr} + \mathcal{N}) \{x \Delta \mathcal{G}_{cr} + kT [x \ln x + (1-x) \ln(1-x)]\}. \tag{8.88}$$

Equilibrium is reached when the energy cost $\Delta \mathcal{G}$ is at a minimum. Under the approximation $x \ll 1$, the total number of particles $n_{cr} + \mathcal{N}$ remains close to the overall number $\mathcal{N}_T$ of molecules present. The equilibrium value of x is the one achieving the minimum of $\Delta \mathcal{G}/(n_{cr} + \mathcal{N}) \simeq \Delta \mathcal{G}/\mathcal{N}_T$. Expressing that $\Delta \mathcal{G}$ must be minimum, that is $\partial [\Delta \mathcal{G}/(n_{cr} + \mathcal{N})]/\partial x = 0$, we find that the equilibrium particle fraction x of critical nuclei is equal to the Boltzmann factor related to the energy barrier $\Delta \mathcal{G}_{cr}$:

$$x = \exp(-\Delta \mathcal{G}_{cr}/kT). \tag{8.89}$$

At equilibrium the volumetric density $\bar{n}_{cr}$ of critical nuclei of the daughter phase can be written in the form

$$\bar{n}_{cr} \equiv \left(\mathcal{N}_A/\overline{V}_1\right) \exp(-\Delta \mathcal{G}_{cr}/kT). \tag{8.90}$$

Because of the low value of kT – let us recall that $k = 1.381 \times 10^{-23}$ J K^{-1} – the value of $\bar{n}_{cr}$ is quite low. For instance, for the transition of liquid water into ice crystals, when adopting the values $\gamma_{12} = \gamma_{LC} = 0.0409$ J m^{-2} and $\Delta S_m = 1.2$ MPa K^{-1}, and for the value of the supersaturation given by Equation (8.62) the one corresponding to the limit $T = -45\,°C$ of water metastability, we obtain the very low value $\bar{n}_{cr} \simeq 2 \times 10^{-26}$ m^{-3}. The extension of the daughter phase is governed by the diffusion and the packing of new molecules on the surface offered by the critical nuclei. Therefore, the extension rate is proportional to this suface and, thus, to $\bar{n}_{cr}$. As an effect, on account of the low value of $\bar{n}_{cr}$ the spontaneous homogeneous nucleation of phase transitions in a free space is rather unlikely.

Heterogeneous nucleation

According to the previous analysis, the lower the energy barrier ΔG_{cr} appearing in the Botlzmann factor of Equation (8.89), the easier the formation of nuclei of the daughter phase. Equation (8.84) for the energy barrier ΔG_{cr} shows that the latter is lowered when the intensity $|\Delta \mu|$ of the supersaturation increases. For instance, according to Equation (8.44), at a given value of the relative humidity h_R, the formation of saltwater droplets from a saline atmosphere is made easier. In contrast, Equation (8.84) for ΔG_{cr} shows that the larger the energy cost induced by the creation of a new interface between the daughter and mother phases, the larger the energy barrier opposing the formation of nuclei of the daughter phase. The spontaneous nucleation of phase transitions will therefore be facilitated when this cost can be lowered,

---

[10] Since the mixing entropy ΔS is defined relative to the total number of particles $n_{cr} + \mathcal{N}$ and not to the total number of moles $n_{cr} \left(4\pi R^3/3\overline{V}_2\right) + \mathcal{N}/\mathcal{N}_A$, the factor affecting ΔS in Equation (8.87) is the Boltzmann constant $k = R/\mathcal{N}_A$ instead of being the constant R for ideal gases.

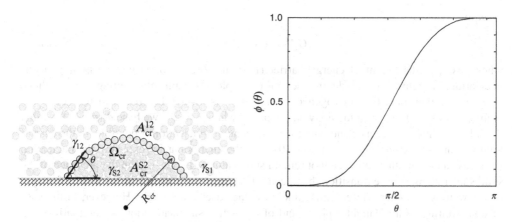

Figure 8.3 Heterogeneous nucleation on a solid substrate and the associated reducing factor $\phi(\theta)$ of the height of the energy barrier opposing the nucleation plotted against the wetting angle θ

as in the presence of impurities or of a solid substrate. The nucleation is then described as heterogeneous. A familiar example of heterogeneous nucleation is given by the trail left behind a jet aircraft. The trail formation is caused by both the solid particles and the water vapor contained in the exhaust from the plane. The extra amount of vapor makes the vapor significantly supersaturated locally, while the solid particles simultaneously trigger its solid condensation.

In order to analyze heterogeneous nucleation let us then consider a nucleus of the daughter phase 2 forming on a plane solid substrate referred to by the subscript S. As shown in Figure 8.3 the actual nucleus forms part of a sphere of radius R. Then let θ be the wetting angle as defined by the Young–Dupré equation

$$\gamma_{S1} = \gamma_{S2} + \gamma_{12} \cos \theta. \tag{8.91}$$

The projected radius of the sphere on the substrate containing one of its parallel lines is $R \sin \theta$. Since the equilibrium Equations (8.17) and (8.85) hold irrespective of the wetting angle θ, the critical radius R_{cr} is still given by Equation (8.84), so that we only have to reconsider the height of the energy barrier ΔG_{cr}^{het} opposing the nucleation. This height ΔG_{cr}^{het} is given by

$$\Delta G_{cr}^{het} = \left(\Omega_{cr}/\overline{V}_2\right) \Delta \mu + \gamma_{12} A_{cr}^{12} + A_{cr}^{S2} (\gamma_{S2} - \gamma_{S1}), \tag{8.92}$$

where, as indicated in Figure 8.3, Ω_{cr} is the volume of the critical nucleus, A_{cr}^{12} is the area of the interface between the critical nucleus and the mother phase 1 and A_{cr}^{S2} is the area of the interface between the critical nucleus and the substrate. Similar calculations to those of Section 6.1.7 show that this volume and these areas can be expressed in the respective forms

$$\Omega_{cr} = 4\pi R_{cr}^3 \phi(\theta)/3; \quad A_{cr}^{12} = 2\pi R_{cr}^2 (1 - \cos \theta); \quad A_{cr}^{S2} = \pi R_{cr}^2 \sin^2 \theta, \tag{8.93}$$

where $\phi(\theta)$ is the function given by Equation (6.64). Using Equation (6.66), we have

$$A_{cr}^{12} - \cos \theta \, A_{cr}^{S2} = 3\Omega_{cr}/R_{cr}. \tag{8.94}$$

Equations (8.84), (8.91) and (8.93) allow us to rewrite Equation (8.92) in the form

$$\Delta G_{cr}^{het} = \gamma_{12} A_{cr}^{12}/3 + A_{cr}^{S2} (\gamma_{S2} - \gamma_{S1})/3 = -\Omega_{cr} \Delta \mu / 2\overline{V}_2, \tag{8.95}$$

or, equivalently,

$$\Delta G_{\text{cr}}^{\text{het}} = \phi(\theta) \Delta G_{\text{cr}}, \tag{8.96}$$

where ΔG_{cr} is the height of energy barrier (Equation (8.84)) opposing the homogeneous nucleation. In Figure 8.3 the reducing factor $\phi(\theta)$ is plotted against the wetting angle θ. Since $\phi(\theta)$ is less than one, the heterogeneous nucleation is always favored with respect to the homogeneous nucleation. In the nonwetting case – when $\theta = \pi$ – we have $\phi(\pi) = 1$, so that the substrate has no influence and only the homogeneous nucleation is possible. In the case of total wetting – when $\theta = 0$ – we have $\phi(0) = 0$: the energy barrier opposing the nucleation vanishes and the mother phase cannot remain supersaturated ($\Delta\mu < 0$) in the presence of the solid substrate; the phase transition then occurs from the formation on the substrate of a layer of the wetting daughter phase, whose thickness increases indefinitely. However, in the case of total wetting of the daughter phase and of a positive supersaturation or, alternatively, in the case of total wetting of the mother phase and of a negative supersaturation, the question arises about the possibility of the coexistence of the daughter phase and the mother phase, with the appropriate phase in the form of a layer lying on the substrate. This is addressed in the following sections.

8.5.2 The precondensed liquid film, the disjoining pressure and the Gibbs adsorption isotherm

Let us consider the effects of a solid wall S on the liquid–vapor phase transition in the case of total wetting of the liquid phase L with regard to its vapor V. According to the results of the previous section, the liquid daughter phase L will form as a liquid film of indefinitely increasing thickness at the surface of the wall. For total wetting, according to Equation (6.39) the spreading coefficient S is positive:

$$S = \gamma_{\text{SV}} - \gamma_{\text{SL}} - \gamma_{\text{VL}} > 0. \tag{8.97}$$

Analogously to Equation (8.82), the free energy cost $\Delta G(e)$, per unit of wall surface, associated with the formation on the wall of a liquid film of thickness e from the supersaturated vapor is

$$\Delta G(e) = \left(e/\overline{V}_{\text{L}}\right) \Delta\mu + W(e) - S, \tag{8.98}$$

where, as in Equation (6.91), $W(e)$ accounts for the effects of the slimness of the film upon the surface energy, and satisfies

$$W(0) = S; \quad W(e \to \infty) = 0. \tag{8.99}$$

Due to attractive intermolecular forces, energy $W(e) - S$ is negative. When the vapor is supersaturated, $\Delta\mu$ is also negative, and both terms of the right-hand side of Equation (8.98) are favorable for the formation of a liquid layer. Since the absolute values of both terms increase as e increases, the formation of a liquid layer with indefinitely increasing thickness is finally favorable regarding the energy cost, as we already concluded at the end of the last section.

When the vapor is undersaturated, $\Delta\mu$ becomes positive. Accordingly, the larger the liquid layer thickness e, the more unfavorable the first term in the right-hand side of Equation (8.98)

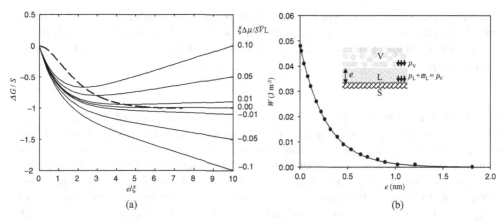

Figure 8.4 As shown in (a), the energy cost ΔG of formation of a precondensed film as a function of the thickness e exhibits a minimum. As the supersaturation varies the minimum is located on the dotted line determined by both Equation (8.100) and the expression relevant for the coupling term $W(e)$. Since ΔG is negative this minimum corresponds to a maximum of energy loss. When adopting $\xi = 2.28$ Å and $S = 50$ mJ m^{-2}, Equation (8.109) for $W(e)$ accurately accounts for the precondensation of argon (mass density $\rho_L = 1.394$ g cm^{-3}) performed at 83 K on a macroporous silica sample (see Pellenq, R.J.M., Coasne, B., Denoyel, R. and Coussy, O. (2009) Simple model for phase transition in confined geometry. 2: capillary condensation/evaporation in cylindrical nanopores. *Langmuir*, **25**, 1393–1402)

to the liquid layer formation. Since it is still the opposite for the attractive energy $W(e) - S$, as shown in Figure 8.4(a) the overall energy cost $\Delta G(e)$ exhibits a minimum, as well as the Gibbs free energy G since ΔG is negative. The formation of a precondensed superheated liquid film with thickness e corresponding to this minimum is now favorable regarding the energy cost. Nullifying the derivative of ΔG with respect to e, we obtain

$$\Delta \mu / \overline{V}_L = -dW/de. \tag{8.100}$$

Substitution of the definition of the disjoining pressure of Equation (6.97) ϖ_L in Equation (8.100) gives the energy balance for the formation of the liquid film in the form

$$\Delta \mu = \varpi_L \overline{V}_L > 0. \tag{8.101}$$

As for the nucleation process, the supersaturation $\Delta \mu$ is the force driving the formation of the precondensed liquid film. In the present case, though, the vapor mother phase is undersaturated, resulting in a positive value of $\Delta \mu$. The chemical potential $\mu_V(p_V, T)$ of the vapor phase is then lower than the chemical potential $\mu_L(p_V, T)$ of the liquid daughter phase subjected to the same thermodynamic pressure p_V which, apparently, does not favor the formation of the condensed daughter phase. Basically, the energy balance of Equation (8.101) shows that, when the vapor is undersaturated, an equilibrium can be established when the energy left over – that is $\Delta \mu > 0$ – exactly supplies the work $\varpi_L de = (\Delta \mu / \overline{V}_L) de$ which the disjoining pressure has to produce against the remaining vapor in the variation de of the thickness of the precondensed liquid film. The related equivalent thermodynamic pressure in the film, p_L,[11] is obtained by substituting Equation (8.101) in Equation (8.17), when taking $1 = V$ and $2 = L$. We return to Equation (6.96) governing the mechanical equilibrium of the V–L interface in the specific form, namely,

$$p_V = p_L + \varpi_L. \qquad (8.102)$$

It is worthwhile noting that the conditions of formation of the precondensed film are the opposites of the ones prevailing in the nucleation process explored in Section 8.5.1. In the nucleation process the supersaturated state of the vapor – associated with $\Delta\mu < 0$ – favors the formation of the critical nucleus, whereas the cost in surface energy opposes it by requiring $p_L > p_V$, as shown by Equation (8.85) when taking $1 = V$ and $2 = L$. The equilibrium of the critical nucleus is then unstable because of the achievement of a maximum state of energy (see Figure 8.3). In contrast, the undersaturated state of the vapor – associated with $\Delta\mu > 0$ – opposes the formation of the precondensed liquid film, whereas the surface energy effects favors it by permitting $p_L < p_V$ according to Equation (8.102), owing to the disjoining pressure that the surface effects induce. The precondensed liquid film is then stable because of the achievement of a minimum state of energy (see Figure 8.4). Remarkably, for low values of the vapor pressure p_V, because of the Kelvin Equation (8.27), the thermodynamic liquid pressure p_L can reach fairly significant negative values related to a tensile state. In order to meet the mechanical equilibrium Equation (8.102) and, therefore, to offset the high negative values of p_L with respect to the small values of the vapor pressure p_V, the disjoining pressure ϖ_L must then have high positive values.[12]

The early formation of a thin precondensed liquid film amounts to vapor adsorption. As a result, the energy change ΔG can be identified as a surface energy change, resulting in a current effective surface energy γ_{SV}^{eff} whose expression is

$$\gamma_{SV}^{eff} = \gamma_{SL} + \gamma_{VL} + (e/\overline{V}_L)\Delta\mu + W(e). \qquad (8.103)$$

From Equation (8.99) it can be checked that $\gamma_{SV}^{eff} = \gamma_{SV}$ as e tends to zero. Let us then consider the equilibrium states successively achieved as the supersaturation $\Delta\mu$ slowly varies, and corresponding to the dashed line in Figure 8.4(a). From Equations (8.100) and (8.103) we then derive

$$d\gamma_{SV}^{eff} = \frac{e}{\overline{V}_L} d\Delta\mu. \qquad (8.104)$$

Since the mother phase is the vapor phase, Equation (8.15) for the supersaturation $\Delta\mu$, the use of Equation (2.28) and of the Gibbs–Duhem Equation (2.18) applied to the vapor, allow us to write

$$d\Delta\mu = d[\mu_L(p_V) - \mu_V(p_V)] = \overline{V}_L dp_V - d\mu_V = (\overline{V}_L/\overline{V}_V - 1) d\mu_V. \qquad (8.105)$$

---

[11] In Sections 2.3.7 and 6.1.10, the equivalent thermodynamic pressure of the phase J of a substance, whatever the confinement conditions to which the phase is subjected, was defined as the equilibrium pressure which the bulk form of the same phase J would have in a large reservoir with the same value of the chemical potential μ_J.

[12] For water, taking $\overline{V}_L = \overline{V}_L^0 = 18\,\text{cm}^3\,\text{mol}^{-1}$, $R = 8.3144\,\text{J}\,\text{K}^{-1}\,\text{mol}^{-1}$, $T = 294\,\text{K}$ and $h_R = 50\,\%$, in Equation (8.33), we obtain $p_L \simeq -93\,\text{MPa}$. With $p_{VS}(T = 294\,\text{K}) \simeq 2.5\,\text{kPa}$ and $p_G = p_V = h_R \times p_{VS} = 1.25\,\text{kPa}$, Equation (8.35) shows that $p_{0V} \simeq p_{VS}$ so that $\varpi_L \simeq -p_L \simeq -93\,\text{MPa}$.

Because of the definition (6.80) illustrated in Figure 6.9 of the number of moles in excess Γ, the number of moles $e/\overline{V}_L$ contained in the liquid layer can be expressed as

$$e/\overline{V}_L = \Gamma + e/\overline{V}_V. \tag{8.106}$$

From Equations (8.104)–(8.106) we return to the Gibbs adsorption isotherm of Equation (6.79) when applied to a single vapor component by taking F = V, that is,

$$d\gamma_{SV}^{eff} = -\Gamma d\mu_V. \tag{8.107}$$

Interestingly, the results above hold irrespective of any specification of the expression of the coupling energy $W(e)$. For instance, as shown by Equation (8.101) the current intensity of the disjoining pressure ϖ_L depends only on the current intensity of the supersaturation $\Delta\mu$. In contrast, the current film thickness e does depend on the expression of the coupling energy $W(e)$. In turn, the determination of $W(e)$ can be carried out from the experimental determination of the thickness e as follows. Substituting Equation (8.39) in Equation (8.100), we obtain

$$W(e) = -\frac{RT}{\overline{V}_L} \int_e^\infty \ln \frac{p_V}{p_{0V}}\, dh. \tag{8.108}$$

Since the vapor pressure p_V can be monitored and the thickness e can be measured, the experimental determination of $W(e)$ is provided by plotting the experimental values of the right-hand side of Equation (8.108) against the liquid layer thickness e. The corresponding results are represented in Figure 8.4(b) for the precondensation of argon on a macroporous silica sample. These experimental data exhibit an accurate exponential dependence of the form

$$W(e) = S \exp\left(-\frac{e}{\xi}\right), \tag{8.109}$$

instead of the form generally used for van der Waals interactions in the disjoining pressure theory. Indeed, according to Equation (6.98), $W(e)$ would have been expected to be proportional to the inverse of the thickness squared, that is $1/e^2$, but such a proportionality fails to account for the experimental data reported in Figure 8.4. Nonetheless, the description of the atom–atom interaction as accounted for by Equation (6.98) assumes that forces of the van der Waals type are the main source of the attractive interaction. This does not apply to the argon/silica system, where the partial charges on the molecules made up of silica create a nonzero permanent electric field close to the surface, and give rise to an attractive so-called polarization or induction interaction. In addition, Equation (6.98) for $W(e)$ does not account for interactions of higher order than the two-body dispersive interactions. In short, intermolecular forces of the van der Waals dispersive type are not the unique forces involved in the formation of a precondensed liquid film and an exponential form for the coupling energy $W(e)$ is more suitable. In Equation (8.109) the characteristic length ξ scales the range of these intermolecular forces and, as shown in Figure 8.4, is found to be 2.78 Å for the argon/silica system. In turn, substituting Equation (8.109) in Equation (8.100), we derive the thickness e of the precondensed liquid in the form

$$e = \xi \ln\left(\frac{S\overline{V}_L}{\xi \Delta\mu}\right). \tag{8.110}$$

This latter relation shows that the divergence of the film thickness e is logarithmic as the supersaturation $\Delta\mu$ tends to zero. It must be pointed out that the previous analysis of the formation of a precondensed film becomes meaningless for temperatures larger than the critical temperature T_{cr}, since the fluid becomes supercritical and the distinction between a liquid phase and a gaseous phase vanishes.

8.5.3 The premelted liquid film and crystallization

Let us now consider the effects of a solid wall S upon the liquid–solid phase transition in the case of total wetting of the liquid phase L with regard to the solid form C of the same substance. The analysis is quite similar to that performed in the previous section except that, unlike the vapor phase, the solid phase C is the daughter phase. The question is then the formation of a supercooled premelted liquid film from a solid crystal phase whose related supersaturation $\Delta\mu$, defined by Equation (8.60), is positive. The free energy cost $\Delta G(e)$ (per unit of wall surface) associated with the formation of a liquid film of thickness e, from the solid phase above the wall is

$$\Delta G(e) = -\left(e/\overline{V}_C\right)\Delta\mu + W(e) - S, \qquad (8.111)$$

where $S = \gamma_{SC} - \gamma_{SL} - \gamma_{CL} > 0$ and where we made the approximation $\overline{V}_L \simeq \overline{V}_C$, while $W(e)$ is again subjected to satisfy the conditions of Equation (8.99). Since the energy $W(e) - S$ due to intermolecular forces is negative, if the supersaturation $\Delta\mu$ related to the solid crystal phase is positive, both terms on the right-hand side of Equation (8.98) are favorable to the formation of a liquid layer. Since the absolute values of both terms increase as e increases, the first term decreases as the thickness e increases, the formation of a liquid layer of indefinitely increasing thickness is favorable regarding the energy cost.

If the supersaturation $\Delta\mu$ related to the solid crystal phase is negative, the larger the liquid layer thickness e, the more unfavorable the first term on the right-hand side of Equation (8.111) to the liquid layer formation. Since it is still the opposite for the attractive energy $W(e) - S$, the overall energy cost $\Delta G(e)$ exhibits a minimum, as shown in Figure 8.4(a), providing the supersaturation is replaced by its opposite $-\Delta\mu$. Since ΔG is negative this minimum accounts for a maximum of energy loss and the formation of a premelted supercooled liquid film with thickness e corresponding to this minimum is now favorable. Nullifying the derivative of ΔG with respect to e, from Equations (6.97) and (8.111) we derive a relation similar to Equation (8.100):

$$-\Delta\mu/\overline{V}_C = -dW/de = \varpi_L. \qquad (8.112)$$

Similarly to Equation (8.101), the energy balance Equation (8.112) shows that the energy left over, that is $-\Delta\mu = -[\mu_C(p_L, T) - \mu_L(p_L, T)] > 0$, provides the work $\varpi_L \overline{V}_C$ that the disjoining pressure has to produce against the remaining crystal, for a mole of the premelted film to form from the crystal. This results in the mechanical equilibrium condition

$$p_C = p_L + \varpi_L, \qquad (8.113)$$

when substituting Equation (8.112) in Equation (8.17) and taking $1 = L$ and $2 = C$.

Why ice skating is possible can now be elucidated, not only from Equation (8.55), but also from Equation (8.112). Skaters can do their sport not because of the very slight lowering of the melting point due to the overpressure they exert on the ice, but because of the premelted thin liquid film forming at the interface between the ice and the atmosphere. Without the need of any overpressure transmitted to the ice by a skater, this premelted liquid film exists because of the lowering of the melting point caused by the disjoining pressure ϖ_L. Indeed, the role played by the solid substrate S is here played by the gaseous atmosphere – subscript G – so that $S = \gamma_{GC} - \gamma_{GL} - \gamma_{CL} > 0$. The mechanical equilibrium of the interface between the liquid film and the atmosphere requires the mechanical pressure of the liquid to be equal to the atmospheric pressure p_{atm}. Since the mechanical equilibrium of the crystal–liquid interface requires the crystal pressure p_C to be equal to the liquid mechanical pressure, we also have $p_C = p_{atm}$. Substitution of Equation (8.113) and $p_C = p_{atm}$ in Equation (8.55) gives

$$\varpi_L \overline{V}_L^0 / \overline{V}_C^0 = (T_m - T) \Delta S_m, \tag{8.114}$$

proving that the disjoining pressure ϖ_L does produce the lowering of the melting point which leads to the formation of the premelted liquid film. In turn, the significant positive values of ϖ_L offset the significant negative values of p_L in order to satisfy the mechanical equilibrium condition $p_L + \varpi_L = p_{atm}$. (For $T_m - T = 10\,\mathrm{K}$ we obtain $\varpi_L = 13.1\,\mathrm{MPa}$ and $p_L \simeq -13\,\mathrm{MPa}$.) The thickness e of the premelted film is given by the combination of Equations (8.109), (8.112) and (8.114):

$$e = \xi \ln \frac{\overline{V}_L^0 S}{\overline{V}_C^0 \xi \Delta S_m (T_m - T)}. \tag{8.115}$$

When adopting $\xi \sim 2.3\,\text{Å}$, as suggested by the experimental data reported in Figure 8.4(b), and the values $\Delta S_m = 1.2\,\mathrm{MPa\,K^{-1}}$, $\overline{V}_L^0 / \overline{V}_C^0 = 1.09$, $\gamma_{GC} \sim \gamma_C = 0.8\,\mathrm{J\,m^{-2}}$, $\gamma_{GL} \sim 0.07\,\mathrm{J\,m^{-2}}$, $\gamma_{CL} \sim 0.4\,\mathrm{J\,m^{-2}}$ and thus $S \sim 0.33\,\mathrm{J\,m^{-2}}$, for $T_m - T > 10\,\mathrm{K}$ we obtain $e > 1.12\,\mathrm{nm}$.[13]

The analysis of the formation of precondensed or premelted liquid films extends to quite similar phase transitions. In the case of the liquid–vapor phase transition where, instead of pure vapor, the mother phase is a gaseous mixture, the approach carried out in the previous section applies provided that the subscript V is replaced by the subscript G in Equations (8.97) and (8.98). In the case of the solute–crystal transition where, instead of a pure liquid, the mother phase is an aqueous solution, the approach developed in this section applies, providing the subscript L refers to the aqueous solution. Indeed, using Equation (8.38) in Equation (8.100), Equation (8.76) in Equation (8.112), we recover Equations (6.96) and (6.97), so that the pressure $p_L + \varpi_L$, where ϖ_L is the disjoining pressure, always applies on the precondensed or premelted liquid film.

The concept of supersaturation $\Delta\mu$ scales the intensity of the driving force ruling a phase transition. The Kelvin Equation (8.27), the Thomson Equation (8.55) or the Correns Equation (8.72), govern the related thermodynamic phase equilibrium. They apply irrespective of the confinement conditions. Phase transitions may occur in unconfined conditions, as in the last section of this chapter when analyzing the nucleation process. Phase transitions may also happen in confined conditions, as is actually the case when they occur in a porous solid. Combining the general laws governing phase transitions and the constitutive equations of

---

[13] For a justification of $\gamma_{GC} \sim \gamma_C$ see the first footnote of Section 6.1.6, and for an assessment of the value of γ_C see the footnote in Section 6.1.4.

unsaturated poroelasticity which we have explored in Chapter 7, we can now address how phase transitions take place in confined conditions and, in particular, understand quantitatively what are the complex mechanical effects they can induce in a deformable porous solid. These are the topics of the next chapter.

Further Reading

Atkins, P. W. (1990) *Physical Chemistry*, 4th edn, Oxford University Press.
Hobbs, P. V. (1974) *Ice Physics*, Clarendon, Oxford.
Lide, D. R. (2001) *Handbook of Chemistry and Physics 2001–2002*, CRC Press.
Lindley, D. (2004) *Degrees Kelvin*, Joseph Henry Press.
Markov, I. V. (2003) *Crystal Growth for Beginners*, 2nd edn, World Scientific.

9

Phase Transition in Porous Solids

An understanding of the mechanical behavior of porous materials subjected to confined phase transitions relates to various issues: cement-based materials in civil engineering, woods in the building industry, plants in botanics, soils in soil science, gels in physical chemistry, vegetables in food engineering, tissues in biomechanics and so on. In civil engineering various concerns are attached to confined phase transitions. Drying shrinkage can induce cracks and enhance the penetration of aggressive agents in concrete structures. Crystallization of sea-salts induced by successive imbibition–drying cycles is recognized as being an important weathering phenomenon in dry environments close to the sea. It often leads to the serious deterioration of porous sedimentary rocks used for building in coastal areas. Furthermore, durability of water-infiltrated materials subjected to frost action is a major concern in cold climates. Ice formation in concrete is the cause of damage worth billions of euros in concrete structures. A better understanding of the mechanics of both the liquid–gas and liquid–solid phase transitions in confined conditions within deformable porous materials is important to improve the resistance of building materials in challenging environmental conditions and, thus, to reduce the maintenance and repair costs.

Before addressing the mechanical effects induced by a phase transition occurring in a porous solid, this chapter first looks into how confined conditions affect a phase transition. We start by showing that the smaller the pore size, the larger the supersaturation has to be for the phase transition to occur, because of the energy cost associated with surface energy effects. For the liquid–solid transition this standard analysis assumes that a spherical stress state always prevails in solid crystals. As we will see the spherical stress state is the most favorable regarding the solid stability. For reasons similar to those examined in Sections 8.5.2 and 8.5.3 governing the formation of precondensed and premelted liquid films on solid substrates, surface energy effects induce the existence of thin liquid films at the surface of the internal walls of porous solids. The premelted liquid film will be shown to play a major role in the mechanics of freezing materials, albeit not significantly affecting the intensity of the pore pressure that is transmitted to the surrounding solid matrix.

The second part of the chapter analyzes and quantifies the effects associated with the drying and freezing of porous materials. This part will finally show the origin of some of the strange phenomena affecting the world of porous solids which we covered in Chapter 1.

Mechanics and Physics of Porous Solids Olivier Coussy
© 2010 John Wiley & Sons, Ltd

9.1 In-pore Phase Transition

In the previous chapter we analyzed how a phase transition can occur in a volume of infinite extent. This section explores how in-pore confinement affects the phase transition and how, in turn, phase transition can be used to assess the pore size distribution of a porous material.

9.1.1 Liquid saturation, pore-entry radius distribution and phase transition

In Section 8.5.1 we concluded that homogeneous nucleation is unlikely.[1] As an effect, because favorable conditions for heterogeneous nucleation are met on the unconfined surface, the phase transition front starts propagating within a porous material from the sample surface. Once the phase transition is triggered, the front progressively enters the accessible pore volume, according to the current value of the supersaturation $\Delta\mu$. The question eventually arises as to how to determine the relation which governs the phase saturation as a function of the supersaturation.

To answer this question let us consider a porous material initially liquid-saturated. As the supersaturation varies, the liquid L can either transform into solid crystals C or into its vapor V. In the latter case air will mix with the vapor to form a gaseous mixture G. In both cases the liquid will be assumed to be the wetting phase — subscript L = W, and the other phase the nonwetting phase (subscript C or L = nW). The progress of the front between the nonwetting and wetting phases has an interface energy cost. This interface energy cost is supplied by the work done by the pressure difference between the two phases. In Section 6.2.1 this was accounted for by the energy balance Equation (6.102), resulting in Equations (6.104) and (6.105) which show that the pressure difference between the two phases is a function of the liquid saturation S_L of the wetting phase. The work that the pressure difference between the two phases can actually produce is supplied by the driving force of the phase transition, namely, the supersaturation. In Section 8.1.3 this was accounted for by the energy balance of Equation (8.17), showing that the pressure difference between the mother and daughter phases is proportional to the supersaturation $\Delta\mu$. Whatever the phase transition, the combination of energy balance Equations (6.104) and (8.17) allows us to conclude that the liquid saturation S_L is a function of the supersaturation $\Delta\mu$:

$$S_L = \mathcal{S}(\Delta\mu). \tag{9.1}$$

It is worthwhile noting that the existence of a relation such as Equation (9.1) has been derived without referring to the pore scale. The supersaturation $\Delta\mu$ depends only on thermodynamic state variables, as shown by the various expressions for $\Delta\mu$ – Equations (8.39), (8.44), (8.62) and (8.77) – related to the phase transitions examined in Chapter 8. As a result, Equation (9.1) appears as a state equation of the porous solid subjected to any phase transition. The function $\mathcal{S}(\Delta\mu)$ relies on the pore scale through the pore access radius which is suited to in-pore phase transition according to the current value of the supersaturation $\Delta\mu$. In the case of the liquid–vapor phase transition, from Equations (6.112), (8.38) and (8.39), the smallest pore access radius $r_{\Delta\mu}$ of the pore volume currently invaded by the gaseous mixture is given by

---

[1] Regarding this point for in-pore crystallization, the reader can usefully refer to Scherer, G. (1993) Freezing gels. *Journal of Non-Crystalline Solids*, **155**, 1–125.

$$r_{\Delta\mu} = \frac{2\gamma_{GL}\cos\theta}{\Delta\mu/\overline{V}_L} = -\frac{2\gamma_{GL}\cos\theta}{(RT/\overline{V}_L)\ln p_V/p_{0V}}. \quad (9.2)$$

Then let $S(r)$ be the pore volume fraction having a pore-entry radius smaller than r, which can be assessed from porosimetry as investigated in Section 6.2.3. As the supersaturation $\Delta\mu > 0$ varies, the fraction $S(r_{\Delta\mu})$ of the pore volume remains filled by the initial liquid in a superheated state. According to Equation (9.2) we can then conclude that the current liquid saturation S_L is a function of the current supersaturation according to

$$S_L = S(r_{\Delta\mu}). \quad (9.3)$$

When the gaseous mixture remains at atmospheric pressure, the equilibrium Equation (9.2) reduces to the Kelvin–Laplace equation:

$$r_{\Delta\mu} = -\frac{2\gamma_{GL}\cos\theta}{(RT/\overline{V}_L)\ln p_V/p_{VS}(T)}. \quad (9.4)$$

Similarly, in the case of the liquid–solid phase transition, from Equations (6.112), (8.61) and (8.62), the smallest pore access radius $r_{\Delta\mu}$ of the pore volume currently invaded by the solid crystals is given by

$$r_{\Delta\mu} = \frac{2\gamma_{CL}\cos\theta}{-\Delta\mu/\overline{V}_C} = \frac{2\gamma_{CL}\cos\theta}{\Delta S_m(T_{0m} - T)}. \quad (9.5)$$

As the supersaturation $\Delta\mu$ varies, the pore volume fraction $S(r_{\Delta\mu})$ remains filled by the initial liquid in a supercooled state associated with $\Delta\mu < 0$. According to Equation (9.5), we can then conclude that the current liquid saturation S_L is again governed by Equation (9.3), but where $r_{\Delta\mu}$ is now given by Equation (9.5) instead of Equation (9.2). When the liquid phase remains at atmospheric pressure, according to Equation (8.56) T_{0m} becomes equal to the melting point T_m and the equilibrium Equation (9.5) reduces to the Gibbs–Thomson equation:

$$r_{\Delta\mu} = \frac{2\gamma_{CL}\cos\theta}{\Delta S_m(T_m - T)}. \quad (9.6)$$

Hence, the general Equation (9.3) combined with either Equation (9.2) or Equation (9.5) provides us with the state function S, whose existence in Equation (9.1) was inferred by macroscopic considerations only.

Letting the liquid be water, in Figure 9.1 $r_{\Delta\mu}$ is plotted against either the relative humidity h_R or the cooling $(T_m - T)$ according to Equation (9.4) or (9.6). These curves provide an alternative experimental means to mercury porosimetry for assessing the pore-entry radius distribution $S(r)$. Indeed, as soon as the liquid water saturation can be measured as a function of either the relative humidity h_R or the cooling $(T_m - T)$ in either a drying or cooling experiment,[2] the pore volume fraction $S(r_{\Delta\mu})$ is known from Equation (9.3), making use of

---

[2] In a drying experiment the current liquid saturation S_L is usually determined by weighing the sample. In a cooling experiment, the current solid saturation S_C can be determined by measuring the current dielectric capacity of the sample, owing to the difference of dielectric constants of liquid water and ice.

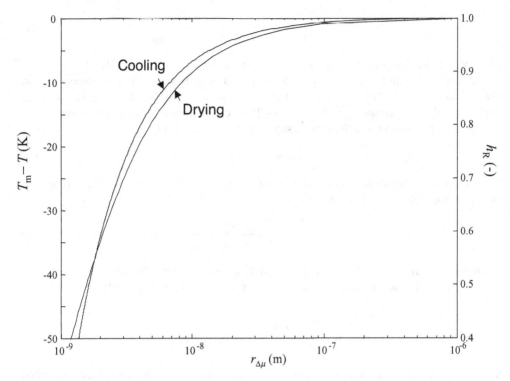

Figure 9.1 The largest pore-entry radius $r_{\Delta\mu}$ associated with the pore volume fraction remaining liquid saturated, when the liquid pressure is maintained at atmospheric pressure, and the temperature is lowered below the melting point ('cooling'); or, alternatively, when the gas pressure is maintained at atmospheric pressure, and the relative humidity is lowered below 100 % ('drying'). The properties are those relative to water: $\overline{V}_L = 18 \text{ cm}^3 \text{ mol}^{-1}$, $\gamma_{CL} = 409 \text{ mJ m}^{-2}$, $\Delta S_m = 1.2 \text{ MPa K}^{-1}$ and $\theta = 0$. For the drying curve the gas is air with $\gamma_{GL} = 73 \text{ mJ m}^{-2}$ and $\theta = 0$, at atmospheric pressure and temperature $T = 293$ K

either Equation (9.4) or (9.6) together with the curves shown in Figure 9.1. Such experimental procedures allow us to explore quite small pores. For instance, Figure 9.1 shows that a value of the relative humidity h_R equal to 90 % allow us to investigate pore-entry radii r having 10 nm as an order of magnitude, whereas the value of the capillary pressure $2\gamma_{GL}/r$ required to investigate the same magnitude of pore-entry radius would be equal to 14.4 MPa.

The evapotranspiration of trees, that is the sum of evaporation and tree transpiration, provides an illustrative example of Equation (9.4). When adopting a zero wetting angle θ in Equation (6.49), the height h of the liquid column which can rise in a capillary tube with radius $r \ll h$ is given by

$$\rho_L g h = 2\gamma_{GL}/r. \qquad (9.7)$$

Adopting $\rho_L = 10^3 \text{ kgm}^{-3}$, $g = 9.81 \text{ ms}^{-2}$, and the value given in the caption of Figure 9.1 for γ_{GL}, from Equation (9.7) we obtain $r = 2.5 \times 10^{-6}$m for $h = 5.9$ m, and $r = 7.5 \times 10^{-9}$m for $h = 1960$ m. The former radius, $r = 2.5 \times 10^{-6}$m, is the radius of the vessels of a tree. The corresponding value of the rise height, $h \sim 6$ m, cannot explain the rise of sap in the

Californian sequoias, whose height may exceed 100m. Nonetheless, it is actually capillarity that makes the sap rise in trees, because the latter radius value, $r = 7.5 \times 10^{-9}$m, is that of the radius of the ultimate duct existing between the cells ending the vessels of the leaves. The associated value $h \sim 2000$ m largely exceeds the height of the trees, and no mechanical equilibrium can be established, resulting in a continuous flow of sap, whose constitutive water ultimately evaporates through the leaves. This roughly depicts the evapotranspiration of trees. In turn, the radius of the ending ducts acts as a pore access radius to the vessel network of the trees. Adopting $R = 8.314 \, \text{JK}^{-1}\text{mol}^{-1}$ and the values given in the caption of Figure 9.1 for the other properties, Equation (9.4) gives $h_R = 86.77\%$ for $r = 7.5 \times 10^{-9}$m. As an effect, when the relative humidity h_R significantly lowers, because of a dry environment, the smallness of the ultimate ducts prevents the tree from excessively drying up. However, the risk of embolism of the tree still exists because of the significant negative values that the water pressure can take within the vessels. With $p_{nW} = p_{atm} = 0.1$ MPa and $r = 7.5 \times 10^{-9}$ m, the Laplace Equation (6.112) gives $p_W = -19.1$ MPa. The liquid water is then in a metastable state with an associated risk of cavitation enhanced by the presence of gas in solute form.

9.1.2 Spherical stress state and stability of in-pore crystallization

The earlier approach to in-pore crystallization, which led to Equation (9.5), implicitly assumed that the crystals are subjected to a spherical stress state and, therefore, that the phase equilibrium is governed by Equation (8.55). This is questionable since the more general phase equilibrium Equation (8.54) allows the equilibrium stress state to be nonspherical and the crystal–liquid interface to be flat. The normal pressure on the crystal side of a flat interface is equal to the liquid pressure on the liquid side of the interface, so that the Laplace Equation (6.112) no longer holds. Let us then perform a stability analysis, in order to determine which one of the two situations depicted in Figure 9.2 is the most favorable regarding energy.

As illustrated in Figure 9.2, at the appropriate temperature a semi-infinite block of the elastic solid phase forms within a cylindrical pore of radius r and of infinite extent in the z direction. As long as equilibrium is achieved, the solid crystal block is subjected to a radial compression applied by the surrounding rigid matrix, while the end face is exposed to liquid water at atmospheric pressure p_{atm}. In the situation depicted in Figure 9.2(a), where the solid–

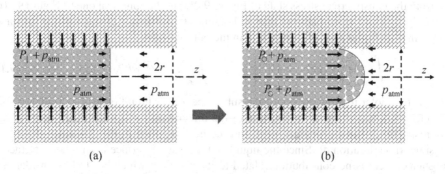

Figure 9.2 Solid crystal block in a cylindrical pore: (a) the end face is flat and the stress state is nonspherical; (b) the end face is spherical and the stress state is spherical

liquid interface is flat, let $P_| + p_\text{atm}$ be the radial compression applied by the surrounding rigid matrix. This results in expressing the stress components σ_{ij} in the crystal block in the form

$$\sigma_{rr} = \sigma_{\theta\theta} = -P_| - p_\text{atm}; \quad \sigma_{zz} = -p_\text{L}; \quad p_\text{L} = p_\text{atm}. \tag{9.8}$$

The expressions in Equation (9.8) for the stress components σ_{ij} and the definitions of Equations (3.30) and (3.31) for the mean stress σ and the deviatoric stress components s_{ij}, respectively, give us

$$-\sigma = 2P_|/3 + p_\text{atm}; \quad s_{rr} = s_{\theta\theta} = -P_|/3; \quad s_{zz} = 2P_|/3. \tag{9.9}$$

Using Equation (9.9), the phase equilibrium Equation (8.54), where we take $p_\text{L} = p_\text{atm}$, gives

$$P_| = \frac{3}{2} \Delta S_\text{m} (T_\text{m} - T). \tag{9.10}$$

Although the stress state of Equation (9.9) is nonspherical, both the mechanical and the phase equilibrium conditions are then satisfied. As a consequence, on the sole basis of thermodynamical equilibrium arguments, there is no need for the end face of the ice block exposed to liquid water to become curved, as in the situation depicted in Figure 9.2(b). In the latter situation, on account of the curvature of the end face and the Laplace Equation (6.112), the stress state in the solid block can now be spherical with

$$\sigma_{rr} = \sigma_{\theta\theta} = \sigma_{zz} = \sigma = -P_\bigcirc - p_\text{atm}; \quad s_{ij} = 0; \quad P_\bigcirc = \frac{2\gamma_\text{CL} \cos\theta}{r}, \tag{9.11}$$

where $P_\bigcirc + p_\text{atm}$ is the radial compression exerted by the surrounding rigid matrix upon the crystal block, and where r is both the half-thickness of the block and the curvature radius of the solid–liquid interface. The previous stress state ensures the mechanical equilibrium of the solid block. Substitution of Equation (9.11) in Equation (8.54), where we take $p_\text{L} = p_\text{atm}$, provides the phase equilibrium condition

$$P_\bigcirc = \Delta S_\text{m} (T_\text{m} - T). \tag{9.12}$$

Therefore, for the same cooling $T_\text{m} - T$ both situations depicted in Figure 9.2 are apparently possible, but not with the same pressure radially applied to the crystal block by the surrounding rigid matrix.

It is therefore necessary to determine whether or not the solid block can spontaneously evolve from the nonspherical stress state of Figure 9.2(a) to the spherical one of Figure 9.2(b). Applying the first and second laws of thermodynamics to isothermal changes of the substance contained in both liquid and solid form within the pore we have

$$\int_\text{pore} de_\text{J}/\overline{V}_\text{J}^0 d\Omega_0 = \delta W + \delta Q; \quad \int_\text{pore} T ds_\text{J}/\overline{V}_\text{J}^0 d\Omega_0 \geq \delta Q, \tag{9.13}$$

where e_J is the molar internal energy of the substance, either liquid (subscript J = L) or solid (subscript J = C), according to the location of the integration point within the pore; δW and δQ are respectively the mechanical work and the heat supplied to the substance during the infinitesimal time duration dt. Since the liquid–solid crystal interface is an inner interface of the substance, there is no contribution related to its evolution which has to be considered in assessing δW and δQ. In addition, because the surrounding solid matrix is rigid, no mechanical work is supplied during a spontaneous evolution of the substance in both liquid and solid form, allowing us to write $\delta W = 0$. The temperature being constant and homogeneous, we also write $dT = 0$ everywhere. Eliminating the heat supply δQ from the two relations in Equation (9.13), and substituting in the resulting equation the relations $\delta W = dT = 0$, as well as the relation $e_J = a_J - Ts_J$, which links the molar internal energy e_J and the molar entropy of phase J to its molar Helmholtz free energy a_J, we derive

$$-\int_{\text{pore}} da_J / \overline{V}_J^0 d\Omega_0 \geq 0. \tag{9.14}$$

Noting that the state of the liquid phase always remains the same, and that the pore has a longitudinal extent sufficiently large to be considered as infinite in both directions, the space integration involves only the current solid crystal block, while the energy associated with the change of curvature of the interface can be neglected with regard to the bulk energy change. Integrating Equation (9.14) between the nonspherical stress state as the initial condition, with $a_C = a_C^|$, and the spherical stress state as the final condition, with $a_C = a_C^\circ$, the spontaneous evolution from the former to the latter state is possible, and the opposite evolution impossible, provided that the stability condition

$$\int_{\text{pore}} a_C^| dV_0 \geq \int_{\text{pore}} a_C^\circ dV_0, \tag{9.15}$$

is fulfilled. Equation (9.15) requires the spherical stress state to relate to a lower state in Helmholtz free energy. Equation (8.9), which gives a_C as a function of μ_C, the detailed expression of Equation (8.46) for μ_C and the elastic constitutive Equations (8.47) give the following expression a_C:

$$a_C = -p_{\text{atm}} \overline{V}_C^0 (\sigma + p_{\text{atm}}) / K_C + \overline{V}_C^0 (\sigma + p_{\text{atm}})^2 / 2K_C$$
$$+ 3\alpha_C \overline{V}_C^0 (T - T_m)(\sigma + p_{\text{atm}}) + \overline{V}_C^0 s_{ij} s_{ji} / 4G_C, \tag{9.16}$$

where the terms depending only on $T - T_m$ and p_{atm} have been omitted, since their expressions are the same whether the stress state is spherical or not, so that they do not play any role in the stability analysis. Using the loading conditions, nonspherical (Equation (9.9)) and spherical (Equation (9.11)), respectively, together with the associated phase equilibrium Equation (9.10) or (9.12), Equation (9.16) for the crystal Helmholtz free energy a_C allows us to show that the stability condition of Equation (9.15) is satisfied because

$$a_C^| - a_C^\circ = \frac{3\overline{V}_C^0}{8G_C} (\Delta S_m)^2 (T_m - T)^2 \geq 0. \tag{9.17}$$

It is worthwhile noting that, since $-(\sigma + p_{\text{atm}})$ has the same value, that is $\Delta S_m (T_m - T)$ in both cases, the difference between $a_C^|$ and a_C° is only due to the absence of a shear contribution to the elastic energy related to the spherical stress state. Thus, although playing no significant role regarding the liquid–solid phase equilibrium Equation (8.54), the shear elastic energy plays a major role in the stability analysis: because the shear stress makes the nonspherical stress state less favorable regarding the free energy cost, the spherical stress state corresponding to a curved end face of the solid block will be achieved spontaneously.

9.1.3 Intermolecular forces and in-pore phase transition

In the previous sections the stress state has always been assumed to be homogeneous in the phases. When one of the two phases in contact within the pore space is the solid crystal phase, it could be argued against the stability argument developed in the previous section that the stress state is not necessarily homogeneous, because only the solid crystals at the liquid–solid interface have to be in thermodynamic equilibrium with the liquid exhibiting the same homogeneous liquid pressure everywhere. As a result, the stress state of the solid crystals a long way from the liquid–solid interface would no longer be governed by the phase equilibrium Equation (8.45) and, therefore, could be nonspherical, involving a shear contribution. Basically, as we already analyzed in Section 8.5.3, because of the surface effects induced by intermolecular forces, there is a film everywhere, with the same equivalent thermodynamic pressure p_L,[3] which remains liquid at the interface between the solid crystals and the surrounding solid. The purpose of this section is to extend the analysis of Section 8.5.3 related to this liquid film to confined conditions.[4]

In view of analyzing the effects of intermolecular forces in confined conditions, let us consider a cylindrical pore of radius R_0, embedded in a porous solid, and initially fully saturated by the liquid in a supercooled state. The free energy cost $\Delta G(R)$, per unit of length of the cylindrical pore axis, associated with the formation of a solid crystal cylindrical core of radius R from the supercooled liquid phase is

$$\Delta G(R) = \left(\pi R^2 / \overline{V}_C\right) \Delta\mu + 2\pi R \gamma_{CL} + 2\pi R_0 \Upsilon(R), \qquad (9.18)$$

with the conditions

$$\Upsilon(R = 0) = 0; \qquad \Upsilon(R = R_0 \to \infty) = S, \qquad (9.19)$$

where we recall that $S = \gamma_{SC} - \gamma_{SL} - \gamma_{CL}$. In Equation (9.18), in addition to the first term representing the bulk contribution, the term $2\pi R \gamma_{CL} + 2\pi R_0 \Upsilon(R)$ accounts for the surface energy change due to the creation of the new C–L interface. In Equation (9.19) the first condition ensures the continuity with the reference state $R = 0$. The second condition ensures that, with respect to the fully liquid-saturated reference state, the surface energy difference associated with the intermolecular forces is $2\pi R_0 (\gamma_{SC} - \gamma_{SL})$, when the liquid has completely transformed into the solid crystals phase, and the pore radius R_0 is sufficiently large in order to neglect any corrective energy associated with the smallness of the pore.

An equilibrium state differing from the initial state may be achieved if the energy cost ΔG can be at a minimum for a nonzero value of the radius R. Since ΔG is negative this minimum corresponds to a maximum of free energy loss. Writing that $d\Delta G / dR = 0$, from Equation (9.18) we obtain

$$-\Delta\mu/\overline{V}_C = \frac{\gamma_{CL}}{R} + \frac{R_0}{R} \frac{d\Upsilon}{dR}. \qquad (9.20)$$

---

[3] In Sections 2.3.7 and 6.1.10, the equivalent thermodynamic pressure of the phase J of a substance, whatever the confinement conditions to which the phase is subjected, has been defined as the equilibrium pressure that the bulk form of the same phase J would have in a large reservoir with the same value of the chemical potential μ_J.

[4] The starting point is inspired by Celestini, F. (1997) Capillary condensation within nanopores of various geometries. *Physics Letters A*, **228**, 94–90. We develop the theory for the liquid–solid phase transition, but the results can be extended to the liquid–vapor phase transition by simply replacing index C by index V, and $\Delta\mu$ by $-\Delta\mu$, the supersaturation $\Delta\mu$ now being positive because it relates to a superheated liquid.

Substituting Equation (8.61) in Equation (9.20), we derive

$$p_C - (p_L + \varpi_L) = \frac{\gamma_{CL}}{R}, \qquad (9.21)$$

where the disjoining pressure ϖ_L is now given by

$$\varpi_L = \frac{R_0}{R} \frac{d\Upsilon}{dR}. \qquad (9.22)$$

Indeed, because of the cylindrical geometry the possible C–L interface has a curvature radius of $1/R$, and Equation (9.21) expresses its mechanical equilibrium. In turn, Equation (9.20) can be rewritten in the form

$$-\Delta\mu/\overline{V}_C = \frac{\gamma_{CL}}{R} + \varpi_L. \qquad (9.23)$$

Interestingly, in a similar way to Equation (8.101), Equation (9.23) shows that the current intensity of the disjoining pressure ϖ_L depends only on the current supersaturation $\Delta\mu$ and the radius of the frozen zone, that is $R \sim R_0$, irrespective of the specification of the expression of the coupling energy $\Upsilon(R)$. In contrast, the current radius of the frozen zone R does depend on the expression of the coupling energy $\Upsilon(R)$.

Based on the experimental results shown in Figure 8.4, a good candidate for the function $\Upsilon(R)$ which meets the conditions of Equation (9.19) is

$$\Upsilon(R) = S\left[\exp\left(-\frac{R_0 - R}{\xi}\right) - \exp\left(-\frac{R_0}{\xi}\right)\right]. \qquad (9.24)$$

The experimental results in Figure 8.4 have shown that the order of magnitude of ξ was 2 Å, so that actual values of R_0 fall beyond the range $R_0/\xi \geq 10$. As a consequence the last term in Equation (9.24) is a small constant quantity and can be dropped with no lack of generality, leading us to adopt for $\Upsilon(R)$:

$$\Upsilon(R) = S \exp\left(-\frac{R_0 - R}{\xi}\right). \qquad (9.25)$$

Substitution of Equation (9.25) in Equation (9.20) gives

$$-\frac{\Delta\mu}{\overline{V}_C} = \frac{\gamma_{CL}}{R} + \frac{R_0 S}{R\xi} \exp\left(-\frac{R_0 - R}{\xi}\right). \qquad (9.26)$$

The actual existence of a radius R, which does satisfy Equation (9.26), depends upon the value of the current supersaturation $\Delta\mu$. The normalized energy cost, that is $\Delta G/2\pi R_0 S$, as a function of R/R_0 is shown in Figure 9.3 for various values of $\Delta\mu$. Since the liquid is in a supercooled state, these values are negative. For small values of the intensity $-\Delta\mu$ of the supersaturation the energy cost ΔG is a strictly increasing function of R and Equation (9.26) has no solution: the energy cost associated with the intermolecular forces is large enough with regard to the energy loss associated with the supersaturation to prevent the liquid–solid phase transition occurring; as a result, the initial equilibrium $R = 0$ remains stable.

The energy cost ΔG will exhibit two other extrema, departing from the one related to the initial state, when the intensity $-\Delta\mu$ of the supersaturation goes beyond the limit value $-\Delta\mu_{\lim}$ such as a radius $R_{\lim} \leq R_0$ satisfying $d\Delta G/dR = 0$ does exist for the first time. In addition to

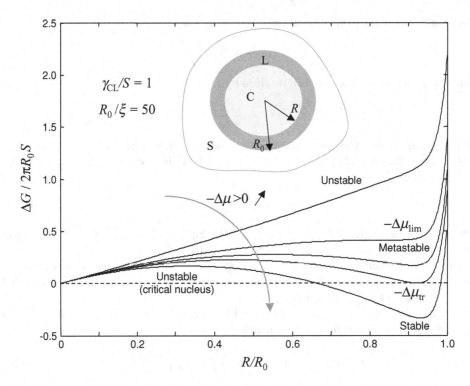

Figure 9.3 Normalized free energy cost $\Delta G/2\pi R_0 S$ plotted against the normalized radius R/R_0: when the intensity $-\Delta\mu$ of the supersaturation $\Delta\mu$ reaches the limit value $-\Delta\mu_{\lim}$, a metastable state exists close to $R/R_0 = 1$; the latter becomes unconditionally stable for supersaturation values beyond the transition value $-\Delta\mu_{tr}$

the condition $d\Delta G/dR = 0$ accounted for by Equation (9.26), the couple $(\Delta\mu_{\lim}, R_{\lim})$ will therefore also have to satisfy the condition $d^2\Delta G/dR^2 = 0$. Making explicit the condition $d^2\Delta G/dR^2 = 0$ with the help of Equation (9.18) where we substitute Equation (9.25), and eliminating the exponential term with the help of the equilibrium Equation (9.26), we derive[5]

$$-\frac{\Delta\mu_{\lim}}{\overline{V}_C} = \frac{\gamma_{CL}}{R_{\lim}}\left(1 + \frac{\xi}{R_{\lim} - \xi}\right). \quad (9.27)$$

As the intensity $-\Delta\mu$ of the supersaturation increases beyond $-\Delta\mu_{\lim}$, the energy cost ΔG now presents both a maximum and a minimum (see Figure 9.3). The maximum, close to $R = 0$, corresponds to the energy barrier of in-pore nucleation. Indeed, neglecting the last term on the right-hand side of Equation (9.18), the value of R corresponding to this maximum is the cylindrical nucleus critical radius obtained by omitting the factor Two in Equation

---

[5] It can be shown that $R_{\lim} = R_0 - \xi \ln\left[SR_0(R_{\lim} - \xi)/\gamma_{CL}\xi^2\right]$. The condition $R_{\lim} \leq R_0$ implies $R_0 \geq \frac{\xi}{2}\left(1 + \sqrt{1 + \frac{4\gamma_{CL}}{S}}\right)$. For lower values of R_0, $R = 0$ would remain the only minimum, and the liquid–solid phase transition could not occur. However, the characteristic length ξ is so small that in practice these values of R_0 are meaningless.

(8.84) for the spherical nucleus critical radius r_{cr}. The existence of a minimum, which does not exist in unconfined conditions (see Figure 8.2), occurs for a value of R close to R_0. For those values of R the thickness $R_0 - R$ of the liquid film becomes sufficiently small and, accordingly, the intensity of the intermolecular forces becomes sufficiently large, for the energy cost ΔG to increase again. The minimum corresponds to a metastable equilibrium as long as the associated value of the energy cost ΔG remains positive, so that the absolute stable equilibrium still remains at $R = 0$. The supersaturation for which the liquid core can transform into a solid is that beyond which the equilibrium state close to R_0 corresponds to an absolute stable equilibrium. The supersaturation $-\Delta \mu_{tr}$ allowing the phase transition of the liquid core to occur is therefore obtained when the associated radius R_{tr} satisfies the equality $\Delta G = 0$. Making explicit the condition $\Delta G = 0$ with the help of Equation (9.18) where we substitute Equation (9.25), and eliminating the exponential term by means of the equilibrium Equation (9.26), we obtain

$$-\frac{\Delta \mu_{tr}}{V_C} = \frac{2\gamma_{CL}}{R_{tr}} \left(1 + \frac{\xi}{R_{tr} - 2\xi}\right), \tag{9.28}$$

where the associated radius R_{tr} has to satisfy[6]

$$R_{tr} = R_0 - \xi \ln \left[\frac{S R_0}{\gamma_{CL} \xi} \left(1 - \frac{2\xi}{R_{tr}}\right)\right]. \tag{9.29}$$

When ξ/R_0 tends to zero, $R_{tr} = R_0$ and Equation (9.28) reduces to Equation (9.5) which we found previously for large pores, and $S > 0$ so that $\theta = 0$. In a confined geometry Equation (9.5) has to be modified according to Equation (9.28) by applying a corrective amplification factor to the usual surface energy γ_{CL}. This can roughly be explained as follows: the surface energy γ_{CL} is due to the interaction energy between molecules and is determined relative to a planar interface; with $S > 0$ the corrective amplification factor accounts for the increase in the interaction energy between adjacent molecules which are more tightly packed owing to the strong curvature imposed by the confined geometry.

As the intensity $-\Delta \mu$ of the supersaturation increases beyond $-\Delta \mu_{tr}$, the liquid film becomes thinner and thinner until it disappears for the supersaturation $-\Delta \mu_{max}$ obtained by taking $R = R_0$ in the equilibrium Equation (9.26). We derive

$$-\frac{\Delta \mu_{max}}{V_C} = \frac{\gamma_{CL}}{R_0} + \frac{S}{\xi}. \tag{9.30}$$

Eventually it can be concluded that, when the intensity $-\Delta \mu$ of the supersaturation lies in the range $-\Delta \mu_{max} > -\Delta \mu > -\Delta \mu_{tr}$, the core of the supercooled liquid transforms into solid crystals, but a thin liquid film of thickness $R_0 - R$ remains unfrozen at the interface between the solid crystals and the solid wall of the pore. The liquid film no longer exists when the intensity $-\Delta \mu$ of the supersaturation goes beyond the value $\Delta \mu_{max}$.

When the intensity $-\Delta \mu$ of the supersaturation lies in the intermediate range $-\Delta \mu_{tr} > -\Delta \mu > -\Delta \mu_{lim}$, the superheated solid crystals may remain solid in a metastable state, provided that the unavoidable thermal fluctuations are not too significant. Owing to these metastable states, as represented in Figure 9.4, a loop of hysteresis is expected to be observed in the plane where the liquid saturation $S_L = 1 - R^2/R_0^2$ is plotted against the intensity

---

[6] The low value of ξ ensures that $R_0/\xi \geq 2 + \gamma_{CL}/S$ for meaningful values of R_0.

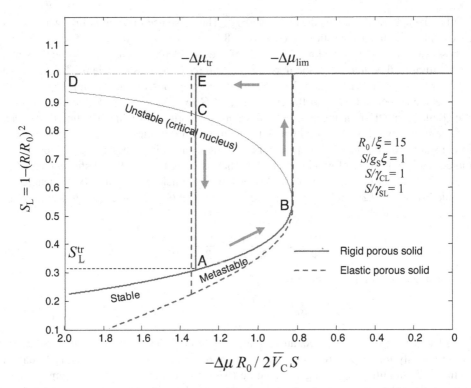

Figure 9.4 The hysteresis loop for phase transition in a confined geometry caused by the presence of metastable states: the effects induced by the elasticity of the surrounding matrix upon in-pore phase transition (dashed lines) are exaggerated regarding the actual values of the dimensionless spreading coefficient $S/g_s\xi \ll 1$; for the rigid case the unstable part of the equilibrium curve is also plotted and according to Equation (9.34) the equilibrium curve has to obey to Maxwell's rule so the area ABC must be equal to the area CDE

$-\Delta\mu$ of the supersaturation. This loop of hysteresis has actually been observed for the liquid–vapor phase transition of argon occurring in a nanoporous silica sample, and is well accounted for by applying the theory developed above to the liquid–vapor phase transition (see Figure 9.5).

In order to derive a property of the equilibrium curve related to this plane, that is irrespective of any particular choice of the function $\Upsilon(R)$ involved in Equation (9.18), we express the energy cost ΔG in the general form

$$\Delta G(S_L) = \pi R_0^2 (1 - S_L) \Delta\mu/\overline{V}_C + \pi R_0^2 w(S_L), \tag{9.31}$$

with $w(1) = 0$. The equilibrium curve has to satisfy $d\Delta G/dS_L = 0$, and therefore is given by the relation

$$\Delta\mu/\overline{V}_C = \frac{dw}{dS_L}. \tag{9.32}$$

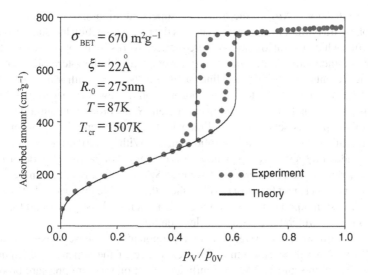

Figure 9.5 The experimental hysteresis loop fitted by the theoretical model for the liquid–vapor transition of argon in a nanoporous silica sample. According to the Kelvin Equation (8.34), the adsorbed amount of argon is plotted against p_V/p_{0V} instead of the supersaturation. The pore radius is derived from the specific surface σ_{BET} and we used $\gamma_{VL} = 0.024 \mathrm{Jm}^{-2}$ and $S = 0.112\ \mathrm{Jm}^{-2}$, in order to assess the reported value of the characteristic length ξ. The hysteresis loop will disappear and the surface energy γ_{VL} vanish as soon as the temperature T is larger than the critical value T_{cr} (for further details see Pellenq, R.J.M., Coasne, B., Denoyel, R. and Coussy, O. (2009) Simple model for phase transition in confined geometry. 2: capillary condensation/evaporation in cylindrical nanopores. *Langmuir*, **25**, 1393–1402)

The transition point S_L^{tr} satisfies $\Delta G \left(S_L^{tr} \right) = 0$ and we obtain

$$- \left(1 - S_L^{tr}\right) \Delta \mu_{tr} / \overline{V}_C = w \left(S_L^{tr} \right). \tag{9.33}$$

These two equations finally combine to give

$$\left(1 - S_L^{tr}\right) \Delta \mu_{tr} = \int_{S_L^{tr}}^{1} \Delta \mu\, \mathrm{d}S_L. \tag{9.34}$$

Conforming with Maxwell's rule illustrated in Figure 8.1, Equation (9.34) states the equality of the areas lying on both sides of the phase transition line $\Delta \mu = \Delta \mu_{tr}$, and delimited by the equilibrium curve (see Figure 9.4).

In the above developments it was assumed that the surrounding solid did not deform. This is only an approximation, since the pressure applying to the solid wall delimiting the pore will cause the surrounding porous solid to deform. According to Equation (9.21) the pressure applied at the interface between the liquid film and the solid is not the crystal pressure p_C, but $p_L + \varpi_L$. The pore deformation caused by $p_L + \varpi_L$ can influence the in-pore phase transition process. This can be analyzed qualitatively as follows. The deformation of the porous solid requires an extra free energy. This extra free energy is made up of both the elastic energy stored in the porous solid and the increase of the interface energy between the porous solid and the liquid due to the increase of the pore radius. The need for this twofold energy will tend to increase the intensity $-\Delta\mu_{tr}$ of the transition supersaturation and, therefore, to delay the liquid–solid phase transition. In contrast, as soon as the phase transition has started, the increase in the pore radius which the disjoining pressure induces works in the opposite direction, since it reduces the confinement, and thus the energy associated with the intermolecular forces. As an effect, as represented by the dashed line in Figure 9.4, the equilibrium curve will have a steeper slope than that related to a rigid porous solid. However, because the disjoining pressure ϖ_L is proportional to the spreading coefficient S, the order of magnitude of the radial deformation of the pore is scaled by the dimensionless coefficient $S/g_S\xi$, where g_S stands for the shear modulus of the surrounding solid matrix.[7] With $S \sim 10\,\text{mJ}\,\text{m}^{-2}$ and, $\xi \sim \text{Å}$, the order of magnitude of $S/g_S\xi \sim 10^8/g_S$ where g_S is expressed in Pascals. The deformation will not significantly reduce the confinement as soon as $S/g_S\xi < 0.1$, resulting in $g_S > 10^3\,\text{MPa}$. Although this condition is met for most elastic solids, the possible influence of the deformation of the surrounding solid upon the in-pore phase transition must be kept in mind for soft elastic materials or materials which have a low tensile strength.

In general, if deformation has no significant effect upon in-pore phase transition, the opposite is not true! So now we must determine if the existence of the supercooled liquid and the associated disjoining pressure do have a significant effect on the pore pressure producing the pore deformation, that is, either p_C or $p_L + \varpi_L$. Substituting Equation (8.61) in Equation (9.5), p_C and p_L are linked to the pore access radius $r_{\Delta\mu}$ according to the relation

$$p_C - p_L = \frac{2\gamma_{CL}\cos\theta}{r_{\Delta\mu}}. \tag{9.35}$$

Equations (9.21) and (9.35) combine to give

$$\varpi_L = \frac{2\gamma_{CL}\cos\theta}{r_{\Delta\mu}} - \frac{\gamma_{CL}}{R}. \tag{9.36}$$

In practice the pore access radius $r_{\Delta\mu}$ is much smaller than R, and Equation (9.36) results in $\varpi_L \simeq 2\gamma_{CL}\cos\theta/r_{\Delta\mu}$ so that Equation (9.35) can be rewritten in the form

$$p_C \simeq p_L + \varpi_L. \tag{9.37}$$

According to Equation (9.37), instead of using the pressure $p_L + \varpi_L$ actually applied to the solid wall of the pore, it is still accurate to adopt the crystal pressure p_C as the pore pressure when determining the deformation induced by the phase transition.

The significance of the premelted liquid film can be summarized as follows. The supercooled liquid film wraps around all the solid crystals. This allows a thermodynamic equilibrium to be established everywhere in the solid phase, at the same homogeneous pressure p_C satisfying the Thomson Equation (8.55), in particular, in the solid crystals that are located a long way from the current interface with the bulk liquid. Except for this major role, the liquid film has no

---

[7] From the standard theory of elasticity it can be shown more precisely that the radial displacement u induced by a pressure p applied to the solid wall of a cylindrical pore embedded in an infinite elastic matrix is given by $u = R_0 p/2g_S$. The elastic energy stored in the solid matrix is equal to $2\pi g_S u^2$. This energy must be added to the energy cost ΔG defined by Equation (9.18), where R_0 also has to be replaced by $R_0 + u$. The minimization of ΔG then has to be worked out with regard to both R and u. This is the minimization which provides the dashed line in Figure 9.4.

9.2 Kinetics and Mechanics of Drying

The drying of materials is a field of investigation in itself. This section is limited to give a first insight into the continuum approach to the drying kinetics and the isothermal drying shrinkage.[8] The isothermal drying of porous materials results from the imbalance between the relative humidity of the surroundings and the higher relative humidity initially prevailing within the porous material. This imbalance results in the evaporation of the inner liquid water and the subsequent evacuation of moisture from the porous material in order to restore the thermodynamic equilibrium.

The moisture exchange with the surroundings can occur through two drying mechanisms. As illustrated in Figure 1.6(a), the first mechanism consists of the evacuation of moisture by means of molecular diffusion of the vapor, through the inner air phase, which is continuously supplied by the inner evaporation of the liquid water phase. Alternatively, as illustrated in Figure 1.6(b), the second mechanism consists of the evaporation of the liquid water phase when arriving at the material surface having been driven there by the capillary pressure gradient induced by the imbalance of relative humidity. The first mechanism can be termed the molecular diffusion drying mechanism and the second mechanism the capillary drying mechanism. The timescale over which these two mechanisms are active, either separately or simultaneously, depends on the transport properties of the porous solid.

During drying, as governed by the Kelvin Equation (8.27), the decrease of the liquid saturation degree simultaneously causes the depressurization of the inner liquid water phase. As a result, when a porous material dries it shrinks, like the sponge of Figure 9.6. While the drying kinetics is governed by the transport phenomena as outlined above, the asymptotic drying shrinkage is only governed by the relative humidity of the surroundings, since, asymptotically, the air pressure recovers the atmospheric pressure value.

In the following sections, through a continuum approach, we determine the transport of moisture which is the most efficient for achieving the drying of a porous material. Using unsaturated poroelasticity, we later on assess the intensity of the drying shrinkage and we examine the stiffening of a porous solid which can be induced by its drying.

9.2.1 Continuum approach to drying kinetics

In order to analyze how the mechanism of drying depends on the transport properties let us then consider an idealized one-dimensional drying experiment. A sample of porous material, lying between $x = d$ and $x = -d$, is in internal thermodynamic equilibrium. The uniform initial liquid saturation degree is S_L^0. At time $t = 0$, the end faces $x = \pm d$ come into contact with the

---

[8] The assumption of isothermal conditions is relevant provided that: (i) the same constant temperature as the initial one is imposed on the external surface of the material; (ii) the characteristic time related to thermal diffusion is much smaller than that associated with fluid transport (for further details see for instance Coussy, O., Eymard, R. and Lassabatère, T. (1998) Constitutive modelling of unsaturated drying deformable materials. *Journal of Engineering Mechanics*, **124**, 658–667).

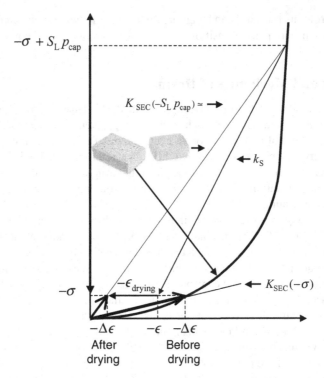

Figure 9.6 Origin of drying stiffening: the change in relative humidity induces a high capillary pressure p_{cap} within the porous solid and causes the partial closure of the cracks and the pores forming the porous network. As a result, the porous solid stiffens, which is contrary to the case represented in Figure 4.2, where the porous solid softens under the tensile stress σ

surrounding atmosphere, whose relative humidity, $h_R^d = p_V^d/p_{VS}(T)$, is lower than the relative humidity, h_R^0, initially prevailing within the porous material. The resulting thermodynamic imbalance enforces the porous material to exchange water vapor with the surroundings. This exchange is carried out by means of the molecular diffusion of vapor through the wet air phase, in order to establish progressively the outer relative humidity h_R^d inside the porous material, from the inner layers close to the end faces. In turn, the liquid water of these layers evaporates simultaneously in order to maintain the liquid–vapor equilibrium. The vapor supplied by the evaporation of the skin layers causes the decrease of the liquid pressure there. As an effect, a gradient in liquid pressure progressively builds up inside the material and causes the liquid of the inner layers to move to the surface ($x = \pm d$) of the material and to evaporate when coming into contact with the surrounding atmosphere. As a result, the saturation degree S_L of the liquid progressively decreases within the material. The whole process stops when the equilibrium saturation associated with the relative humidity h_R^d of the surroundings is attained. At the same time as the previous moisture exchange with the surroundings, a molecular diffusion of dry air occurs in the opposite direction to the molecular diffusion of vapor, that is, from the surrounding inside the porous solid. This gives rise to an inner buildup of gas pressure. Later on, this gas pressure will vanish because of the overall advective transport which, in response to any gas pressure increase, will enforce the vapor–dry air mixture to leave the porous material. To actually know the timescale over which these complex coupled transports occur requires a further analysis which can only be carried out by writing down the equations governing the drying of a porous continuum.

To this end, let us start by expressing the mass conservation laws. In a similar way to Equation (5.16), let $\mathring{m}_{L\to V}$ be the rate of liquid mass transforming into vapor. When neglecting both the velocity and the deformation of the porous solid, the one-dimensional continuity equations associated with the liquid L, the vapor V and the dry air A are given by

$$\frac{\partial (\phi_0 S_L \rho_L)}{\partial t} + \frac{\partial}{\partial x}(\rho_L q_L) = -\mathring{m}_{L\to V}; \tag{9.38a}$$

$$\frac{\partial (\phi_0 (1 - S_L) \rho_V)}{\partial t} + \frac{\partial}{\partial x}(\rho_V v_V) = \mathring{m}_{L\to V}; \tag{9.38b}$$

$$\frac{\partial (\phi_0 (1 - S_L) \rho_A)}{\partial t} + \frac{\partial}{\partial x}(\rho_A v_A) = 0, \tag{9.38c}$$

respectively. We now extend the transport laws we found in saturated conditions to unsaturated conditions. This is achieved by replacing in Equation (5.22) the constant porosity $\phi = \phi_0$ by the partial porosity $\phi_0 (1 - S_L)$ related to the wetted air, and the tortuosity τ by the partial tortuosity $\alpha(S_L)\tau$, where the function $\alpha(S_L)$ has to satisfy $\alpha(0) = 1$ and $\alpha(1) = 0$. The one-dimensional transport laws can then be expressed in the form

$$v_V = q_G - \phi_0 (1 - S_L) \alpha(S_L) \tau D_0 \frac{p_{atm}}{p_V} \frac{\partial}{\partial x}\left(\frac{p_V}{p_G}\right); \tag{9.39a}$$

$$v_A = q_G + \phi_0 (1 - S_L) \alpha(S_L) \tau D_0 \frac{p_{atm}}{p_A} \frac{\partial}{\partial x}\left(\frac{p_V}{p_G}\right). \tag{9.39b}$$

In these expressions q_G is the volume flow vector of the vapor–dry air mixture. Indeed, from Equation (9.39) we obtain the average relation

$$\rho_G q_G = \rho_V v_V + \rho_A v_A. \tag{9.40}$$

However, note that Equation (9.40) holds irrespective of any specification of the transport laws. Indeed, it could have been derived straight from Equations (5.11) and (5.15) and the relation $\overline{N}_J = \rho_J/\rho_G$ stemming from Equation (2.38). The ideal gas state Equation (2.23), successively applied to dry air A and vapor V with $p_G = p_A + p_V$, the mass balance Equations (9.38b) and (9.38c) and Equation (9.40) combine to give us

$$\frac{\partial (\phi_0 (1 - S_L) \rho_G)}{\partial t} + \frac{\partial}{\partial x}(\rho_G q_G) = \frac{RT}{M_W}\mathring{m}_{L\to V}, \tag{9.41}$$

where M_W is the water molar mass. The volume flow vectors q_G and q_L of the gaseous mixture and the liquid water are governed by Darcy's law of Equation (6.128). For the one-dimensional experiment under consideration, not taking into account body forces, we write

$$q_G = -\frac{\varkappa k_{rG}(S_L)}{\eta_G}\frac{\partial p_G}{\partial x}; \quad q_L = -\frac{\varkappa k_{rL}(S_L)}{\eta_L}\frac{\partial p_L}{\partial x}. \tag{9.42}$$

Here, the air is the nonwetting phase nW, while the liquid water is the wetting phase W. Combining Equation (6.106) where, for the sake of simplicity, we adopt $p_{entry} = 0$, with

Equation (6.136), we then write

$$p_L = p_G - \pi_0 \Pi(S_L). \tag{9.43}$$

We now assume an incompressible flow for the liquid so that the liquid mass density ρ_L is a constant. Substituting the liquid Darcy's law (Equation (9.42)) in Equation (9.38a) and using Equation (9.43), we derive

$$\frac{\partial (\phi_0 S_L)}{\partial t} - \frac{\pi_0 \varkappa}{\eta_L} \frac{\partial}{\partial x} \left(D_L(S_L) \frac{\partial S_L}{\partial x} + \frac{k_{rL}(S_L)}{\pi_0} \frac{\partial p_G}{\partial x} \right) = -\mathring{m}_{L\to V}/\rho_L, \tag{9.44}$$

where $D_L(S_L)$ is the dimensionless liquid diffusion coefficient defined by

$$D_L(S_L) = -k_{rL}(S_L) \frac{d\Pi}{dS_L}. \tag{9.45}$$

Substituting the Darcy's law applied to the gas, that is, Equation (9.41), in Equation (9.42), and combining the resulting equation with Equation (9.44), we derive the equation governing the advective transport of the vapor–dry air mixture:

$$\frac{\partial ((1 - S_L) p_G)}{\partial t} - \frac{\varkappa}{2\phi_0 \eta_G} \frac{\partial}{\partial x} \left(k_{rG}(S_L) \frac{\partial}{\partial x} p_G^2 \right) \tag{9.46}$$
$$+ \frac{\rho_L RT}{M_W} \left[\frac{\partial S_L}{\partial t} - \frac{\varkappa \pi_0}{\phi_0 \eta_L} \frac{\partial}{\partial x} \left(D_L(S_L) \frac{\partial S_L}{\partial x} + \frac{k_{rL}(S_L)}{\pi_0} \frac{\partial p_G}{\partial x} \right) \right] = 0.$$

In a consistent way, this equation reduces to Equation (5.37) in the absence of liquid water, that is, for $p_G = p$ and $S_L = 0$. The equation governing the transport of moisture in both the liquid and vapor forms is obtained by substituting in Equation (9.38b) the vapor transport law of Equation (9.39a), Darcy's law related to q_G (Equation (9.42)), and the relation $\rho_V = p_V M_W / RT$, and subsequently adding the resulting equation to Equation (9.44). We then obtain

$$\frac{\partial [(1 - S_L) p_V]}{\partial t} - \frac{\varkappa}{\phi_0 \eta_G} \frac{\partial}{\partial x} \left(p_V k_{rG}(S_L) \frac{\partial p_G}{\partial x} \right)$$
$$- \tau D_0 p_{\text{atm}} \frac{\partial}{\partial x} \left[(1 - S_L) \alpha(S_L) \frac{\partial}{\partial x} \left(\frac{p_V}{p_G} \right) \right]$$
$$+ \frac{\rho_L RT}{M_W} \left[\frac{\partial S_L}{\partial t} - \frac{\pi_0 \varkappa}{\phi_0 \eta_L} \frac{\partial}{\partial x} \left(D_L(S_L) \frac{\partial S_L}{\partial x} + \frac{k_{rL}(S_L)}{\pi_0} \frac{\partial p_G}{\partial x} \right) \right] = 0. \tag{9.47}$$

Of the three thermodynamic variables p_V, S_L and p_G, which are involved in the two previous diffusion equations, only two are actually independent. Indeed, the two differential Equations (9.46) and (9.47) must be solved while also ensuring the thermodynamic equilibrium between the liquid and vapor phases of water. This equilibrium is governed by the Kelvin Equation (8.27), which can be written in the form

$$-\pi_0 \Pi(S_L) + p_G - p_{\text{atm}} = \frac{\rho_L RT}{M_W} \ln \frac{p_V}{p_{VS}(T)}. \tag{9.48}$$

To complete the set of equations, when considering only half the sample, that is, $0 \leq x \leq d$, we add the boundary and symmetry conditions

$$x = d: \; p_G = p_{atm}; \; p_V = p_V^d; \quad x = 0: \; \frac{\partial S_L}{\partial x} = \frac{\partial p_V}{\partial x} = \frac{\partial p_G}{\partial x} = 0, \quad (9.49)$$

while the initial conditions are expressed in the form

$$t = 0: \quad p_G = p_{atm}; \; p_V = p_V^0. \quad (9.50)$$

Because of the initial thermodynamic equilibrium between the vapor and liquid phases, the relation governing the uniform initial saturation S_L^0 is provided by combining the Kelvin Equation (9.48) and the initial conditions of Equation (9.50) into

$$-\pi_0 \Pi \left(S_L^0 \right) = \frac{\rho_L RT}{M_W} \ln \frac{p_V^0}{p_{VS}(T)}. \quad (9.51)$$

Similarly, the Kelvin Equation (9.48) and the first two boundary conditions of Equation (9.49) give the relation governing the value S_L^d of the liquid saturation prevailing on the sample border $x = \pm d$:

$$-\pi_0 \Pi \left(S_L^d \right) = \frac{\rho_L RT}{M_W} \ln \frac{p_V^d}{p_{VS}(T)}. \quad (9.52)$$

The set of Equations (9.46)–(9.52) capitalizes the continuum approach to the isothermal drying of any porous material. If the external relative humidity, that is $p_V^d/p_{VS}(T)$, is lower than that which can be associated by Equation (9.4) with the smallest pore-entry radius $r_{\Delta\mu}$ of the pore volume invaded by the gaseous mixture, the surface ($x = \pm d$) of the material is fully dried ($S_L^d = 0$) and a drying front starts propagating within the internal layers. This special case will not be examined further.

9.2.2 Drying asymptotics

An inspection of the set of Equations (9.46) and (9.47) leads us to identify three characteristic times τ_F, τ_D^G and τ_D^L, which are defined by

$$\tau_F = \frac{d^2}{\tau D_0}; \quad \tau_D^G = \frac{\phi_0 \eta_G d^2}{p_{atm} \varkappa}; \quad \tau_D^L = \frac{\phi_0 \eta_L d^2}{\pi_0 \varkappa}, \quad (9.53)$$

respectively. The characteristic time τ_F scales the rate at which the vapor and the dry air are transported within the porous material by means of the molecular – Fickian – diffusion, while the characteristic times τ_D^G and τ_D^L scale the rate at which the advective – Darcean – transport of the gaseous mixture and the liquid water phase, respectively, take place.

Let us now introduce the dimensionless quantities

$$\bar{t} = t/\tau_F; \quad \bar{x} = x/d; \quad \overline{p}_G = p_G/p_{atm}; \quad \overline{p}_V = p_V/p_{atm}; \quad \overline{\omega} = \frac{\rho_L RT}{p_{atm} M_W}. \quad (9.54)$$

Interestingly, the pressure $\overline{\omega} p_{atm}$ is the pressure $\rho_L RT/M_W$ which the water vapor would have at the considered temperature if its density was the liquid water density. As a result, adopting $T = 293\,\text{K}$ we obtain the quite large value 1335.6 for the dimensionless pressure coefficient

$\overline{\omega}$. Through the introduction of the dimensionless time $\bar{t}$ defined in Equation (9.54), the time is now scaled by the characteristic molecular diffusion time τ_F. Making use of Equation (9.54) in the set of Equations (9.46) and (9.47), the equations governing the advective transport of the vapor–dry air mixture and the transport of moisture in both the liquid and vapor forms can be rewritten in the dimensionless form as

$$\frac{\partial \left((1 - S_L)\overline{p}_G\right)}{\partial \bar{t}} - \frac{\tau_F}{2\tau_D^G} \frac{\partial}{\partial \bar{x}} \left(k_{rG}(S_L) \frac{\partial \overline{p}_G^2}{\partial \bar{x}}\right) + \qquad (9.55)$$

$$\overline{\omega}\left[\frac{\partial S_L}{\partial \bar{t}} - \frac{\tau_F}{\tau_D^L} \frac{\partial}{\partial \bar{x}} \left(D_L(S_L) \frac{\partial S_L}{\partial \bar{x}} + k_{rL}(S_L) \frac{p_{atm}}{\pi_0} \frac{\partial \overline{p}_G}{\partial \bar{x}}\right)\right] = 0$$

and

$$\frac{\partial \left[(1 - S_L)\overline{p}_V\right]}{\partial \bar{t}} - \frac{\tau_F}{\tau_D^G} \frac{\partial}{\partial \bar{x}}\left(\overline{p}_V k_{rG}(S_L) \frac{\partial \overline{p}_G}{\partial \bar{x}}\right) - \frac{\partial}{\partial \bar{x}}\left[(1 - S_L)\alpha(S_L) \frac{\partial}{\partial \bar{x}}\left(\frac{\overline{p}_V}{\overline{p}_G}\right)\right]$$

$$+\overline{\omega}\left[\frac{\partial S_L}{\partial \bar{t}} - \frac{\tau_F}{\tau_D^L} \frac{\partial}{\partial \bar{x}} \left(D_L(S_L) \frac{\partial S_L}{\partial \bar{x}} + k_{rL}(S_L) \frac{p_{atm}}{\pi_0} \frac{\partial \overline{p}_G}{\partial \bar{x}}\right)\right] = 0, \qquad (9.56)$$

respectively.

The timescale over which the coupled transports successively occur is governed by the relative range of the characteristic times τ_F, τ_D^G and τ_D^L. We are now going to explore the drying mechanisms for quite permeable materials, where the advective transport of the gaseous mixture occurs much faster than the vapor molecular diffusion, and for weakly permeable materials where this is the opposite. Let us then assess the range of the values of the intrinsic permeability related to the two cases.

When adopting $T = 293$ K, Equation (5.21) produces $D_0 = 2.48 \times 10^{-5} \mathrm{m^2 s^{-1}}$. Adopting in addition the values $\eta_G = 1.8 \times 10^{-5}$ kg m^{-1}s^{-1} and $p_{atm} = 101\,325$ Pa, from their definitions given in Equation (9.53) the ratio of the characteristic times, τ^F and τ_D^G, which scale the two transport processes are assessed in the form

$$\frac{\tau^F}{\tau_D^G} = \frac{p_{atm} \varkappa}{\eta_G \phi_0 \tau D_0} = \frac{\varkappa/\phi_0 \tau}{4.41 \times 10^{-15} \mathrm{m^2}}, \qquad (9.57)$$

so that we can write:

for quite permeable materials: $\tau_D^G/\tau^F \ll 1$; $\varkappa/\phi_0\tau \gg 4.41 \times 10^{-15}$ m^2; (9.58a)
for weakly permeable materials: $\tau_D^G/\tau^F \gg 1$; $\varkappa/\phi_0\tau \ll 4.41 \times 10^{-15}$ m^2. (9.58b)

Taking $\eta_L = 10^{-3}$ kg m^{-1} s^{-1} for the liquid viscosity, we also obtain

$$\tau_D^G/\tau_D^L = \eta_G \pi_0 / \eta_L p_{atm} = \pi_0/5.6\,\mathrm{MPa}. \qquad (9.59)$$

The value π_0 stands for a reference capillary pressure value related to the whole distribution of pore-entry radii. When we substitute $p_{nW} - p_W = p_G - p_L = 5.6$ MPa in the Laplace Equation (6.58), together with a zero wetting angle $\theta = 0$ and the air–water interfacial energy $\gamma_{nWW} = \gamma_{GL} = 73$ mJ m^{-2}, we obtain the value $r = 26$ nm for the pore-entry radius associated with π_0. Equation (5.31) has shown that the intrinsic permeability $\varkappa$ is proportional to the square ℓ^2 of a characteristic length related to the geometry of the porous network where the flow takes place. If we adopt $\ell \equiv 26$ nm, the order of magnitude of the associated value of the intrinsic permeability $\varkappa$ is therefore 7×10^{-16} m^2. Owing to Equations (9.58) and (9.59), this value of $\varkappa$ allows us to add the following conditions

$$\text{for quite permeable materials:} \quad \tau_D^G/\tau_D^L \ll 1; \qquad (9.60a)$$
$$\text{for weakly permeable materials:} \quad \tau_D^G \sim \tau_D^L. \qquad (9.60b)$$

Equation (9.60a) means that, because of a sufficiently high value of the intrinsic permeability $\varkappa$, the advective transport of the gaseous mixture occurs much faster than the advective transport of liquid water. In contrast, because of a sufficiently low value of the intrinsic permeability $\varkappa$, Equation (9.58b) means that the vapor molecular diffusion occurs much faster than the advective transport of the gaseous mixture. Because of the much lower value of the liquid water viscosity compared with that of the gaseous mixture, it could have been expected that the advective transport of liquid water always occurs much slower than that of the gaseous mixture irrespective of the value of the intrinsic permeability $\varkappa$. Equation (9.60a) reveals that this is only true for quite permeable materials. As shown by Equation (9.60b), for weakly permeable materials the advective transport of liquid water occurs at the same rate as the advective transport of gaseous mixture. This is because the large difference in viscosity values between the liquid and the gas is offset by the large value of the capillary pressure driving the liquid flow, which can be associated with a low value of the intrinsic permeability. With the help of Equations (9.58) and (9.60), we can now perform an asymptotic analysis of the dimensionless transport Equations (9.55) and (9.56) governing the drying kinetics.

Drying of quite permeable materials

For quite permeable materials, Equation (9.58a) ensures that the molecular diffusion of vapor and dry air occurs much slower than the advective transport of the vapor–dry air mixture. In addition, because of Equation (9.60a), for quite permeable materials we have $\tau_F/\tau_D^G \gg \tau_F/\tau_D^L$. Retaining only the term having the main order of magnitude with respect to $\tau^F/\tau_D^G \gg 1$, when adopting $\bar{t} \sim 1$ the dimensionless Equation (9.55) governing the transport of the air–vapor mixture as a whole reduces to

$$\left(\tau^F/\tau_D^G\right) \times \frac{\partial}{\partial \bar{x}} \left[k_{rG}(S_L) \frac{\partial \overline{p_G^2}}{\partial \bar{x}} \right] = 0. \qquad (9.61)$$

Because of the boundary and symmetry conditions of Equation (9.49), the transport Equation (9.61) integrates to give the simple equality $p_G = p_{atm}$. Taking advantage of this equality, and noting that the approximation $p_V \ll \rho_L RT/M_W$ allows us to neglect the first term of Equation (9.56), the dimensionless moisture diffusion Equation (9.56) and the vapor–liquid equilibrium Equation (9.48) combine to give

$$t \gg \tau_D^G: \quad \frac{\partial S_L}{\partial t} - \frac{\pi_0 \varkappa}{\phi_0 \eta_L} \frac{\partial}{\partial x} \left[D_V(S_L) \frac{\partial S_L}{\partial x} + D_L(S_L) \frac{\partial S_L}{\partial x} \right] = 0. \qquad (9.62)$$

In the diffusion Equation (9.62), $D_V(S_L)$ is a dimensionless vapor diffusion coefficient, whose expression is

$$D_V(S_L) = \frac{\eta_L \tau_D^G}{\eta_G \tau^F} \frac{p_{VS}(T)(1-S_L)\alpha(S_L)}{p_{atm}\overline{\omega}^2 \, k_{rL}(S_L)} \exp\left[-\frac{\pi_0}{p_{atm}\overline{\omega}}\Pi(S_L)\right] \times D_L(S_L). \quad (9.63)$$

The moisture diffusion Equation (9.62) must be solved with the initial condition $S_L(x, t=0) = S_L^0$ and the boundary condition $S_L(x=d, t) = S_L^d$, where S_L^0 and S_L^d are given by Equations (9.51) and (9.52). Furthermore, if we adopt $T = 293$ K, with $p_{VS}(T) = 2333$ Pa, we find

$$(\eta_L/\eta_G)(p_{VS}/p_{atm}\overline{\omega}^2) = 7.17 \times 10^{-7}. \quad (9.64)$$

In spite of the large value of the viscosity ratio η_L/η_G, the ratio D_V/D_L remains small because of the large value of $\overline{\omega} \sim 10^3$. The inequality $D_V \ll D_L$ ensures that the advective transport of water in liquid form is more favorable than the transport in gaseous form by means of confined molecular diffusion. Bearing in mind the meaning of $\overline{\omega}$ reported above after its definition of Equation (9.54), this is because of the large volume which the water would occupy in gaseous form after evaporation, when compared with the volume occupied by the same amount of water in liquid form. As a result of the inequality $D_V \ll D_L$, the moisture diffusion Equation (9.62) accurately reduces to

$$t > \tau_D^L \gg \tau_D^G : \quad \frac{\partial S_L}{\partial t} - \frac{\pi_0 \varkappa}{\phi_0 \eta_L} \frac{\partial}{\partial x}\left[D_L(S_L)\frac{\partial S_L}{\partial x}\right] = 0. \quad (9.65)$$

According to the approximate analysis above, the mechanism of drying of quite permeable materials satisfying Equation (9.60a) can be summarized as follows. Owing to the large gas permeability, as captured by Equation (9.61) the vapor–dry air mixture cannot sustain any overpressure with respect to the initial atmospheric pressure. Indeed, any buildup of the gas pressure vanishes because of the high speed of the advective Darcean flow compared with the slow speed of both the molecular diffusion and the liquid advective flow. During the period $t > \tau_D^L \gg \tau_D^G$, because of the large value of $\overline{\omega}$, the evaporation of the inner liquid water and its subsequent transportation towards the material surface through molecular diffusion is not favorable since the confinement does not facilitate the expansion of the liquid transforming to its vapor. For $t \gg \tau_D^G$ this results in a drying mechanism where the moisture content is governed by the single diffusion Equation (9.65), where the moisture transport is achieved almost exclusively by the advective transport of water in liquid form. In turn, this liquid water transportation makes the liquid pressure decrease sufficiently in order to meet the thermodynamic vapor–liquid equilibrium condition at any time. In short the capillary drying mechanism sketched out in Figure 1.6(b) prevails.

Drying of weakly permeable materials

For weakly permeable materials, Equation (9.58b) ensures that the transport of vapor mainly occurs through molecular diffusion. Besides, according to Equation (9.60b), the advective transport of both the liquid water and the gaseous mixture occurs at the same rate. This already allows us to conclude that the drying process of weakly permeable materials will be achieved in two steps which are well separated in time. Indeed, the molecular diffusion is accounted for by the third term in the moisture transport Equation (9.56). This term, according to Equation (9.58b), is now much larger than the second term accounting for the contribution of the advective transport of the gaseous mixture to the vapor transport. As a result, in Equation (9.56) we can cancel the first and second terms with respect to the other terms because of the large value of both τ_D^G/τ_F and $\bar{\omega}$, as well as the last term because of the approximation $p_{atm} \ll \pi_0$. Therefore, for weakly permeable materials the moisture transport Equation (9.56) reduces to

$$-\frac{\partial}{\partial \bar{x}}\left[(1 - S_L)\alpha(S_L)\frac{\partial}{\partial \bar{x}}\left(\frac{\bar{p}_V}{\bar{p}_G}\right)\right] + \bar{\omega}\left[\frac{\partial S_L}{\partial \bar{t}} - \frac{\tau_F}{\tau_D^L}\frac{\partial}{\partial \bar{x}}\left(D_L(S_L)\frac{\partial S_L}{\partial \bar{x}}\right)\right] = 0. \quad (9.66)$$

After this preamble, let us now carry out an asymptotic analysis of the drying process again. For $\bar{t} \sim 1$, that is for $t \sim \tau^F \ll \tau_D^G \sim \tau_D^L$, when retaining only the term having the main order of magnitude with respect to $\tau^F/\tau_D^G \sim \tau^F/\tau_D^L \ll 1$, the transport terms vanish in the transport Equation (9.55) governing the advective transport of the vapor–dry air mixture. The latter reduces to

$$t \ll \tau_D^G \sim \tau_D^L : \quad \frac{\partial\left[(1 - S_L)\bar{p}_G\right]}{\partial \bar{t}} = -\bar{\omega}\frac{\partial S_L}{\partial \bar{t}}. \quad (9.67)$$

For times $t \sim \tau^F \ll \tau_D^G \sim \tau_D^L$, the advective transport of both the liquid water and the gaseous mixture is not active, so that the change in the liquid saturation S_L is only due to the inner evaporation. The inner evaporation then acts as a source term for the overall gas pressure, which is accounted for by the right-hand side of Equation (9.67). Taking into account the initial conditions of Equations (9.50) and (9.51), the transport Equation (9.67) integrates in the form

$$\frac{p_G}{p_{atm}} = 1 + (1 - \bar{\omega})\frac{S_L - S_L^0}{1 - S_L}. \quad (9.68)$$

The steady state related to the molecular diffusion, that is for large values of the dimensionless time $\bar{t}$, is obtained by integrating Equation (9.66), having equated the time derivative $\partial S_L/\partial \bar{t}$ to zero and neglected the last term because of the approximation $\tau_F/\tau_D^L \sim \tau_F/\tau_D^G \ll 1$ associated with weakly permeable materials. Owing to the symmetry condition of Equation (9.49) at $x = 0$, the integration of Equation (9.66) shows that, once the steady state has been established, the ratio p_V/p_G no longer depends on x. The corresponding constant value of p_V/p_G is provided by taking advantage of the boundary conditions at $x = d$ in Equation (9.49):

$$\bar{t} \gg 1 : \quad \frac{p_V}{p_G} = \frac{p_V^d}{p_{atm}}. \quad (9.69)$$

The asymptotic values of the liquid saturation and the gas pressure, S_L^∞ and p_G^∞, which relate to the end of the first step of the drying process, that is for times $t \sim \tau_D^G \sim \tau_D^L$, must simultaneously satisfy Equations (9.48), (9.68) and (9.69). This provides us with S_L^∞ as the solution of the nonlinear equation

$$-\frac{\pi_0}{p_{atm}}\Pi\left(S_L^\infty\right) + (1 - \bar{\omega})\frac{S_L^\infty - S_L^0}{1 - S_L^\infty} = \bar{\omega}\ln\left\{\frac{p_V^d}{p_{VS}(T)}\left[1 + (1 - \bar{\omega})\frac{S_L^\infty - S_L^0}{1 - S_L^\infty}\right]\right\}. \quad (9.70)$$

In order to assess the drop in liquid water saturation, and the maximum excess of gas pressure which can be expected at the end of the first drying step, we choose the capillary curve given by Equation (6.107) where we adopt $m = 0.4$ and $\pi_0 = 10$ MPa. In addition, we take $T = 293$ K, with $p_{VS}(T) = 2333$ Pa and $h_R^0 = p_V^0/p_{VS}(T) = 87\%$, while, for the outer relative humidity, we choose $h_R^d = p_V^d/p_{VS}(T) = 50\%$. Recalling that $\rho_L = 10^3$ kg m^{-3}, $R = 8.314$ J mol^{-1} K^{-1} and $M_W = 18$ g, and using successively Equations (9.51), (9.68), (9.69) and (9.70), we numerically obtain

$$S_L^0 = 0.58165; \quad S_L^\infty = 0.58141; \quad \frac{p_G^\infty}{p_{atm}} = 1.7846. \tag{9.71}$$

Although the difference between the initial saturation S_L^0 and the asymptotic saturation S_L^∞ turns out to be negligible, it is worthwhile noting that the asymptotic gas pressure is nearly twice the initial atmospheric pressure. This is due to the fact that the vapor occupies a much larger volume than the same amount of liquid water prior to its evaporation, resulting in a significant source term for the overall gas pressure. This is accounted for by the large value reported above for the pressure coefficient $\overline{\omega} \sim 10^3$, which scales the source term in the right-hand side of Equation (9.67).

According to the above analysis, in response to the lowering of the relative humidity of the surrounding, for weakly permeable materials the period $t \ll \tau_D^G$ is dominated by the molecular diffusion of water vapor. The vapor molecular diffusion tends to establish in the inner layers the vapor molar fraction prevailing in the surroundings, but not yet the outer relative humidity. Indeed, the value of the asymptotic liquid water saturation S_L^∞ provided by Equation (9.71) does not match the value S_L^d of the saturation at the border $x = d$ given by Equation (9.52). For instance, when using the same experimental data as in Equation (9.71), Equation (9.52) for S_L^d gives

$$S_L^d = 0.223. \tag{9.72}$$

Basically, the boundary condition of Equation (9.54) at $x = d$ is mainly met with the help of the advective liquid transport. The latter starts at once within the skin layer close to the end surfaces at $x = d$, where the capillary pressure gradient concentrates at very early times. To analyze this further step of the drying process, we now scale the time by the characteristic time τ_D^L according to

$$\tilde{t} = t/\tau_D^L = \overline{t} \times \tau_F/\tau_D^L. \tag{9.73}$$

Substituting Equation (9.73) in Equation (9.66), the dimensionless moisture transport equation reads

$$-\frac{\tau_D^L}{\tau_F}\frac{\partial}{\partial \overline{x}}\left[(1 - S_L)\alpha(S_L)\frac{\partial}{\partial \overline{x}}\left(\frac{\overline{p}_V}{\overline{p}_G}\right)\right] + \overline{\omega}\left[\frac{\partial S_L}{\partial \tilde{t}} - \frac{\partial}{\partial \overline{x}}\left(D_L(S_L)\frac{\partial S_L}{\partial \overline{x}}\right)\right] = 0. \tag{9.74}$$

Because of Equation (9.58b) governing the permeability range for weakly permeable materials, the first term in Equation (9.74) is the leading term and must be equated to zero, allowing us to recover the asymptotic regime of Equation (9.69) of the earlier drying step.[9]

---

[9] Because of the large value of $\overline{\omega}$, the larger the value of τ_D^L/τ_F and, therefore, the more weakly permeable the material, the more accurate is the analysis. A refined quantitative analysis requires a numerical treatment of the set of Equations (9.46)–(9.51), confirming the analysis for usual weakly permeable materials (see Mainguy, M., Coussy, O. and Baroghel-Bouny, V. (2001) The role of air pressure in the drying of weakly permeable materials. *Journal of Engineering Mechanics*, **127**, 582–592).

Indeed, the molecular Fickian diffusion operates at a timescale which is much smaller than the one related to the advective Darcean liquid transport. Therefore, when the second step of the drying process becomes significantly active within the material, the molecular diffusion within the porous material has already made the vapor molar fraction $x_V = p_V/p_G$ uniform down to the outer value p_V^d/p_{atm}. Equating to zero the first term in Equation (9.74), in the ultimate drying step of weakly permeable materials, the moisture content is finally governed for times $t > \tau_D^L \sim \tau_D^G$ by the transport Equation (9.65) to which we now add the initial and boundary conditions

$$\bar{x} = 1: \ S_L = S_L^d; \ \bar{x} = 0: \ \frac{\partial S_L}{\partial \bar{x}} = 0; \ \tilde{t} = 0: \ S_L = S_L^\infty \simeq S_L^0. \tag{9.75}$$

Therefore, in the later drying step of weakly permeable materials, the liquid saturation is still governed by the same water liquid diffusion Equation (9.65) which we found for the quite permeable materials. In this later drying step the gradient of capillary pressure drives the liquid water towards the end faces of the sample, where liquid water finally evaporates. Paradoxically, for weakly permeable porous materials, this capillary drying step sketched out in Figure 1.6(b) is therefore also the efficient step of drying, since Equation (9.71) has shown that the first drying step causes a negligible evaporation. For weakly permeable material the main difference lies in the existence of an initial period where a gas pressure builds up, which does not exist for quite permeable materials. These general conclusions will have to be partly revisited in situations close to either complete saturation, that is for $S_L = 1$, or desaturation, that is for $S_L = 0$, since then the gas permeability k_{rG} and the liquid permeability k_{rL} respectively vanish.

9.2.3 Drying mechanics

Drying shrinkage

During drying, when the relative humidity of the surroundings drops below the initial relative humidity prevailing within the porous material, the resulting thermodynamic imbalance enforces the porous material to exchange water with the surroundings, in order to establish the outer relative humidity within the porous material. This causes the liquid saturation degree to decrease, resulting in the depressurization of the inner liquid water with respect to the atmospheric pressure. A drying shrinkage of the material ensues. While the drying kinetics is governed by the transport phenomena explored in Section 9.2.2, the asymptotic drying shrinkage is governed by the outer relative humidity only.

In order to assess the intensity of the drying shrinkage, letting the fluids $1 = G$ and $2 = L$ be the gas formed from the wetted air and the liquid water, respectively, and making use of the pore isodeformation assumption of Equation (7.42), we first rewrite the constitutive Equation (7.31a) of unsaturated porous solid in the form

$$\sigma = K\epsilon - b(S_G p_G + S_L p_L), \tag{9.76}$$

where σ, p_G and p_L stand for the excess of the considered stress or pressure with respect to the atmospheric pressure p_{atm}. During a free drying experiment, both the stress excess σ and the gas pressure excess p_G remain equal to zero. In addition, the liquid pressure excess p_L identifies with the pressure excess of the wetting fluid with respect to the pressure of the nonwetting fluid. According to Equation (6.105), the liquid pressure excess p_L then identifies with the opposite of the capillary pressure, p_{cap}, and therefore depends only on the saturation $S_W = S_L$ related to liquid water. Thus taking $\sigma = p_G \equiv 0$ and $p_L \equiv -p_{cap}$ in Equation (9.76), we can express the volumetric drying shrinkage ϵ_{drying} in the form

$$\epsilon_{drying} = -\frac{b}{K} S_L p_{cap}(S_L). \tag{9.77}$$

In addition, the capillary pressure p_{cap} is provided by the Kelvin Equation (8.33):

$$p_{cap} = -\frac{\rho_L RT}{M_W} \ln h_R, \tag{9.78}$$

where h_R is the relative humidity imposed by the surroundings. In Section 6.2 we explored how the pore volume fraction, $1 - S(r)$, having a pore-entry radius greater than r and the capillary pressure curve, $p_{cap}(S_L)$, were related. As a result, from the knowledge of the outer relative humidity h_R, the pore volume fraction, $1 - S(r)$, and the factor b/K, an assessment of the volumetric drying shrinkage ϵ_{drying} of a porous solid is provided by Equations (9.77) and (9.78).

Drying stiffening

Substituting Equation (9.77) in Equation (9.76), we can rewrite the constitutive Equation (9.76) in the form

$$\sigma = K \Delta\epsilon; \quad \Delta\epsilon := \epsilon - \epsilon_{drying}, \tag{9.79}$$

where $-\Delta\epsilon$ accounts for the volumetric contraction resulting from the application of the volumetric compressive stress $-\sigma$. In linear poroelasticity the bulk modulus K is constant and Equation (9.79) predicts that any compressive stress $-\sigma$ will produce the same contraction $-\Delta\epsilon$, whether it is applied before or after a drying period. This absence of stiffening caused by drying contradicts familiar observations which can be made for soft porous materials: as illustrated in Figure 9.6, a sponge is much stiffer after drying than before drying.

The stiffening induced by drying originates from the partial closure of the porous space due to the compressive stress induced within the solid matrix by the capillary pressure p_{cap}. To account for this drying stiffening, we can extend to unsaturated conditions the results which we obtained in Sections 4.1.3 and 7.4.1, when exploring nonlinear poroelasticity. Indeed, according to Equation (9.76), the average pressure $S_G p_G + S_L p_L$ in unsaturated conditions plays the same role as the one played by the pore pressure p in saturated conditions. Replacing the Terzaghi effective stress, $\sigma' = \sigma + p$ in Equation (4.34a), by its unsaturated analogous $\sigma' = \sigma + S_G p_G + S_L p_L$, we can extend Equation (9.76) in the form

$$\sigma = K_{SEC}(\sigma')\epsilon - b_{SEC}(\sigma')(S_G p_G + S_L p_L), \tag{9.80}$$

where K_{SEC} and b_{SEC} are the secant poroelastic properties which we defined in Section 4.1.3. During drying where $p_G = 0$ we have $\sigma + S_G p_G + S_L p_L = \sigma - S_L p_{cap}$. Because of the high intensity of the capillary pressure induced by drying,[10] we can use the approximation

---

[10] When taking $h_R = 50\%$ and $T = 293$ K, the Kelvin Equation (9.78) provides us with the value $p_{cap} = 93.8$ MPa.

$\sigma - S_L p_{\text{cap}} \simeq -S_L p_{\text{cap}}$. We then extend Equations (9.77) and (9.79) in the form

$$\epsilon_{\text{drying}} \simeq -\frac{b_{\text{SEC}}\left(-S_L p_{\text{cap}}\right)}{K_{\text{SEC}}\left(-S_L p_{\text{cap}}\right)} S_L p_{\text{cap}}(S_L), \qquad (9.81)$$

and

$$\sigma \simeq K_{\text{SEC}}\left(-S_L p_{\text{cap}}\right) \Delta\epsilon; \quad \Delta\epsilon := \epsilon - \epsilon_{\text{drying}}. \qquad (9.82)$$

As illustrated in Figure 9.6 which is an extension of Figure 4.2 to unsaturated conditions, Equation (9.82) now does account for the drying stiffening. On account of the Kelvin Equation (9.78), the change in relative humidity induces a high capillary pressure within the porous solid, resulting in a value of the secant bulk modulus K_{SEC} much larger after drying, where $K_{\text{SEC}} \simeq K_{\text{SEC}}\left(-S_L p_{\text{cap}}\right)$, than before drying, where $K_{\text{SEC}} = K_{\text{SEC}}(-\sigma)$. Therefore, the contraction $-\Delta\epsilon$ recorded for the same volumetric compressive stress $-\sigma$ is much smaller when the stress is applied after drying rather than before drying.

9.3 Mechanics of Confined Crystallization

The durability of water-infiltrated materials subjected to frost action is a major concern in cold climates. At first sight it might be thought that the mechanics of freezing porous materials is similar to that of a sealed water-filled bottle when subjected to frost action. As illustrated in Figure 1.1(a) the failure of the bottle results from the pressure buildup caused by the undrained 9 % expansion of the liquid water solidifying within the single large pore formed by the inside of the bottle. Unexpectedly the mechanics of freezing porous materials is not that simple. As reported in Figure 1.1(b) a porous solid saturated by benzene, which unlike water contracts when solidifying, still expands slightly when subjected to temperatures below the melting point of benzene. Furthermore a porous solid, initially saturated with liquid water, but presenting air voids, can shrink when it is subjected to temperatures below the melting point of water. In Section 9.1.1 we showed that the temperature at which the saturating liquid solidifies depended upon the entry radius of the pore it filled. Accordingly, at a given temperature below the melting point only a part of an initially liquid-saturated porous solid exhibiting a pore size distribution will be frozen. Combining this result with unsaturated poroelasticity, this section investigates the mechanics of porous solids subjected to confined crystallization.

9.3.1 Cryogenic swelling

The main cause of the unexpected expansion reported in Figure 1.1 of a porous solid initially saturated by liquid benzene and subjected to freezing temperatures is the cryosuction process. This process can roughly be described as follows. When the temperature is continuously lowered below the melting point, benzene is present in the pore space in both solid and liquid forms. In order to ensure the thermodynamic equilibrium between the solid and liquid phases of benzene, liquid benzene is constantly driven from the remaining unfrozen solution towards the already crystallized sites. The unexpected expansion is finally due to the increase in the crystal pressure caused by the further freezing of the extra liquid entering the pore.

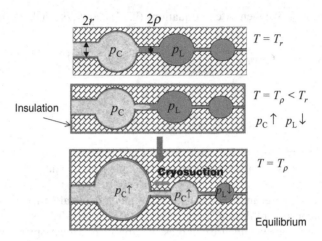

Figure 9.7 The cryosuction process which induces the unexpected expansion observed for a cement paste initially saturated by liquid benzene and subjected to freezing

More precisely let us consider the pore shown in Figure 9.7, whose access is constituted of two cylindrical channels. Their respective radii are r and ρ, with $r > \rho$. The liquid contained in this pore solidifies at the temperature T_r given by the Gibbs–Thomson Equation (9.5). Assuming a zero wetting angle we take $\theta = 0$ in Equation (9.5) and we obtain $T_r = T_{0m} - 2\gamma_{CL}/r\Delta S_m$. As the value of the temperature drops down to $T_\rho = T_{0m} - 2\gamma_{CL}/\rho\Delta S_m < T_r$, because of the equilibrium Equation (8.57) the pressure difference $p_C - p_L$ must increase. This pressure increase is required to offset the free energy difference between the solid phase and the liquid phase caused by the lower entropy of the former and this increase is partly achieved by increasing the crystal pressure p_C. In the already solidified pores this crystal pressure increase can only be achieved by some extra liquid entering the pore and freezing. The entry within the pore of this extra liquid is made possible by the premelted liquid film lying between the crystals and the solid walls of the pore, whose existence was analyzed in Section 9.1.3. This premelted liquid film also exists in the entry channels. As a result, the extra liquid can even enter into the pores whose entry channels are already frozen, like the pores of entry radii r and ρ (Figure 9.7) in the temperature range $T < T_\rho < T_r$. The extra liquid is driven from the supercooled liquid towards the already solidified pore. The actual driving force of this phenomenon known as cryosuction is the chemical potential difference which exists between the supercooled liquid some distance from the already solidified pore, and the liquid directly in contact with the crystals already formed. At a given temperature as, for instance, at temperature T_ρ in Figure 9.7, cryosuction stops when equilibrium is achieved everywhere, that is, as soon as the crystal pressure meets the condition of Equation (8.57). For sufficiently low cooling rates this equilibrium can be assumed to be achieved at any time. As the temperature is continuously lowered below T_m, the resulting continuous increase in the crystal pressure p_C applied to the solid walls of all the already frozen pores causes the cryogenic swelling of the porous solid.

Since a bulk liquid phase cannot exert any tensile stress on the pore solid walls, the contraction of benzene when it solidifies does not affect the overall deformation of the porous

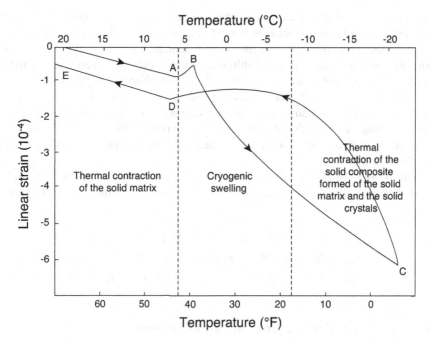

Figure 9.8 As reported in Figure 1.1, a sample which is initially saturated by liquid benzene exhibits an unexpected expansion when its temperature is lowered below the benzene melting point; this expansion is caused by the cryosuction driving the liquid benzene towards the already solidified pores when the temperature decreases below the melting point

solid. This contraction only gives some more volume to the air unavoidably left in the sample after sealing. For similar reasons, the overall deformation is not affected by the thermal contraction of liquid benzene because the latter is much more significant than the thermal contraction of the porous volume enclosed by the solid walls. As a result the pressure of liquid benzene remains close to atmospheric pressure, so that T_{0m} remains close to the melting point T_m of benzene.

Three periods can be identified in the experimental data reported in Figure 9.8 concerning the continuous cooling of a cement paste initially saturated by liquid benzene. In the first period, which starts at the initial temperature above the melting point T_m, the deformation is caused by the thermal contraction of the solid matrix. In the second period, which starts at the melting point T_m, the liquid progressively transforms into solid crystals, resulting in the unexpected cryogenic swelling. In the third and last period, which starts after the liquid has been completely transformed into solid crystals, the deformation is caused by the thermal contraction of the solid composite formed from the solid matrix and the solid crystals.

9.3.2 The hydraulic pressure

The analysis above has qualitatively explained at the pore scale the origin of the cryogenic swelling due to cryosuction. Even though cryosuction and the related cryogenic swelling play a significant role in the deformation of water-infiltrated porous solids upon the frost action, the other major factor is obviously the 9 % density change of liquid water transforming into ice. Since the water filling a pore freezes at a cooling depending upon the size of the pore-entry, the swelling ability of water-infiltrated materials exhibiting a pore size distribution also depends on a so-called hydraulic pressure effect. According to this effect, the expansion of liquid water transforming into ice causes a significant expulsion of liquid water from the freezing sites. This expulsion causes a pressure buildup in the solution still remaining unfrozen and accommodating this extra liquid water. Unsaturated poroelasticity, which we explored in Chapter 7, can help us to quantify the various effects causing the overall deformation. This is the topic of this section.

In all that follows, we take the melting point T_m as the initial reference temperature, and the atmospheric pressure p_{atm} as the zero reference pressure. State Equations (7.62) then give the relations

$$\sigma = K\epsilon - b_C p_C - b_L p_L - 3\alpha_S K (T - T_m); \qquad (9.83a)$$
$$\varphi_C = b_C \epsilon + p_C/N_C + p_L/N_{CL} - 3\alpha_{\varphi_C} (T - T_m); \qquad (9.83b)$$
$$\varphi_L = b_L \epsilon + p_C/N_{CL} + p_L/N_{LL} - 3\alpha_{\varphi_L} (T - T_m); \qquad (9.83c)$$
$$s_{ij} = 2G e_{ij}. \qquad (9.83d)$$

In addition, according to Equation (7.63), we have the relations

$$\alpha_{\varphi_C} = \alpha_S (b_C - \phi_0 S_C); \quad \alpha_{\varphi_L} = \alpha_S (b_L - \phi_0 S_L). \qquad (9.84)$$

Considering a stress-free experiment, the overall volumetric strain ϵ is given by taking $\sigma = 0$ in Equation (9.83a). If, in addition, we assume pore volumetric isodeformation, making use of Equation (7.42) we obtain

$$\epsilon = b \frac{S_L p_L + S_C p_C}{K} - 3\alpha_S (T_m - T). \qquad (9.85)$$

In the cases considered below, the pore pressures p_J of the liquid and solid phases of the same substance – when they are present simultaneously – are linked by the Thomson Equation (8.55). Since the atmospheric pressure is taken as the zero reference pressure, we rewrite Equation (8.55) in the form

$$p_C - p_L + \left(1 - \overline{V}_L^0/\overline{V}_C^0\right) p_L = \Delta S_m (T_m - T). \qquad (9.86)$$

As a result, only the crystal saturation $S_C = 1 - S_L$ and the liquid pressure p_L remain unknown in Equation (9.85). They have to be determined by conditions specific to the problem under consideration.

Then let n be the overall density of moles of the same substance currently contained in the porous space in both solid and liquid form. Mole density n depends on the current partial porosity ϕ_J related to each phase J. Assuming infinitesimal transformations, so that $\left(\overline{V}_J - \overline{V}_J^0\right)/\overline{V}_J^0 \ll 1$ and $\varphi_J \ll 1$, and retaining only the main order terms with respect to $\left|1 - \overline{V}_L^0/\overline{V}_C^0\right|$ ($\simeq 0.09$ for water) $\ll 1$, from the definition of Equation (7.17) for the

Lagrangian saturation S_J, the mole density n can be written in the form

$$n = \phi_C/\overline{V}_C + \phi_L/\overline{V}_L \simeq n_0 + \phi_0 S_C \left(1/\overline{V}_C^0 - 1/\overline{V}_L^0\right)$$
$$+ \phi_0 S_L \left(1/\overline{V}_L - 1/\overline{V}_L^0\right) + \phi_0 S_C \left(1/\overline{V}_C - 1/\overline{V}_C^0\right) + (\varphi_C + \varphi_L)/\overline{V}_L^0, \quad (9.87)$$

where $n_0 = \phi_0/\overline{V}_L^0$ stands for the initial number of moles. The constitutive Equations (8.47) and (8.49) of liquid and solid water provide us with the relation

$$1/\overline{V}_J \simeq 1/\overline{V}_J^0 \left[1 + p_J/K_J - 3\alpha_J (T - T_m)\right]. \quad (9.88)$$

Substituting Equation (9.88) in Equation (9.87), together with the constitutive Equations (9.83b) and (9.83c) of the porous solid, while making use of Equations (7.42), (7.39) and (9.84), we derive

$$n - n_0 = \Delta n_1 + \Delta n_2 + \Delta n_3, \quad (9.89)$$

where

$$\Delta n_1 = \left(1/\overline{V}_C^0 - 1/\overline{V}_L^0\right) \phi_0 S_C; \quad (9.90a)$$

$$\Delta n_2 = \left[\frac{b\sigma}{K} + S_C p_C \left(\frac{\phi_0}{K_C} + \frac{1}{N} + \frac{b^2}{K}\right) + S_L p_L \left(\frac{\phi_0}{K_L} + \frac{1}{N} + \frac{b^2}{K}\right)\right]/\overline{V}_L^0; \quad (9.90b)$$

$$\Delta n_3 = 3\phi_0 (\alpha_S - S_C \alpha_C - S_L \alpha_L)(T - T_m)/\overline{V}_L^0, \quad (9.90c)$$

where the modulus N can be expressed with the help of Equation (4.18).

Let us now restrict the analysis to water. The bulk moduli of water in liquid and solid forms are $K_L = 2200$ MPa and $K_C = 8800$ MPa, respectively. In addition, for cohesive materials such as cement-based materials or rocks, the bulk moduli k_S and K of respectively the solid matrix and the porous solid satisfy the inequalities $k_S > K > 20\,000$ MPa. Since the value of the melting entropy of water is $\Delta S_m = 1.2$ MPa K^{-1}, while the order of magnitude of the thermal dilation coefficient α_J is 10^{-5} K^{-1}, the contributions to the mole change $n - n_0$, which are weighted by either a factor $\Delta S_m (T_m - T)$ divided by a bulk modulus or by a factor $\alpha_J (T_m - T)$, can be disregarded. Making use of both this approximation and the approximation $\phi_0/K_L \gg 1/N + b^2/K$ when putting Equation (9.86) in Equation (9.90b), from Equation (9.89) we derive

$$n - n_0 \simeq b\sigma/K\overline{V}_L^0 + \phi_0 (S_L/K_L + S_C/K_C) p_L/\overline{V}_L^0 - \left(1/\overline{V}_L^0 - 1/\overline{V}_C^0\right) \phi_0 S_C. \quad (9.91)$$

The mole balance Equation (9.91) holds irrespective of the problem under consideration. In the right-hand side of Equation (9.91), the first term accounts for the mole change made possible by the change in pore volume due to the external stress. This change in pore volume is positive for a tensile stress, that is for $\sigma > 0$, and negative for a compressive stress, that is for $\sigma < 0$. According to Equation (9.88) and to the above approximations about the compressibility properties, the volume decrease of the molar volume of water phase J is close to p_L/K_J. The volume decrease which the current volume of phase J undergoes under its pressure is therefore $\phi_0 S_J p_L/K_J$ per unit of overall volume. As a result, in the right-hand side of Equation (9.91) the second term accounts for the mole change made possible by the decrease in volume of the current in-pore phases caused by their deformation. In contrast, the increase of volume due to the phase change opposes an increase of the mole density n, which is accounted for by the last term.

For a sealed sample[11] the number $nd\Omega_0$ of water moles remains constant, so that $n-n_0 = 0$. Considering a stress-free experiment, we have $\sigma = 0$. Substitution of both these conditions in Equation (9.91) gives

$$p_L = S_C \left(1 - \overline{V}_L^0/\overline{V}_C^0\right) \frac{K_L K_C}{S_L K_C + S_C K_L}. \tag{9.92}$$

In turn, substituting Equation (9.92) in Equation (9.86), we find that $p_C \simeq p_L$. An assessment of the overall volumetric strain ϵ induced by the freezing process is provided by substituting the approximation $S_L p_L + S_C p_C \simeq p_L$ and Equation (9.92) for p_L in Equation (9.85), where the thermal strain can be neglected because of the low values of the thermal dilation coefficients reported above. As a conclusion, in contrast to the case examined in the previous section, as soon as the liquid, such as water, expands significantly when transforming into a sealed porous solid, the volumetric strain of an initially water-saturated sample subjected to freezing is mainly caused by the buildup of the hydraulic pressure p_L resulting from the liquid water expelled from the freezing sites. This analysis might lead one to think that the pressure difference $p_C - p_L$ between the liquid and the solid permitted by the Thomson Equation (9.86) does not actually have any influence. In fact, it does have a significant influence since we are now going to derive how the Thomson equation governs the yet unknown crystal saturation S_C and, therefore, the liquid pressure p_L through Equation (9.92). Keeping in mind that the atmospheric pressure has been taken as the zero reference pressure, we substitute Equation (9.92) in Equation (8.56) to derive the shift in the melting point due to the liquid pressurization:

$$T_m - T_{0m} = S_C \frac{K_L K_C \left(1 - \overline{V}_L^0/\overline{V}_C^0\right)^2}{(S_L K_C + S_C K_L)\Delta S_m}. \tag{9.93}$$

Substitution of Equation (9.93) in Equation (9.5) gives the current pore-access radius $r_{\Delta \mu}$, whose substitution in Equation (9.3) gives us the relation linking S_C and the current cooling $T_m - T$:

$$S_C = 1 - S_L = 1 - S(r)$$

$$T_m - T = [1 - S(r)] \frac{K_L K_C \left(1 - \overline{V}_L^0/\overline{V}_C^0\right)^2}{(S(r) K_C + [1 - S(r)] K_L)\Delta S_m} + \frac{2\gamma_{CL} \cos\theta}{r \Delta S_m}, \tag{9.94}$$

where $1 - S(r)$ is the cumulative volume fraction of the pore volume having a pore-entry radius greater than r. Equation (9.94) gives the crystal saturation degree S_C as a function of the cooling $T_m - T$, in a parametric form with respect to the current pore-entry radius allowing the liquid water to transform into ice. This is illustrated in Figure 9.9 for the cement paste whose function $1 - S(r)$ was reported in Figure 1.2. Figure 9.9 shows the crystal saturation S_C and the hydraulic pressure $S_L p_L + S_C p_C \simeq p_L$ plotted against the cooling $T_m - T$. In order to assess the shifts resulting from the liquid pressure, the crystal saturation S_C is also plotted against the cooling $T_m - T$ when assuming no liquid pressurization ($p_L = 0$).

---

[11] The sample may be naturally sealed by the ice forming at $T = T_m$ on its external surfaces and preventing liquid water from escaping during freezing.

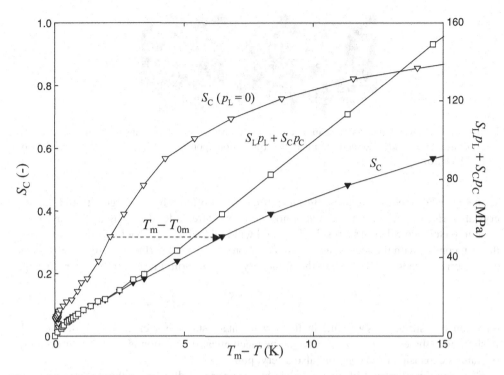

Figure 9.9 The high hydraulic pressure induced by the confinement causes a shift of the melting point by several K, which significantly limits the extent of the frozen zone; this shift and the induced limitation of the solid water saturation are shown by comparing the upper curve, where $p_L = 0$, with the lower curve related to confined conditions causing the indicated pore overpressure $S_L p_L + S_C p_C$

The results above are related to an initially liquid-saturated porous solid, so that there is no possibility for the liquid water from the freezing sites to escape from the pores. As shown in Figure 9.9 the intensity of the liquid pressure is so great that usual porous solids cannot sustain it, leading to an irreversible damage of the material. In contrast, if liquid water can escape from the pores, the liquid pressure buildup will ultimately vanish. This is the topic of the next section.

9.3.3 Freezing and air voids

The discovery that a distributed network of small air voids in cement-based materials enhanced their resistance to frost was accidentally observed by T. C. Powers.[12] Indeed, air voids appropriately spaced within cement-based materials have long been observed to limit the expansion. These air voids act as expansion reservoirs and limit the pressure buildup in the still unfrozen solution. When the liquid enters an air void whose pressure is atmospheric it is no longer confined

---

[12] See Powers, T. C (1949) The air requirement of frost-resistant concrete. *Highway Research Board, PCA Bulletin*, **33**, 184–211. Treval Clifford Powers (1900–1997) is nowadays recognized as the father of the modern science of cement-based materials.

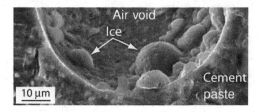

Figure 9.10 Air voids incorporated in a cement paste operate as both expansion reservoirs and cryopumps; as a result, they efficiently mitigate the frost action upon cement-based materials (courtesy of Paulo Monteiro)

so that it instantaneously freezes. Because of the space offered by the large air void, the ice crystals formed in this way remain at atmospheric pressure, at least as long as the air void is not completely ice-filled. As a result, the liquid within the immediate vicinity of the void and, thus, in contact with the ice already crystallized on the air void surface where $p_C = 0$, must depressurize because it has to meet the liquid–crystal equilibrium Equation (9.86), requiring

$$p_L = - \left(\overline{V}_C^0 / \overline{V}_L^0 \right) \Delta S_m (T_m - T) \qquad (9.95)$$

on the air void surface. As a result, the liquid water at a distance is sucked towards the air void. In short, the efficiency of air voids against the mechanical frost action is twofold: they mainly operate as expansion reservoirs but also as cryopumps.

This twofold efficiency of air voids has been confirmed directly in the experiment[13] reported in Figures 9.10 and 9.11. After subjecting a sample of water-saturated cement paste to a continuous temperature decrease down to some degrees below 0 °C, the cooling was suddenly stopped and the temperature was later on maintained at a constant value. Figure 9.10 illustrates how liquid water was observed to enter the air void and solidify into ice. Figure 9.11 shows how, in spite of stopping the temperature decrease, ice was still observed to continue filling up the air void progressively, whereas the void was simultaneously shrinking. The delayed nature of the void shrinkage regarding the stopping of the temperature decrease shows that the process involved a coupling between the flow through the porous solid towards the void and the deformation of the latter. This delayed nature also shows that the intensity of the hydraulic pressure depends on the cooling rates compared with the ability of the porous material to drain the extra liquid water expelled from the freezing sites towards the nearby air voids. Let us examine further how the intensity of all the above depicted phenomena can be quantitatively assessed.

We now consider a water–air infiltrated material, whose air voids are much larger than the water-filled pores of the surrounding porous matrix. A representative sample of such a porous material consists of a spherical air void of radius R, embedded in a spherical shell of porous material of outer radius $R + L$, so that the distance L scales the spacing between two adjacent air voids. This sample of porous material is subjected to a uniform cooling $T_m - T > 0$. On the outer radius $\varrho = R + L$, the radial stress $\sigma_{\varrho\varrho}$ is zero, and there is no liquid flow. In addition, as previously explained, at $\varrho = R$ the liquid immediately in contact with the ice already formed in the air void depressurizes at the liquid pressure given by Equation (9.95). Collecting these

---

[13] See Mehta and Monteiro (2006) referred to in the Further Reading Section.

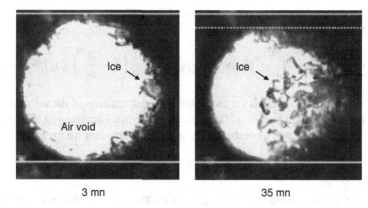

Figure 9.11 In spite of stopping the temperature decrease, ice is still observed to continue filling up the air void, while the void is observed to shrink simultaneously (courtesy of Paulo Monteiro)

mechanical and hydraulic boundary conditions, we write

$$\sigma_{\varrho\varrho}|_{\varrho=R} = 0; \quad \sigma_{\varrho\varrho}|_{\varrho=R+L} = 0; \quad (9.96a)$$

$$p_L|_{\varrho=R} = -\left(\overline{V}_C^0/\overline{V}_L^0\right) \Delta S_m (T_m - T); \quad \frac{\partial p_L}{\partial \varrho}\bigg|_{\varrho=R+L} = 0. \quad (9.96b)$$

In the presence of an air void the stress field is no longer uniform as it was in the previous section. In the absence of body forces the overall momentum Equation (3.35) reads

$$\nabla \cdot \underline{\underline{\sigma}} = 0. \quad (9.97)$$

In the momentum balance Equation (9.97) the strain tensor $\underline{\underline{\varepsilon}}$ can be expressed with the help of the displacement $\underline{\xi}$ by using Equation (3.14). According to the analysis of the previous section, we anticipate that the thermal terms are negligible in the poroelastic constitutive Equation (9.83a). So combining Equations (3.14), (9.83a), (9.83d) and (9.97), while still assuming the pore volumetric isodeformation, namely $b_J = bS_J$, we extend the Navier Equation (5.53) to unsaturated conditions in the form

$$\left(K + \frac{4}{3}G\right)\nabla\epsilon - G\nabla \times (\nabla \times \underline{\xi}) - b\nabla(S_L p_L + S_C p_C) = 0. \quad (9.98)$$

Because of the problem symmetry, the only nonzero displacement component is the radial displacement $u(\varrho, t)$, so that the volumetric strain can be expressed in the form

$$\epsilon = \frac{1}{\varrho^2}\frac{\partial}{\partial \varrho}\left(\varrho^2 u\right). \quad (9.99)$$

Because the displacement is purely radial, it is irrotational, resulting in $\nabla \times \underline{\xi} = 0$. The differential Equation (9.98) can therefore be integrated according to

$$\epsilon = b\frac{S_L p_L + S_C p_C}{K + 4G/3} + f(t), \quad (9.100)$$

where $f(t)$ is a still unknown integration function of time. We put Equation (9.99) in Equation (9.100) and integrate the resulting equation:

$$\frac{u(\varrho,t)}{\varrho} = \frac{b}{(K+4G/3)\varrho^3}\int_R^\varrho \rho^2(S_L p_L + S_C p_C)d\rho + \frac{1}{3}\left(1 - \frac{R^3}{\varrho^3}\right)f(t) + \frac{R^3}{3\varrho^3}\epsilon_{\text{void}}(t), \quad (9.101)$$

where $\epsilon_{\text{void}}(t) = 3u(R,t)/R$ stands for the volumetric deformation of the air void. Since the expression of the radial strain is $\varepsilon_{\varrho\varrho} = \partial u/\partial\varrho$, the constitutive Equations (9.83a) and (9.83d) and Equations (9.99) and (9.100), combine to give us the expression for the radial stress $\sigma_{\varrho\varrho}$:

$$\sigma_{\varrho\varrho} = (K+4G/3)f(t) - 4G\frac{u}{\varrho}. \quad (9.102)$$

The mechanical boundary conditions of Equation (9.96a) and the displacement and stress expressions of Equations (9.101) and (9.102) allow us to determine simultaneously the time function $f(t)$ and the air void volumetric deformation $\epsilon_{\text{void}}(t)$ in the form

$$f(t) = \frac{4}{3}\frac{G}{K+4G/3}\epsilon_{\text{void}}(t) \quad (9.103)$$

and

$$\epsilon_{\text{void}}(t) = \frac{b\langle S_L p_L + S_C p_C\rangle}{K}, \quad (9.104)$$

where $\langle S_L p_L + S_C p_C\rangle$ stands for the mean pore pressure:

$$\langle S_L p_L + S_C p_C\rangle = \frac{1}{(4\pi/3)[(R+L)^3 - R^3]}\int_R^{R+L} 4\pi\rho^2(S_L p_L + S_C p_C)d\rho. \quad (9.105)$$

The time history for the deformation depends upon the history of the pore overpressure field $S_L p_L + S_C p_C$ and, eventually, on the liquid overpressure p_L because of the Thomson Equation (9.86). In turn, the determination of p_L requires us to analyze the coupling between the liquid flow through the porous solid and the deformation of the latter. Since ice flows very slowly when compared with the usual cooling timescales, the flow involves only liquid water. When using liquid Darcy's law of Equation (9.42), the water continuity equation, which expresses the mole conservation of water, whatever its solid or liquid form, can finally be written in the form

$$\frac{dn}{dt} = -\frac{1}{\overline{V}_L^0}\nabla\cdot\underline{q}_L = \frac{1}{\overline{V}_L^0}\frac{\varkappa}{\eta_L}\frac{1}{\varrho^2}\frac{\partial}{\partial\varrho}\left[k_{rL}(S_L)\varrho^2\frac{\partial p_L}{\partial\varrho}\right]. \quad (9.106)$$

The water mole density n is expressed by Equation (9.91). Because of the poor compressibility of the porous solid compared with that of water, an analysis of the order of magnitude of the various contributions to n (similar to that carried out in the previous section) leads us to neglect the term $b\sigma/K\overline{V}_L^0$ in Equation (9.91). When disregarding this term, the diffusion equation governing p_L is derived by substituting Equation (9.91) in Equation (9.106):

$$\frac{\partial}{\partial t}\left(\frac{S_L K_C + S_C K_L}{K_C K_L}p_L\right) = \frac{\partial}{\partial t}\left[\left(1 - \frac{\overline{V}_L^0}{\overline{V}_C^0}\right)S_C\right] + \frac{\varkappa}{\phi_0\eta_L}\frac{1}{\varrho^2}\frac{\partial}{\partial\varrho}\left[k_{rL}(S_L)\varrho^2\frac{\partial p_L}{\partial\varrho}\right]. \quad (9.107)$$

The diffusion Equation (9.107) has now to be associated with the hydraulic boundary conditions of Equation (9.96b). The diffusion Equation (9.107) shows that the time history of the liquid pressure is governed by the competition between the source term resulting from the liquid expelled from the freezing sites during the liquid–solid transition – the first term on the right-hand side of Equation (9.107) – and the capacity of the liquid flow to evacuate this liquid source towards the air void – the second term on the right-hand side of Equation (9.107).

However, in order to determine the intensity of the liquid pressure from Equation (9.107), we now have to assess the source term and, therefore, determine how the crystal saturation degree, S_C, is governed. Again bearing in mind that the atmospheric pressure has been taken as the zero reference pressure, the Thomson Equation (8.56) gives the local shift $T_m - T_{0m}$ of the melting point in the form

$$T_m - T_{0m} = \left(1 - \overline{V}_L^0/\overline{V}_C^0\right) \frac{p_L}{\Delta S_m}. \tag{9.108}$$

Substitution of Equation (9.108) in Equation (9.5) provides the local current pore-access radius $r_{\Delta\mu}$, whose substitution in Equation (9.3) gives us the relation relating the local crystal saturation degree $S_C(\varrho, t)$ to both the current cooling $T_m - T$ and the local liquid pressure $p_L(\varrho, t)$. The relation is expressed in a parametric form with regard to the current local pore-entry radius $r(\varrho, t)$:

$$S_C = 1 - S(r); \quad T_m - T = \left(1 - \overline{V}_L^0/\overline{V}_C^0\right) \frac{p_L}{\Delta S_m} + \frac{2\gamma_{CL} \cos\theta}{r \Delta S_m}. \tag{9.109}$$

We can now quantify the effect of the cooling rate on the deformation of a freezing porous solid containing air voids. First let $\overset{o}{T}$ be the cooling rate such that

$$T_m - T = \overset{o}{T} t. \tag{9.110}$$

The time-dependent liquid pressure field, $p_L(\varrho, t)$, can be determined by solving the set of Equations (9.107), (9.109) and (9.110), while the mean pore pressure $\langle S_L p_L + S_C p_C \rangle$ is known from its expression given in Equation (9.105) and from Equation (9.86) linking p_C and p_L. With the aim of determining $p_L(\varrho, t)$ numerically we select the cooling rate $\overset{o}{T} = 3\,\mathrm{K\,h^{-1}}$ corresponding to *in situ* values. For the pore volume fraction $1 - S(r)$ involved in Equation (9.109), as in the previous section we choose the one reported in Figure 1.2 and related to a cement paste whose porosity is $\phi_0 = 0.34$. In addition to the values of the water properties recalled in the caption of Figure 9.1, together with the values previously given for the bulk moduli K_L and K_C, we take for the value of the water viscosity $\eta_L \simeq 1.79 \times 10^{-3}\,\mathrm{Pa\,s}$. We also take the expression provided by Equation (6.131) and $m = 0.55$ for the relative permeability k_{rL} as a function of the liquid saturation S_L (see Figure 6.12). The value retained for the air void radius is $R = 50\,\mu\mathrm{m}$, while the spacing length is $L = 300\,\mu\mathrm{m}$. Using these values, we finally find that the rate of the liquid flow towards the air void governed by the diffusion Equation (9.107) is scaled by the characteristic time τ given by

$$\tau = \frac{\phi_0 \eta_L L^2}{\varkappa K_L} \simeq \frac{2.5 \times 10^{-20}}{\varkappa\,(\mathrm{m}^2)}\,(\mathrm{s}). \tag{9.111}$$

Figure 9.12 shows the mean pressure $\langle S_L p_L + S_C p_C \rangle$ plotted against the cooling $T_m - T$ for various values of the intrinsic permeability $\varkappa$. For an infinitely permeable material or,

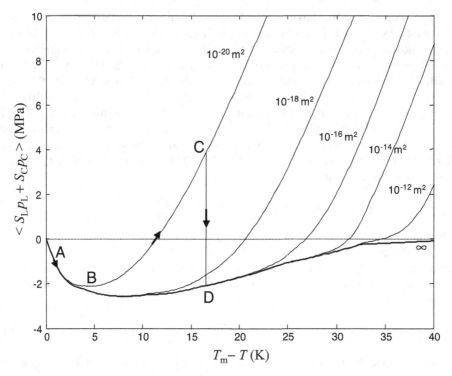

Figure 9.12 Mean pore pressure plotted against cooling for a cooling rate equal to 3 K h^{-1} and for various values of the intrinsic permeability $\varkappa$ ranging from 10^{-20} m^2 to infinity

equivalently, for an infinitely slow cooling, the thermodynamic equilibrium is achieved at any time in the shell surrounding the void. Accordingly, the crystal pressure is zero everywhere in the shell, and the mean pressure reduces to the liquid pressure term $S_L p_L$. Because of the Thomson equation, the liquid is depressurized, with $p_L < 0$, resulting in the material shrinkage, with $\epsilon_{\text{void}}(t) < 0$ in Equation (9.104). As the cooling carries on increasing and, therefore, ice carries on forming, the liquid saturation S_L decreases to zero. As a result, at some stage of the cooling, the mean pressure $S_L p_L$ reaches a minimum and goes back to zero. Simultaneously, the air void stops shrinking and starts expanding to recover its original size.

If the intrinsic permeability $\varkappa$ decreases or, equivalently, the cooling rate $\overset{\circ}{T}$ increases, there is competition between the pore pressure buildup, which is associated with the freezing of the liquid water within the porous shell, and the flow of liquid water towards the air void, that tends to make the pore pressure buildup vanish. Nevertheless, usual permeability values are larger than 10^{-20} m^2, and the values given in Equation (9.111) for the characteristic time τ scaling the flow rate remain quite small, when compared with the usual *in situ* cooling rates ~ 3 K h^{-1}. As a result, as shown in Figure 9.12, even for small values of the intrinsic permeability, when the cooling process begins the mean pressure curve remains confused with that related to an infinite permeability or, equivalently, to infinitely slow coolings. However, as the cooling goes on increasing and liquid water progressively solidifies, the liquid saturation goes to zero and the liquid permeability $\varkappa k_{rL}(S_L)$ significantly decreases. This permeability decrease prevents the liquid expelled from the freezing sites from flowing towards the air void. Conditions similar to the undrained ones which we have examined in the previous section progressively establish within the shell, from the outer limit $\varrho = R + L$ towards the air void surface $\varrho = R$, ultimately resulting in the increase of the mean pore pressure. As expected and shown in Figure 9.12, the lower the intrinsic permeability $\varkappa$, the sooner the mean pressure curve starts deviating from the curves related to an infinitely slow cooling rate – point A – and finally starts increasing – point B. If the cooling now suddenly stops, an efficient liquid flow can progressively reestablish towards the air void. Therefore, as represented in Figure 9.12 by the path CD, the mean pore pressure will again decrease and ultimately reach the value corresponding to an infinitely permeable porous solid or, equivalently, an infinitely slow cooling. This mean pore pressure decrease will result in the observed shrinkage of Figure 9.11.

In practice the appropriate spacing factor L for mitigating the frost action on cement-based materials has to meet the condition that the initial volume of the void is capable of welcoming all the liquid water expelled from the adjacent freezing sites because of the ice–liquid density difference. Accordingly, there will be no in-void pressure buildup in the long term. The zone of influence of a void is the spherical shell with inner radius R, outer radius $R + L$ and volume $V_{\text{shell}} = (4\pi/3)\left[(R+L)^3 - R^3\right]$. At equilibrium related to the cooling $T_{\text{m}} - T$ the volume change due to the solidification of water within the shell is $\left(\overline{V}_{\text{C}}^0/\overline{V}_{\text{L}}^0 - 1\right)\phi_0 S_{\text{C}}(T_{\text{m}} - T) V_{\text{shell}}$ where $S_{\text{C}}(T_{\text{m}} - T)$ is given by the curve such as the one shown in Figure 9.9. As a result there will be no pressure buildup within the void in the long term if this volume change is less than $4\pi R^3/3$.

9.3.4 Weathering and the crystallization of sea-salts

The crystallization of sea-salts induced by successive imbibition–drying cycles constitutes a major weathering phenomenon in dry environments close to the sea. As illustrated in Figure 1.5 it often leads to the serious deterioration of porous sedimentary rocks used for building in coastal areas. The successive imbibitions of a stone by a fresh salted solution increase the supersaturation of the current solution, resulting in the progressive in-pore crystallization of sea-salts. The drying period following each imbibition enhances the salt concentration of the residual solution, so that the crystal can also grow while drying. We now address the salt crystallization in a porous solid and the associated mechanical effects originating from a supersaturated salted solution, which initially saturates the stone and whose salt content is known. Increasing the salt content of this saturating solution amounts to accounting for the increase of the salt content of the solution filling the porosity of the stone after each new imbibition period.

A comparison of the Correns Equation (8.74) with the Thomson Equation (8.57) shows that, regarding salt crystallization, the quantity $(RT/\overline{V}_{\text{C}}) \ln[x/x_{\text{0sat}}(T)]$ plays the same role as the one played by $\Delta S_{\text{m}}(T_{\text{0m}} - T)$ in the freezing process. Equation (9.5) can then be adapted for salt in-pore crystallization in the form

$$r_{\Delta\mu} = \frac{2\gamma_{\text{CL}} \cos\theta}{(RT/\overline{V}_{\text{C}}) \ln[x/x_{\text{0sat}}(T)]}, \quad (9.112)$$

whose substitution in Equation (9.3) gives the liquid saturation S_L as a function of the current solute molar fraction x. In the previous relation x stood for the molar fraction of the residual solution after the crystallization has occurred. Because of the conservation of the overall salt mass, x is linked to the initial molar fraction x_0 related to the saturating salted solution before crystallization through the relation

$$\frac{S_C}{\overline{V}_C} + \frac{x S_L}{(1-x)\overline{V}_W + x\overline{V}_C} = \frac{x_0}{(1-x_0)\overline{V}_W + x_0\overline{V}_C}. \quad (9.113)$$

In Equation (9.113) $\overline{V}_W$ is the molar volume of pure water; $\overline{V}_C$ stands for both the molar volume of the salt crystals and for the molar volume of salt in pure solute form prior to the mixing with the solvent. This amounts to assuming a zero dilation coefficient δ in Equation (8.68). As a result, Equation (8.73) shows that $x_{0\mathrm{sat}} = x_{\mathrm{sat}}$: so there is no change in the solubility when the crystal and the solution are subjected to the same pressure. Neglecting the variations of $\overline{V}_C$ with the crystal deformation, and recalling that $S_C + S_L = 1$, the molar fraction x of the residual solution and the solid saturation S_C can be determined as a function of x_0 by solving the set of Equations (9.3), (9.112) and (9.113). Once x has been determined, assuming that the solution pressure p_L remains at the zero reference atmospheric pressure, the crystal pressure p_C is then known directly from the Correns Equation (8.72). As x_0 increases p_C also increases, even in the sites where salt crystallization can occur for a lower value than the current value of x_0. As a result, as x_0 increases, a greater volume of crystal has to form in each pore, in order constantly to adjust the crystal pressure to the value required by the current value of x_0, according to both the salt mass conservation expressed by Equation (9.113) and the Correns Equation (8.72).[14] This is quite similar to the phenomenon governing the cryogenic swelling examined in Section 9.3.1.

As x_0 increases, the crystal pressure p_C applied to the internal solid walls delimiting the pores generates an increasing tensile stress within the solid matrix of the stone. This tensile stress, when it exceeds the tensile strength of the solid matrix, causes the stone failure. With the aim of determining the molar fraction x_0 causing the failure, we can proceed as we did in Section 7.3.3. Assuming that the stone is stress-free, the free energy a_S^* reduces to the sole contribution of the internal crystal pressure p_C, and can finally be expressed in the form

$$a_S^* = \frac{1}{2} p_C \varphi_C = \frac{1}{2K}\left(b_C^2 + \frac{K}{N_{CC}}\right) p_C^2. \quad (9.114)$$

Adapting the approach of Section 7.3.3, the stone failure can be determined from Equations (7.47) and (7.49), and we obtain

$$\varpi = \frac{RT}{\overline{V}_C} \ln \frac{x}{x_{\mathrm{sat}}(T)} \sqrt{\frac{1}{2}\left(b_C^2 + \frac{K}{N_{CC}}\right)} = \varpi_{\mathrm{cr}}. \quad (9.115)$$

In Equation (9.115) ϖ scales the intensity of the tensile stress which the crystal pressure p_C induces in the solid matrix, while ϖ_{cr} stands for the tensile strength of the solid matrix. Adopting the function $S(r)$ reported in Figure 6.11 for a Berea sandstone, in Figure 9.13

---

[14] This greater crystallized volume is made possible by the pore volume change due to the deformation. Its order of magnitude is therefore much smaller than the size of the pore so that it does not have to be taken into account in Equation (9.113) accounting for the conservation of the overall salt mass.

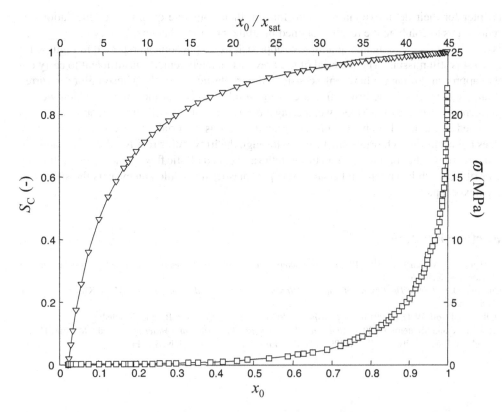

Figure 9.13 Solid crystal saturation S_C (upper curve) and solid matrix tensile stress ϖ (lower curve) plotted against the salt molar fraction x_0 of the invading saturating solution before crystallization; (the data used for the computation is that reported for a Berea sandstone in Figure 6.11)

the crystal saturation degree S_C and the tensile stress intensity ϖ are plotted against x_0. The values of the properties related to the salt are those of a sodium sulfate salt,[15] namely $\overline{V}_C = 220 \, \text{cm}^3 \, \text{mol}^{-1}$, $x_{sat} = 2.235 \times 10^{-2}$ and $\gamma_{CL} = 100 \, \text{mJ} \, \text{m}^{-2}$. The poroelastic properties of the Berea sandstone are $\phi_0 = 0.228$, $b = 0.817$ and $K = 5.3 \times 10^3 \, \text{MPa}$. To assess the values of b_C and N_{CC} we used the pore isodeformation Equation (7.42) and Equation (7.46), with $k_S = 28.9 \times 10^3 \, \text{MPa}$ and $g_S = 1.54 \times 10^3 \, \text{MPa}$. The increase of x_0 accounts for that of the salt molar fraction of the saturating solution at the end of a new imbibition. Knowing the value ϖ_{cr} of the tensile strength of the solid matrix, and the value of the salt molar fraction of the invading solution, the critical number of imbibition–drying cycles which the Berea sandstone could sustain can then be calculated from Figure 6.11.

The attentive reader will have noticed that many of the intriguing phenomena depicted in the first chapter of this book devoted to the strange world of porous solids have by now been investigated in detail. For most of them, the same attentive reader has had to wait for this

---

[15] These values are those of mirabilite which is a typical sodium sulfate. For details see Flatt, R. (2002) Salt damage in porous materials: how high supersaturations are generated. *Journal of Crystal Growth*, **242**, 435–454.

chapter for their definitive clarification. Indeed, their complete quantitative elucidation has required us to combine the results obtained in all the previous chapters. However, the journey is not over. A convincing and quantitative approach to the phenomena reported in Figures 1.3 and 1.4 is still missing. The material of the previous chapters remains insufficient to carry out this approach. So far we have only considered poroelastic materials, whose failure occurred dramatically through the brittle fracture of the solid matrix as in the analysis which we just performed related to the stone weathering due to sea-salt crystallization. The phenomena involved in Figures 1.3 and 1.4 involve granular porous materials, whose failure obviously does not occur through brittle fracture, but through the irreversible relative sliding of the solid grains forming the solid matrix and through the associated plastic flow of the porous solid they constitute. Such behavior can be said to be poroplastic. Its detailed analysis is the subject of the next chapter.

Further Reading

Brinker, J. C. and Scherer, G. W. (1990) *Sol-gel Science : the Physics and Chemistry of Sol/gel Processing,* Elsevier Science & Technology.

Cushman, J. H. (1997) *The Physics of Fluids in Hierarchical Porous Media: Angstroms to Miles,* Kluwer Academic Publishers.

Hall, C. and Hoff, W. D. (2000) *Water Transport in Brick, Stone and Concrete,* Brunner-Routledge.

Mehta, P. K. and Monteiro, P. (2006) *Concrete, Microstructure, Properties, and Materials,* 3rd edn, McGraw Hill.

Petrenko, V. F. and Whitworth, R. W. (1999) *Physics of Ice,* 1st edn, Oxford University Press.

10

The Poroplastic Solid

In the introduction to his book *The Mathematical Theory of Plasticity*, whose first edition was published in 1950 when he was then still in his 28th year, Rodney Hill (born in 1921), one of the pioneers of classical plasticity, wrote: 'Theory of plasticity is the name given to the mathematical study of stress and strain in plastically deformed solids, especially metals (...). At the present time metals are the only plastic materials for which there is enough data to warrant the construction of a *general* theory. For this reason the theory is related specifically to the properties of metals, though it may apply to other potentially plastic materials (e.g. ice, clay, rock).'

Plasticity is the property which a material has to exhibit permanent strains having been loaded beyond some stress threshold, and later on unloaded. The question 'when does plastic deformation occur?' can be answered by specifying the yield locus delimiting the elastic domain in the appropriate stress space. If the celebrated names of Charles Augustin Coulomb (1736–1806), Henri Édouard Tresca (1814–1885) and Richard von Mises (1883–1953) were early on associated with the yield loci related to soils and metals, these yield loci can constitute only a part of the constitutive equations of plasticity. For a given plastic material the plastic yield locus has to be completed by the plastic flow rule which answers the question 'how does plastic strain occur?', and provides the plastic strain rate as a function of stress. At the time when Hill published his masterly book, the origin of plasticity for metals had been progressively known as resulting from the irreversible relative sliding of crystalline planes, enhanced by the motion of dislocations. To elaborate the corresponding plastic flow rule, Hill precisely formulated the elegant principle of maximum plastic work, which can be interpreted as a principle of maximum production of entropy.

The introduction to his book concerning the general applicability of the theory of plasticity to materials other than metals was premonitory. In the late 1940s and early 1950s, Kenneth Harry Roscoe (1914–1970) of Cambridge University developed testing techniques to study the yielding of sand and clays under shear conditions. The data he obtained, and further triaxial tests performed at Imperial College, led Roscoe, Schofield (born in 1930) and Wroth (1929–1991) to publish a seminal paper in 1958 introducing the critical state concept.[1] This concept states that soils, if they are continuously sheared, will ultimately come into a well-defined critical state where they – plastically – flow as a frictional fluid. In the critical state shear distortions occur without any further change in the stress and specific volume. Plasticity of soils results from the irreversible sliding of so many particles forming the solid skeleton that the physical connection with the microscopic scale is not as straightforward as for metals. Basically, right from its formulation, the critical state concept, which avoids going down to the solid particle scale, was going to open the way to intensive research into soil plasticity. However, the presence of water in soil porosity requires us to involve, rather than the stress itself, the effective stress concept, which we have already encountered in Chapters 3, 4 and 7 devoted to saturated and unsaturated poroelastic solids. Nowadays, the extension of the effective stress concept to plastic solids such as soils is still debated for unsaturated conditions.

This chapter is an invitation to explore the extension of poroelasticity to poroplasticity, by applying the concepts we have just mentioned. Plasticity is particularly relevant to account for the constitutive equations of metals and soils. However, we believe that the general theory of plasticity can prove to be useful for addressing mechanical behaviors involving irreversible strains whatever their origin, providing the *in situ* material behavior shows some ductility before its possible ultimate brittle collapse. For instance, modern ultra high-performance concrete behaves that way, even if the microscopic processes to which plasticity can be attributed are not fully identified.

10.1 Basic Concepts of Plasticity

10.1.1 Plastic loading function and flow rule

Plasticity is the ability of materials to undergo permanent strains after removing the load to which they were subjected. Plastic deformation occurs when the stress reaches a threshold. The simplest plastic model consists of representing this threshold by a friction element as shown in Figure 10.1(a): the friction strength C or tangential cohesion restricts the admissible stresses τ to a domain $\mathcal{D}$ expressed in the form of a scalar plastic loading function $f(\tau)$:

$$\tau \in \mathcal{D}: \quad f(\tau) = |\tau| - C \leq 0. \tag{10.1}$$

When the applied stress intensity $|\tau|$ is below the friction strength C, the friction element stays at rest and its irreversible sliding or plastic strain γ^p cannot occur. We then write

$$f(\tau) < 0: \quad d\gamma^p = 0. \tag{10.2}$$

In turn, when the stress reaches the stress plastic threshold C, the friction element can slide. Since a friction element cannot store any free energy, the displacement γ reduces to the plastic strain γ^p while, according to the second law of thermodynamics, the whole infinitesimal plastic work $\tau d\gamma^p$ given to the system is transformed into heat. This is expressed in the form

$$\delta W^p = \tau d\gamma^p \geq 0, \tag{10.3}$$

---

[1] Roscoe, K. H., Schofield, A. N. and Wroth, C. P. (1958) On the Yielding of Soils. *Géotechnique*, **8**, 22–53.

The Poroplastic Solid

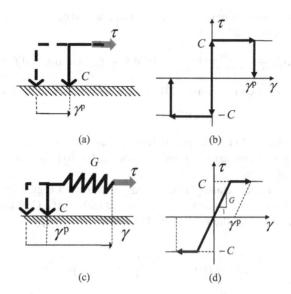

Figure 10.1 (a) The friction element (left) and (b) its constitutive law in the (γ, τ) plane; (c) the elastoplastic element and (d) its constitutive law in the (γ, τ) plane. In both cases the domain D_E of admissible stresses, here $|\tau| - C < 0$, remains unaffected by a previous loading – plasticity is said to be ideal

where δW^P is the so-called plastic work. The intensity $|d\gamma^P|$ of the infinitesimal current plastic strain $d\gamma^P$ remains undetermined. Only its direction – that of the applied stress τ – is fixed. Noting that $\partial f / \partial \tau = \text{sign}(\tau)$, this allows us to write

$$f(\tau) = 0: \quad d\gamma^P = d\Lambda \frac{\partial f}{\partial \tau}; \qquad d\Lambda \geq 0, \tag{10.4}$$

where $d\Lambda$ is the nonnegative plastic multiplier. Note that the occurrence of the plastic strain is not related to a timescale associated with the plastic strain, but to the stress history or chronology. This is why we prefer reasoning in terms of infinitesimal increments like $d\gamma^P$, rather than in rates like $d\gamma^P/dt$.

The cumulative permanent or plastic strain γ^P is irreversible. As illustrated in Figure 10.1(b), when the applied stress τ decreases below the stress strength C, the friction element stops and the displacement γ^P remains. For the plastic strain to occur, the intensity τ of the applied stress must not only be on the plastic loading surface defined by $f(\tau) = 0$, but must remain constantly equal to the stress strength C. This is accounted for through the plastic consistency condition

$$df = \frac{\partial f}{\partial \tau} d\tau = 0. \tag{10.5}$$

In contrast, when the intensity of the applied stress is on the plastic loading surface, but leaves it to reenter the interior of the domain of admissible stresses, so that $df < 0$, the increment $d\gamma^P$ becomes zero. We write

$$f(\tau) = 0; \quad df < 0: \quad d\gamma^P = 0. \tag{10.6}$$

Interestingly, the constitutive Equations (10.2), (10.4) and (10.6) of the friction element can be summarized in the condensed form

$$d\gamma^p = d\Lambda \frac{\partial f}{\partial \tau}; \quad f \leq 0; \quad d\Lambda \geq 0; \quad d\Lambda \times f = 0; \quad d\Lambda \times df = 0, \qquad (10.7)$$

while the plastic work defined in Equation (10.3) can be expressed in the form

$$\delta W^p = C \left| d\gamma^p \right|. \qquad (10.8)$$

The first part of Equation (10.7) is the plastic flow rule and the four other conditions, which represent the plastic loading–unloading conditions of the friction element, are often referred to as Kuhn–Tucker conditions.

To elaborate on the simplest elastoplastic system, as shown in Figure 10.1(c), the friction element can be assembled with a linear spring of constant stiffness G. Considering the possible existence of an initial stress τ_0, the isothermal Helmholtz free energy a_S of the elastoplastic system is given by

$$a_S = \tau_0 \left(\gamma - \gamma^p \right) + \frac{1}{2} G \left(\gamma - \gamma^p \right)^2. \qquad (10.9)$$

The second law of thermodynamics applied to the system leads to the inequality

$$\tau d\gamma - da_S \geq 0. \qquad (10.10)$$

When no plastic flow occurs, as for instance in an unloading process, we have $d\gamma^p = 0$. There is no dissipation, so that the inequality of Equation (10.10) reduces to an equality. Substituting Equation (10.9) in Equation (10.10), we obtain

$$\tau = \frac{\partial a_S}{\partial \gamma} = \tau_0 + G \left(\gamma - \gamma^p \right). \qquad (10.11)$$

In turn, substituting state Equation (10.11) in Equation (10.10), we recover the inequality of Equation (10.3), so that the flow rule of Equation (10.7) still applies for the elastoplastic system.

10.1.2 The principle of maximum plastic work

Equation (10.3) for the plastic work δW^p designates the stress τ as the driving force of the plastic rate $d\gamma^p$. Accordingly, the flow rule has to provide us with the relation linking τ and $d\gamma^p$. For the friction element of Figure 10.1(a) this relation, given by Equation (10.7), can be derived from simple considerations. In preparation for its extension to complex systems, the flow rule of Equation (10.7) can be shown to derive from a more general principle, namely, the principle of maximum plastic work.

Any plastically admissible stress τ must lie within – or on the boundary of – the domain $\mathcal{D}$ of admissible stresses which is defined by $f(\tau) \leq 0$ and includes the zero stress, so that $f(\tau = 0) \leq 0$. The principle of maximum plastic work can be expressed as follows. Compared with any plastically admissible stress τ^*, the actual stress τ associated with the plastic displacement $d\gamma^p$ produces the maximum plastic work:

$$\forall \tau^* : f(\tau^*) \leq 0; \quad (\tau - \tau^*) d\gamma^p \geq 0. \qquad (10.12)$$

Since the plastic work δW^p here identifies with the dissipation, that is, the work which is spontaneously transformed into heat, the principle of maximum plastic work turns out to be a principle of maximum production of entropy. Therefore, this principle can be understood as a principle of noneconomy of matter: among all possible plastically admissible stresses τ^*, the actual stress τ associated with the plastic strain $d\gamma^p$ is the one that maximizes the dissipation (into heat) of the supplied mechanical energy $\tau d\gamma$. While somewhat obvious for the friction element, later on we will see that this principle is a suitable concept to extend the flow rule of Equation (10.7) to complex plastic systems and materials. In fact, the principle of maximum plastic work stated in Equation (10.12) will be shown to combine the two key elements of the plasticity model: the normality of flow rule and the convexity of the domain $\mathcal{D}$ of admissible stresses.

To illustrate this consider for a while that τ lies within $\mathcal{D}$, so that $f(\tau) < 0$. Since $\tau - \tau^*$ may be positive or negative, the only possibility for satisfying Equation (10.12) in this case is $d\gamma^p = 0$. Hence, Equation (10.12) includes Equation (10.2). Next, since the zero stress $\tau^* = 0$ is plastically admissible, we can take $\tau^* = 0$ in Equation (10.12). This allows us to recover the positivity of the plastic work δW^p, as stated in Equation (10.3).

As a result of Equation (10.3), when τ lies on the boundary of $\mathcal{D}$ (that is when $f(\tau) = 0$) a nonzero value of $d\gamma^p$ must take the sign of the stress τ. If τ is positive, say $\tau = \tau^+ > 0$, the nonzero value of $d\gamma^p$ is thus positive. The principle of maximum plastic work expressed in Equation (10.12) then requires $\tau^* \leq \tau^+$ when τ^* belongs to the interior $f(\tau^*) \leq 0$ of domain $\mathcal{D}$. When τ is negative, say $\tau = \tau^-$, we must similarly have $\tau^- \leq \tau^*$, so that the domain $\mathcal{D}$ of admissible stresses must finally be a segment $\tau^- \leq \tau \leq \tau^+$. As a result, when the stress lies on the boundary of $\mathcal{D}$ (that is when either $\tau = \tau^+$ or $\tau = \tau^-$) the plastic strain $d\gamma^p$ must point in the outward direction to the segment $\tau^- \leq \tau \leq \tau^+$, so that we retrieve the condition of Equation (10.4).

The inequality of Equation (10.12) requires the increment $\tau - \tau^* = d\tau$ to satisfy $d\tau\, d\gamma^p \geq 0$. Substituting Equation (10.4) in $d\tau\, d\gamma^p \geq 0$, we obtain $d\Lambda = 0$ as soon as df (given by Equation (10.5)) is negative, that is, as soon as an unloading process occurs. This is nothing other than the condition of Equation (10.6). In short, collecting the results, the unloading–loading conditions and the plastic flow rule of Equation (10.7) can therefore be derived from the principle of maximum plastic work expressed in Equation (10.12).

It must finally be pointed out that the expression for the intensity of the plastic strain $d\gamma^p$ provided by the plastic flow rule of Equation (10.7) is irrespective of the particular choice adopted for the plastic loading function f defining the domain $\mathcal{D}$ of admissible stresses. If two distinct plastic loading functions were chosen for defining the same domain $\mathcal{D}$, the plastic multiplier would adjust in order that the flow rule of Equation (10.4) provides the same value for $d\gamma^p$ whatever the chosen loading function.

The principle of maximum plastic work is a founding principle of plasticity models. It combines the two pillars of plasticity, namely, the normality of the flow rule and the convexity of the domain $\mathcal{D}$ of admissible stresses. The normality of the flow rule stipulates that the plastic strain must point in the outward normal direction to the domain of admissible stresses. Here, this normality was accounted for in Equation (10.7) by the proportionality of $d\gamma^p$ to $\partial f/\partial \tau$ and the positivity of the plastic multiplier $d\Lambda$. When the principle of maximum plastic work applies, the knowledge of the plastic loading function f is enough to define the flow rule, which is then described as associated (with the plastic loading function). The convexity of a stress domain means that, whenever two stress states lie in the domain including its boundary, the whole segment joining the two stress states must lie within the domain. Here, the convexity of the domain of admissible stresses means that the domain has to be a segment of a line.

By taking $\tau^* = 0$ in Equation (10.12), we showed that the principle of maximum plastic work ensures the plastic work to be positive. Basically, the principle of maximum plastic work also ensures both the stability and the uniqueness of any equilibrium stress state. In preparation for showing the stability property, let us consider the system constituted from any three-dimensional assembly of $i = 1$ to n elastoplastic elements such as the one represented in Figure 10.1(c). Further, consider an equilibrium state of this system subjected to any external loading which we do not need to specify. In this equilibrium state the stress applying to the ith element is denoted by $\tau_0^{(i)}$. Still subjected to the same external loading, unavoidable internal fluctuations may modify the stress applying to the ith element from $\tau_0^{(i)}$ to $\tau^{(i)}$. The stability issue consists of exploring if fluctuations $\tau^{(i)} - \tau_0^{(i)}$ will spontaneously amplify or vanish. Since the stresses $\tau_0^{(i)}$ and $\tau^{(i)}$ are in equilibrium with the same external loading, the stresses $\tau^{(i)} - \tau_0^{(i)}$ are self-stress. If $\gamma^{(i)}$ denotes the strain associated with $\tau^{(i)}$, the overall infinitesimal strain work of the system related to the infinitesimal strain $d\gamma^{(i)}$ and the self-stress $\tau^{(i)} - \tau_0^{(i)}$ must be zero, giving

$$\sum_{i=1}^{n} \left(\tau^{(i)} - \tau_0^{(i)} \right) d\gamma^{(i)} = 0. \tag{10.13}$$

Using the constitutive Equation (10.11) adapted for the ith element subjected to the initial stress $\tau_0^{(i)}$ and whose spring stiffness is G_i, from Equation (10.13) we obtain

$$d \left[\sum_{i=1}^{n} \frac{\left(\tau^{(i)} - \tau_0^{(i)} \right)^2}{2G_i} \right] = -\sum_{i=1}^{n} \left(\tau^{(i)} - \tau_0^{(i)} \right) d\gamma^{p(i)}. \tag{10.14}$$

Because the stress field $\tau_0^{(i)}$ is plastically admissible, the principle of maximum plastic work can be applied to each element i by letting $\tau = \tau^{(i)}$, $\tau^* = \tau_0^{(i)}$ and $d\gamma^p = d\gamma^{p(i)}$, in Equation (10.12). From Equations (10.12) and (10.14), we obtain

$$d \left[\sum_{i=1}^{n} \frac{\left(\tau^{(i)} - \tau_0^{(i)} \right)^2}{2G_i} \right] \leq 0. \tag{10.15}$$

The stability of the stress sate $\tau_0^{(i)}$ follows. Indeed, in the inequality of Equation (10.15) the function in parenthesis is a positive function of its arguments $\tau^{(i)} - \tau_0^{(i)}$, and the inequality states that it cannot increase further.

The uniqueness of the equilibrium stress state can be shown in a similar way. Let $(\tau_I^{(i)}, \gamma_I^{(i)})$ and $(\tau_{II}^{(i)}, \gamma_{II}^{(i)})$ be two possible stress solutions for the same history of external loading, which starts from a zero loading so that there is no initial stress. Since $\tau_I^{(i)} - \tau_{II}^{(i)}$ are self-stresses, the overall infinitesimal strain work of the system related to the infinitesimal strain $d\gamma_I^{(i)} - d\gamma_{II}^{(i)}$

The Poroplastic Solid

and the self-stresses $\tau_I^{(i)} - \tau_{II}^{(i)}$ must be zero, giving

$$\sum_{i=1}^{n} \left(\tau_I^{(i)} - \tau_{II}^{(i)}\right)\left(d\gamma_I^{(i)} - d\gamma_{II}^{(i)}\right) = 0. \tag{10.16}$$

Use of the constitutive Equation (10.11), with a zero initial stress, in Equation (10.16) gives us the equality

$$d\left[\sum_{i=1}^{n} \frac{\left(\tau_I^{(i)} - \tau_{II}^{(i)}\right)^2}{2G_i}\right] = -\sum_{i=1}^{n} \left(\tau_I^{(i)} - \tau_{II}^{(i)}\right)\left(d\gamma_I^{p(i)} - d\gamma_{II}^{p(i)}\right). \tag{10.17}$$

Both possible solutions, $\tau_I^{(i)}$ and $\tau_{II}^{(i)}$, are required to lie within the invariant domain $\mathcal{D}$ of admissible stresses. Applying the inequality of Equation (10.12) to both possible solutions successively, we obtain

$$\sum_{i=1}^{n} \left(\tau_I^{(i)} - \tau_{II}^{(i)}\right) d\gamma_I^{p(i)} \geq 0; \quad \sum_{i=1}^{n} \left(\tau_{II}^{(i)} - \tau_I^{(i)}\right) d\gamma_{II}^{p(i)} \geq 0, \tag{10.18}$$

so that the right-hand side of Equation (10.17) turns out to be negative:

$$d\left[\sum_{i=1}^{n} \frac{\left(\tau_I^{(i)} - \tau_{II}^{(i)}\right)^2}{2G_i}\right] \leq 0. \tag{10.19}$$

In Equation (10.19) the function in parenthesis is a positive function of its arguments $\tau_I^{(i)} - \tau_{II}^{(i)}$. It scales the distance relative to the energy between the two possible solutions $\tau_I^{(i)}$ and $\tau_{II}^{(i)}$. Accordingly, the inequality of Equation (10.19) states that the distance between two possible solutions can only decrease. As a result, since both of them are required to meet the same zero initial conditions, the distance, which is zero at time $t = 0$, remains zero as time passes – the stress solution is unique. Because the intensity of the plastic multiplier $d\Lambda$ in the flow rule of Equation (10.7) remains undetermined, the same conclusion does not hold for the deformation $\gamma^{(i)}$. However, as soon as there is a coupling of the plastic constitutive equation with a viscous phenomenon, the rate $d\Lambda/dt$ becomes governed by the characteristic time of the viscous phenomenon, and the deformation solution can be shown to be unique. When the viscous phenomena relate to those depicted in Section 4.4, the coupling operates at the material scale, resulting in an overall viscoplastic constitutive equation. When the viscous phenomenon relates to the viscous flow of the fluid within the porous solid constituting a structure, as for instance the dam of Figure 10.14, the coupling operates at the structure scale. However, even in the absence of any viscous phenomenon, the plastic multiplier is eventually governed by the elastic zone which curbs the expansion of the plastic zone before the final collapse of the structure. This causes the overall ductility of a structure made locally from a nonductile elastoplastic material. Nevertheless, this phenomenon – also called hardening – can also operate at the material scale. This is the topic of the next section.

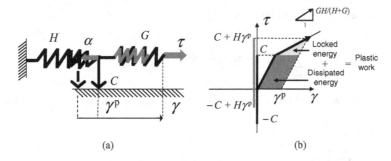

Figure 10.2 Kinematic hardening elastoplastic element

10.1.3 Hardening plasticity

For the elastoplastic element represented in Figure 10.1 the plastic irreversible strain occurs with no change of the current strength. After a complete unloading process restoring the zero stress $\tau = 0$, the subsequent plastic strain will occur when the intensity of the stress again reaches the same threshold C. Indeed, the plastic loading function f given in Equation (10.3) has remained unaffected by the previous plastic strain, so that the system has no memory of the earlier loading history. The plasticity is then described as ideal.

This is no longer the case for the hardening elastoplastic element of Figure 10.2. For this element the friction element of strength C is subjected both to the external load τ and to the hardening force α sustained by the hardening element. Instead of Equation (10.3), the plastic loading function of the hardening friction element is given by

$$\tau \in \mathcal{D}: \quad f(\tau, \alpha) = |\tau - \alpha| - C \le 0. \tag{10.20}$$

For the hardening elastoplastic element of Figure 10.2, the hardening element is made up of a linear spring of stiffness constant H. As a result, the hardening force α is linked to the plastic strain γ^p by the relation

$$\alpha = H\gamma^p, \tag{10.21}$$

or, equivalently,

$$\alpha = \frac{da_H}{d\gamma^p}; \quad a_H = \frac{1}{2} H \left(\gamma^p\right)^2. \tag{10.22}$$

The energy a_H is the elastic energy which is stored in the hardening element. Although this energy is not dissipated into heat by friction, it is also not recovered during an unloading process where the stress τ is removed, since it becomes trapped in the hardening element. For this reason the energy a_H is termed the locked energy. Substitution of Equation (10.21) in Equation (10.20) gives

$$\tau \in \mathcal{D}: \quad -C + H\gamma^p \le \tau \le C + H\gamma^p. \tag{10.23}$$

As shown by Equation (10.23), and illustrated in Figure 10.2(b), during plastic loading the domain of admissible stresses $\mathcal{D}$ is translated according to the algebraic quantity $H\gamma^p$. This specific hardening is called kinematic hardening.

Alternatively to the plastic loading function given by Equation (10.20), we can write

$$\tau \in \mathcal{D}: \quad f(\tau, \alpha) = |\tau| - C - \alpha \leq 0, \tag{10.24}$$

where the hardening force α is now linked to the plastic strain γ^p by the relation

$$\alpha = H\beta = \frac{da_H}{d\beta}; \quad \beta = \int_0^{\gamma^p} |d\gamma^p|; \quad a_H = \frac{1}{2}H\beta^2. \tag{10.25}$$

Substitution of Equation (10.25) in Equation (10.24) gives

$$\tau \in \mathcal{D}: \quad -C - H\beta \leq \tau \leq C + H\beta. \tag{10.26}$$

According to the definition of Equation (10.25), the hardening variable β is a positive monotonic increasing function. Therefore, the inequalities of Equation (10.26) show that, during plastic loading whatever the sign of τ is, the domain $\mathcal{D}$ of admissible stresses continuously expands symmetrically from both sides of the zero stress $\tau = 0$. Again, this hardening is specific and is called isotropic hardening.

To encompass more general hardening systems, instead of Equation (10.21), we now write

$$\alpha = h(\beta), \tag{10.27}$$

where β stands for the hardening state variable associated with the hardening force α. The hardening state variable β evolves only when the plastic strain γ^p does, although β is equal to γ^p only as a special case. Similarly to Equation (10.22), from Equation (10.27) we alternatively write

$$\alpha = \frac{da_H}{d\beta}, \quad a_H = \int_0^\beta h(c)\,dc, \tag{10.28}$$

so that $\alpha d\beta$ accounts for the infinitesimal variation da_H of the locked energy a_H, which is a quadratic function of β only as a special case. Instead of Equation (10.9) for the overall free energy a_S, when adopting a zero initial stress we now have

$$a_S = \frac{1}{2}G(\gamma - \gamma^p)^2 + a_H. \tag{10.29}$$

Considering elastic changes, where $d\gamma^p = 0$ so that both $d\beta$ and da_H are also zero, substitution of Equation (10.29) in Equation (10.10) first allows us to retrieve the state Equation (10.11) providing τ_0 is zero. Requiring the state equation to be continuous, Equation (10.11) must still hold during plastic changes. Substituting Equation (10.29) in Equation (10.10) and taking into account Equation (10.11), we derive the inequality

$$\tau d\gamma^p - da_H = \tau d\gamma^p - \alpha d\beta \geq 0. \tag{10.30}$$

In a way consistent with the above definition for the locked energy a_H, the inequality of Equation (10.30) shows that, because of the trapping of the energy da_H, only a part of the plastic work $\delta W^p = \tau d\gamma^p$ is transformed into heat. This is illustrated in Figure 10.2(b) for the hardening elasto-plastic element represented in Figure 10.2(a).

During the plastic loading of a hardening elastoplastic system, we may still write that a positive plastic multiplier $d\Lambda$ does exist such that the plastic increment $d\gamma^p$ is given by the flow rule of Equation (10.4). However, in a way consistent with Equation (10.30) the current domain $\mathcal{D}$ of admissible stresses now depends on the hardening force α, so that it is defined by $f(\tau,\alpha) \leq 0$. As a result the consistency condition stated in Equation (10.5) has to be extended in the form

$$df = \frac{\partial f}{\partial \tau} d\tau + \frac{\partial f}{\partial \alpha} d\alpha = 0. \tag{10.31}$$

Using Equation (10.27), from the consistency condition of Equation (10.31) we obtain

$$\frac{\partial f}{\partial \tau} d\tau + \frac{\partial f}{\partial \alpha} \frac{dh}{d\beta} d\beta = 0. \tag{10.32}$$

We said above that, owing to the very definition of the hardening variable β, its increment $d\beta$ can be nonzero providing only the plastic strain $d\gamma^p$ and, therefore, the plastic multiplier $d\Lambda$, are nonzero. This allows us to infer that, whatever the hardening system under consideration, the increment $d\beta$ must be proportional to $d\Lambda$. Since $d\beta$ and $d\Lambda$ are proportional, the consistency condition of Equation (10.32) allows us to write $d\Lambda$ in the form

$$d\Lambda = \frac{1}{H} \frac{\partial f}{\partial \tau} d\tau, \tag{10.33}$$

where H is the hardening modulus, whose expression is

$$H = -\frac{\partial f}{\partial \alpha} \frac{dh}{d\beta} \frac{d\beta}{d\Lambda}. \tag{10.34}$$

In Equation (10.34) $d\beta/d\Lambda$ stands for the proportionality factor which exists between $d\beta$ and $d\Lambda$, and not for the derivative of β with respect to some variable Λ. For instance, in the hardening model of Equation (10.25), we have $d\beta = |\partial f/\partial \tau| d\Lambda$. For this isotropic hardening system defined by Equations (10.24) and (10.25), or for the kinematic hardening system of Figure 10.2 defined by Equations (10.20) and (10.21), it can be checked that the hardening modulus H defined by Equation (10.33) is actually the stiffness constant H of the hardening element.

According to Equation (10.33), since the plastic multiplier $d\Lambda$ is positive, provided only that the hardening modulus H is positive, plastic loading does actually occur when $(\partial f/\partial \tau) d\tau$ is positive. This condition means that the increment $d\tau$ must point in the direction of the outward normal to the domain $\mathcal{D}$ of admissible stresses. When the modulus H becomes zero, since the plastic multiplier $d\Lambda$ cannot be infinite, $\partial f/\partial \tau$ must become zero in Equation (10.33): the value of $d\Lambda$ then remains undetermined and ideal plasticity is recovered. When the modulus H becomes negative, plastic loading occurs when $(\partial f/\partial \tau) d\tau$ is negative so that the increment $d\tau$ must now point in the direction of the inward normal to the domain $\mathcal{D}$ of admissible stresses. In this case, as illustrated in Figure 10.3, the current strength decreases and, instead of hardening plasticity, we have softening plasticity. However, as also illustrated in Figure 10.3, if $(\partial f/\partial \tau) d\tau$ is negative, the response of the softening system, instead of being plastic, might be elastic. In stress-monitored experiments performed on softening systems, the response in strain is therefore not unique, and plastic loading is not guaranteed. As a result, the softening branch can only be shown in strain-monitored experiments. For real systems and materials, the relation linking the hardening force α and the hardening state variable β is in general not linear, resulting in a hardening modulus H that is not a constant and a stress–strain relationship which is not linear. As illustrated in Figure 10.3, from positive (hardening

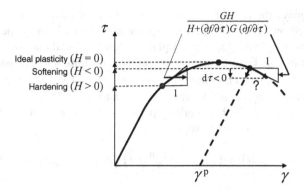

Figure 10.3 Hardening plasticity, ideal plasticity and softening plasticity; in the case of softening plasticity the same negative stress increment $d\tau$ may result in either plastic loading or elastic unloading

plasticity), the hardening modulus H may become zero (ideal plasticity) and ultimately negative (softening plasticity).

An instructive example of nonmonotonic hardening plasticity is provided by the irreversible phenomenon of capillary hysteresis which we explored in Section 6.2.4. Looking first at Figure 10.4, reproduced from Figure 6.13, the current line segment defined by $p_{\text{cap}}^{\text{IMB}} \leq p_{\text{cap}} \leq p_{\text{cap}}^{\text{DRA}}$,

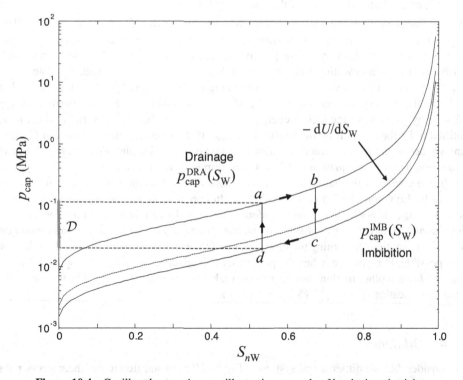

Figure 10.4 Capillary hysteresis as an illustrative example of hardening plasticity

as *ad* or *bc*, then plays the role of the current domain $\mathcal{D}$ of admissible capillary pressure p_{cap}. The role played by stress τ in the (γ, τ)-plane of Figure 10.2 may then be devoted to the capillary pressure p_{cap} in the $(S_{n\text{W}}, p_{\text{cap}})$-plane, where we recall that $S_{n\text{W}} = 1 - S_{\text{W}}$ is the nonwetting fluid saturation. For a given porous solid both the imbibition pressure $p_{\text{cap}}^{\text{IMB}}$ and the drainage pressure $p_{\text{cap}}^{\text{DRA}}$ are experimental functions, $p_{\text{cap}}^{\text{IMB}}(S_{n\text{W}})$ and $p_{\text{cap}}^{\text{DRA}}(S_{n\text{W}})$, of the same variable $S_{n\text{W}}$. Therefore, they are not independent. As a result, the plastic loading function f defining the domain $\mathcal{D}$ of admissible capillary pressures can be written as a function of p_{cap} and of a unique hardening force α according to

$$p_{\text{cap}} \in \mathcal{D}: \quad f(p_{\text{cap}}, \alpha) = \left(p_{\text{cap}} - \varpi_{\text{cap}}^{\text{IMB}}(\alpha)\right)\left(p_{\text{cap}} - \varpi_{\text{cap}}^{\text{DRA}}(\alpha)\right) \leq 0, \quad (10.35)$$

with

$$\alpha = h(S_{n\text{W}}), \quad (10.36)$$

where the function h has to match the experimental data $p_{\text{cap}}^{\text{IMB}}(S_{n\text{W}})$ and $p_{\text{cap}}^{\text{DRA}}(S_{n\text{W}})$ according to

$$\varpi_{\text{cap}}^{\text{IMB}}[h(S_{n\text{W}})] = p_{\text{cap}}^{\text{IMB}}(S_{n\text{W}}); \quad \varpi_{\text{cap}}^{\text{DRA}}[h(S_{n\text{W}})] = p_{\text{cap}}^{\text{DRA}}(S_{n\text{W}}). \quad (10.37)$$

At this stage of the reinterpretation of capillary hysteresis as an example of hardening plasticity, the conditions of Equation (10.37) do not lead to a unique choice of the triplet $\left(h, \varpi_{\text{cap}}^{\text{IMB}}, \varpi_{\text{cap}}^{\text{DRA}}\right)$ and, therefore, of the hardening force α.

As the capillary pressure moves within the current domain $\mathcal{D}$ of admissible capillary pressure, there is no change of the saturation $S_{n\text{W}} = 1 - S_{\text{W}}$ of the nonwetting fluid. In this simplified approach to capillary hysteresis it is thus assumed that there is no 'elastic' or reversible contribution to the nonwetting fluid saturation $S_{n\text{W}} = 1 - S_{\text{W}}$. As a result, the role above played by γ^p is now devoted to $1 - S_{\text{W}}$. Comparing Equation (6.117) with Equation (10.30), we obtain the correspondence $\tau \text{d}\gamma^p \equiv p_{\text{cap}} \text{d} S_{n\text{W}} = -p_{\text{cap}} \text{d} S_{\text{W}}$, while the interface energy $U(S_{n\text{W}})$ plays the role of the locked energy $a_H(\beta)$, so that the hardening variable β is finally identified to be the nonwetting fluid saturation $S_{n\text{W}}$. If the interface energy function $U(S_{n\text{W}})$ is known, according to Equation (10.27), where we take $\beta = S_{n\text{W}}$ and $a_H = U$, the choice of the function $h(S_{n\text{W}})$ in Equation (10.37) is then imposed by the relation $h = \text{d}U/\text{d}S_{n\text{W}}$. As a result, the choice of the hardening variable α associated with the hardening state variable $\beta = S_{n\text{W}}$ by Equation (10.36) becomes unique. In contrast, as long as the interface energy function $U(S_{n\text{W}})$ is not known, a unique identification of the capillary hardening force α, that is consistent regarding energy considerations, cannot actually be achieved. It is only with new pieces of information concerning the morphology of the pore space that a unique identification of α is possible. For instance, when the pore space can be represented by large spherical pores connected to each other by thin capillary tubes, making use of Equation (6.123), we obtain the unique identification $\alpha = \frac{3}{2} p_{\text{cap}}^{\text{IMB}}(S_{n\text{W}}) = h(S_{n\text{W}})$.

10.1.4 Dilatancy

Now consider the two-dimensional system of Figure 10.5. In addition to the shear stress τ the system is subjected to the compressive stress $-\sigma$. The domain $\mathcal{D}$ of admissible stresses now

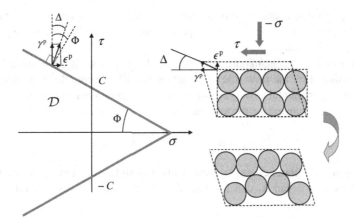

Figure 10.5 When being submitted to the combined action of a shear stress τ and a compressive stress $-\sigma$, the solid grains of an initially well-packed system become less tangled up, so that dilatancy is observed. According to the second law of thermodynamics, the dilatancy angle Δ scaling the expansion cannot be greater than the friction angle Φ associated with the loading function $f(\tau, \sigma)$

has to be defined in the (σ, τ)-plane. Figure 10.5 shows such a domain, which is convex and defined by the Coulomb plastic loading function

$$(\sigma, \tau) \in \mathcal{D}: \quad f(\sigma, \tau) = |\tau| + \sigma \tan \Phi - C \leq 0, \qquad (10.38)$$

where Φ is termed the friction angle. According to Equation (10.38), the larger the intensity $-\sigma$ of the compressive stress, the larger the shear stress τ required to cause the plastic flow of the system. If the system obeys the principle of maximum plastic work, the normality rule governing the plastic strain increments $d\gamma^p$ and $d\epsilon^p$ associated with stresses τ and σ leads us to write

$$d\gamma^p = d\Lambda \frac{\partial f}{\partial \tau}; \quad d\epsilon^p = d\Lambda \frac{\partial f}{\partial \sigma}. \qquad (10.39)$$

Substitution of Equation (10.38) in Equation (10.39) provides the flow rule in the explicit form

$$|d\gamma^p| = d\Lambda; \quad d\epsilon^p = d\Lambda \tan \Phi. \qquad (10.40)$$

The flow rule of Equation (10.40) states that the plastic distortion $|d\gamma^p|$ is accompanied by a plastic expansion $d\epsilon^p > 0$ irrespective of the sign of the normal applied stress σ. For systems formed of solid grains this can be explained as shown in Figure 10.5: when undergoing the distortion $|d\gamma^p|$ induced by the shear stress τ the solid grains become less tangled up so that, in turn, dilatancy is observed. This is the dilatancy phenomenon mentioned in the introduction, at the origin of the drying out of the wet sand around the footprint of a walker on a beach and illustrated in Figure 1.4.

For a two-dimensional system, the expression of the plastic work δW^p extends in the form

$$\delta W^p = \sigma d\epsilon^p + \tau d\gamma^p. \qquad (10.41)$$

When substituting Equation (10.40) in Equation (10.41), and using the plastic loading condition $f = 0$ where f is given by Equation (10.38), the same expression of Equation (10.8) is retrieved for δW^p. More generally, it can be shown that, whenever the principle of maximum plastic work applies, the plastic work δW^p can be expressed as a function involving the plastic strain increments $d\epsilon^p$ and $d\gamma^p$ only.

In practice, for most systems, in particular the ones formed of solid grains as in Figure 10.5, the dilatancy predicted by the flow rule of Equation (10.40) is generally too important when compared to the experimental observations. Instead of Equation (10.40), the experimental observations can be accounted for through the modified flow rule

$$\left|d\gamma^p\right| = d\Lambda; \quad d\epsilon^p = d\Lambda \tan \Delta, \tag{10.42}$$

where Δ is termed the dilatancy angle, now distinct from the friction angle Φ. The related dilatancy coefficient, $\delta = \tan \Delta$, can be determined experimentally as the ratio $d\epsilon^p / |d\gamma^p|$ since we have

$$d\epsilon^p = \delta \left|d\gamma^p\right|. \tag{10.43}$$

When the principle of maximum plastic work applies, as in Equation (10.39), the flow rule is associated with the plastic loading function f. The potential defining the flow rule is then described as associated and equal to f. When the dilatancy angle Δ differs from the friction angle Φ, this is no longer the case. According to usual practice, a nonassociated potential g can then be introduced to express the flow rule in the form

$$d\gamma^p = d\Lambda \frac{\partial g}{\partial \tau}; \quad d\epsilon^p = d\Lambda \frac{\partial g}{\partial \sigma}. \tag{10.44}$$

The flow rule of Equation (10.42) can be expressed as Equation (10.44), when identifying the nonassociated potential g as

$$g(\sigma, \tau) = |\tau| + \sigma \tan \Delta. \tag{10.45}$$

When the principle of maximum plastic work does not apply, the plastic strains do not achieve a maximum production of entropy. Accordingly, the positivity of the plastic work is not guaranteed and has to be checked. Substituting Equation (10.42) in Equation (10.41), while using the plastic loading condition $f(\tau, \sigma) = |\tau| + \sigma \tan \Phi - C = 0$, we derive

$$\delta W^p = [\sigma (\tan \Delta - \tan \Phi) + C] \left|d\gamma^p\right|. \tag{10.46}$$

In agreement with a previous comment, the expression of Equation (10.46) for the plastic work δW^p shows that this is only when the dilatancy angle Δ is equal to the friction angle Φ, and therefore when the principle of maximum plastic work does apply, that the plastic work can be expressed as a function of the plastic strain only. The domain $\mathcal{D}$ of admissible stresses includes an infinite compressive stress, that is an infinite negative value for σ, while the value of σ cannot exceed $C \cot \Phi$. As a result, irrespective of the value of cohesion C, the plastic work δW^p, whose expression is given by Equation (10.46), will always be positive providing the dilatancy angle Δ is smaller than the friction angle Φ:

$$\Delta \leq \Phi. \tag{10.47}$$

10.1.5 Three-dimensional plasticity

Consider now a three-dimensional representative volume element of a solid material exhibiting a plastic behavior. Having been subjected to an external loading and subsequently completely unloaded, this volume element undergoes a permanent or plastic deformation. Noting σ_{ij} the components of the current stress tensor, and ε_{ij}^p the components of the plastic strain tensor, the three-dimensional plastic work δW^p is expressed in the form

$$\delta W^p = \sigma_{ij} d\varepsilon_{ij}^p. \tag{10.48}$$

Extending the definition introduced in the previous section, the dilatancy coefficient δ quantifies the magnitude of the distortion compared with the volumetric strain. To exploit this definition we can first introduce the plastic volumetric strain ϵ^p capturing the irreversible volume change:

$$d\epsilon^p = d\varepsilon_{ii}^p. \tag{10.49}$$

A three-dimensional mean plastic distortion γ^p is then introduced through the definition of its variation $d\gamma^p$:

$$d\gamma^p = |d\gamma^p| = \sqrt{\frac{1}{2} d\gamma_{ij}^p d\gamma_{ji}^p}; \quad \gamma_{ij}^p = 2e_{ij}^p = 2\left(\varepsilon_{ij}^p - \frac{1}{3}\epsilon^p \delta_{ij}\right). \tag{10.50}$$

When $i \neq j$ the plastic distortion, $\gamma_{ij}^p = 2e_{ij}^p = 2\varepsilon_{ij}^p$, accounts for the irreversible change undergone by the angle made between material directions $\underline{e}_i$ and $\underline{e}_j$ that were normal prior to the deformation. As a result, the extended three-dimensional definition of Equation (10.50) for $d\gamma^p$ reduces to its two-dimensional definition illustrated in Figure 10.5. The latter represents for instance the plane $(\underline{e}_1, \underline{e}_2)$ when the nonzero components of the plastic strain tensor reduce to $\varepsilon_{11}^p = \varepsilon_{22}^p = \varepsilon_{33}^p = \epsilon^p$, and $\varepsilon_{12}^p = \gamma_{12}^p/2$ so that $d\gamma^p = d|\gamma_{12}^p|$. Irrespective of any specific model, the three-dimensional definition of the dilatancy coefficient δ is then provided by Equation (10.43) and the definitions of Equations (10.49) and (10.50).

The one-dimensional statement of Equation (10.12) for the principle of maximum plastic work extends to three dimensions in the form

$$\forall \sigma_{ij}^* : f\left(\sigma_{ij}^*\right) \leq 0; \quad \left(\sigma_{ij} - \sigma_{ij}^*\right) d\varepsilon_{ij}^p \geq 0. \tag{10.51}$$

The normality of the flow rule follows from the principle of maximum plastic work expressed in Equation (10.51) and allows us to write

$$d\varepsilon_{ij}^p = d\Lambda \frac{\partial f}{\partial \sigma_{ij}}. \tag{10.52}$$

For an isotropic material, since no direction is favored, the plastic loading function f depends upon only the first three invariants of the stress tensor. Restricting ourselves to the first two invariants, this amounts to writing

$$f\left(\sigma_{ij}\right) = f\left(\sigma, \tau\right), \tag{10.53}$$

where $\sigma = \frac{1}{3}\sigma_{ii}$ is the mean stress and τ denotes the shear stress defined by

$$\tau = \sqrt{\frac{1}{2} s_{ij} s_{ji}}, \qquad (10.54)$$

where s_{ij} stands for the components of the deviatoric stress tensor $\underline{\underline{s}}$ defined by Equation (3.31). Substitution of Equation (10.53) in Equation (10.52), and use of the definitions of Equations (3.30), (3.31) and (10.54) give us the flow rule

$$d\varepsilon_{ij}^p = d\Lambda \left(\frac{1}{3} \frac{\partial f}{\partial \sigma} \delta_{ij} + \frac{\partial f}{\partial \tau} \frac{s_{ij}}{2\tau} \right). \qquad (10.55)$$

A nonassociated flow rule will be obtained by replacing the plastic loading function f in Equation (10.55) by a nonassociated potential g. The results of the previous section can then be extended by using the plastic loading function f whose expression is given by Equation (10.38), or by using the nonassociated potential g whose expression is given by Equation (10.45).

10.1.6 Limit analysis and stability of dry sandpiles

Limit analysis aims to determine the critical loading causing the collapse of structures without having to determine the behavior prior to failure. This determination can be carried out by analyzing the maximum rate of dissipated energy which the structure may set against the work rate of applied forces in the likely modes of plastic collapse. Consider then a possible candidate $\underline{V}$ for the velocity field describing the plastic collapse of the structure Ω under consideration. Restricting ourselves to the case where the body forces $\underline{f}$ reduce to gravity forces $\underline{g} = g\underline{e}_z$, the work rate dW/dt of applied forces associated with the velocity field $\underline{V}$ is expressed in the form

$$\frac{dW}{dt} = \int_{\partial \Omega} \underline{V} \cdot \underline{\underline{T}} dA + \int_{\Omega} \underline{V} \cdot \underline{g} d\Omega, \qquad (10.56)$$

where $\underline{T}$ are the surface forces applying on the structure border $\partial \Omega$. According to the mechanical energy balance we derived in Section 3.3.1, the work rate dW/dt can be equated to the overall strain work rate associated with any stress field $\underline{\underline{\sigma}}$ in mechanical equilibrium with the applied forces, that is, satisfying both Equations (3.29) and (3.35). This amounts to writing that the mechanical energy balance

$$\frac{dW}{dt} = \int_{\Omega} \frac{dW}{dt} d\Omega = \int_{\Omega} \sigma_{ij} \frac{d\varepsilon_{ij}}{dt} d\Omega \qquad (10.57)$$

must hold whatever the velocity field $\underline{V}$.

At this stage it is crucial to point out that the energy balance of Equation (10.57) does not require the stress $\underline{\underline{\sigma}}$ and the strain rate $d\varepsilon_{ij}/dt$ to be linked by the constitutive equations of the material forming the structure. However, if the ultimate strength of the constitutive material forming the structure may be accounted for by a plastic loading function $f(\sigma_{ij})$, for any velocity field $\underline{V}$ associated with the strain rate $d\varepsilon_{ij}/dt$, we can determine the maximum of the strain work rate $\sigma_{ij} d\varepsilon_{ij}/dt$ with regard to all the admissible stresses satisfying the condition $f(\sigma_{ij}) < 0$. For a particular velocity field $\underline{V}$ let us then assume that this maximum, when integrated over the whole structure, becomes less than the work rate dW/dt associated with the applied forces. For this particular velocity field $\underline{V}$ this means that we cannot exhibit any stress field $\underline{\underline{\sigma}}$, both admissible and in mechanical equilibrium with the applied forces, which can fulfill the mechanical energy balance of Equation (10.57). As a result, since Equation (10.57) has to be fulfilled whatever the velocity field $\underline{V}$, the collapse of the structure will unavoidably occur. This statement can be expressed in the form of the kinematical theorem of limit analysis, reading

$$\exists \underline{V}: \sup_{f(\sigma_{ij}) \leq 0} \int_{\Omega} \sigma_{ij} \frac{d\varepsilon_{ij}}{dt} d\Omega < \frac{dW}{dt} \implies \text{collapse of the structure.} \qquad (10.58)$$

As emphasized above, there is no restriction imposed on the velocity field $\underline{V}$ and, therefore, on the strain rate $d\varepsilon_{ij}/dt$. This allows us to explore the statement of Equation (10.58) by choosing the velocity field $\underline{V}$ such that the associated strain rate $d\varepsilon_{ij}$ is normal to the domain $f(\sigma_{ij}) < 0$ of admissible stresses. The strain work $\sigma_{ij}d\varepsilon_{ij}$ may then be equated to the plastic work δW^p related to the ideal plastic material associated with the plastic loading function $f(\sigma_{ij})$ and verifying the principle of maximum plastic work. For instance, if the plastic loading function is the one given by Equation (10.38), taking $\Delta = \Phi$ in Equation (10.46) to ensure the normality of the flow rule, we derive

$$\sup_{f(\sigma_{ij}) \leq 0, \Delta = \Phi} \int_{\Omega} \sigma_{ij} \frac{d\varepsilon_{ij}}{dt} d\Omega = \int_{\Omega} C \frac{|d\gamma^p|}{dt} d\Omega. \qquad (10.59)$$

By way of illustration let us analyze the stability of a sandpile in its virgin state under the action of only gravity forces. The cohesion of sand is due to the packing of the solid grains. The virgin state of sand can be viewed as the one prevailing prior to any external packing. The sand is then loose and cohesionless so that its strength is only governed by the friction existing between the grains. Taking $C = 0$ in Equation (10.59), the dissipated energy that a sandpile has to oppose in its virgin state to the work rate of external forces is shown to be zero, as soon as the collapse mechanism corresponds to a normal plastic flow rule. This is the case of the collapse mechanism shown in Figure 10.6. In this mechanism the block ABC moves rigidly

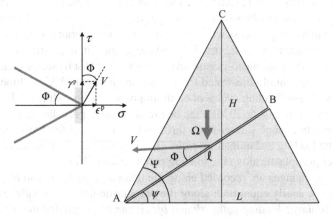

Figure 10.6 Possible collapse mechanism for a dry sandpile

with a velocity of constant intensity V, and whose angle made with the line AB is the friction angle Φ. Then letting Ω be the weight of the block ABC, and ψ the angle made by the line AB with the horizontal line, the work rate dW/dt of the gravity forces can be written in the form

$$\frac{dW}{dt} = \Omega V \sin(\psi - \Phi). \tag{10.60}$$

Combining Equations (10.58) and (10.60) leads to the statement

$$\exists \psi : \sin(\psi - \Phi) > 0 \implies \text{collapse of the dry sandpile.} \tag{10.61}$$

As soon as the angle Ψ defining the slope of the sandpile is greater than Φ, the inequality in Equation (10.61) is achieved for any value of ψ satisfying $\Psi > \psi > \Phi$. It can therefore be concluded that, in the absence of any cohesion, the slope angle Ψ cannot exceed the friction angle Φ without the sandpile collapsing.

10.2 From Plasticity to Poroplasticity

10.2.1 The poroplastic solid

As a natural extension of plasticity, poroplasticity is the ability of porous materials to undergo permanent strains and permanent changes in porosity and, consequently, permanent changes in fluid mass content. In order to account for both permanent or plastic strains ε_{ij}^p and permanent or plastic changes of porosity φ^p, when assuming both a zero stress and a zero pore pressure reference state, the isotropic state equations of elasticity (Equation (4.6)) can be extended in the form

$$\sigma = K(\epsilon - \epsilon^p) - bp \tag{10.62a}$$

$$\varphi - \varphi^p = b(\epsilon - \epsilon^p) + p/N \tag{10.62b}$$

$$s_{ij} = 2G\left(e_{ij} - e_{ij}^p\right). \tag{10.62c}$$

According to the state Equations (10.62), the strain and porosity changes $\varepsilon_{ij} - \varepsilon_{ij}^p$ and $\varphi - \varphi^p$ are the elastic contributions to the strain and changes of porosity which can be recovered at the end of a complete unloading process where the stress σ_{ij} and the pore pressure p recover their zero values. The permanent or plastic strain ε_{ij}^p and the permanent or plastic change of porosity φ^p are then recorded. The poroelasto-plastic behavior accounted for by state Equations (10.62a) and (10.62b) is experimentally illustrated for a limestone in Figure 10.7. A limestone sample is subjected to loading–unloading cycles of confining pressure $-\sigma$, while the fluid pressure is maintained at zero. Up to $-\sigma \simeq 20$ MPa the behavior remains poroelastic: the same strain ϵ and the same porosity change φ (measured through the change of fluid volume content) are recorded along the loading and unloading paths Oa and aO. Beyond this initial threshold the behavior becomes poroplastic along the loading paths ab and bb', since irreversible or plastic strains and porosity changes are recorded along the unloading paths bc and $b'c'$. Since ϵ and φ are observed to be nearly equal both along the poroelastic unloading paths aO, bc or $b'c'$ and along the poroplastic loading paths ab and bb', it can be concluded that $\epsilon^p = \varphi^p$, so that the solid matrix is plastically incompressible.

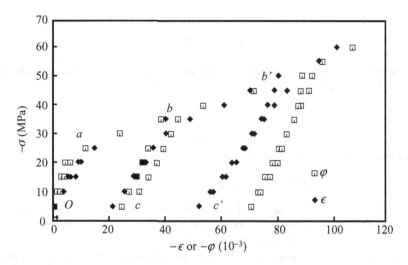

Figure 10.7 Experimental evidence of the poroplastic behavior of a limestone (courtesy of Frédéric Skoczylas)

The usual expression of Equation (10.48) for the plastic work rate for plastic solids extends to poroplastic solids in the form

$$\delta W^p = \sigma_{ij} d\varepsilon_{ij}^p + p d\varphi^p. \tag{10.63}$$

As $d\varepsilon_{ij}^p$ is the plastic strain increment associated with stress σ_{ij}, $d\varphi^p$ is the plastic increment of porosity associated with the pore pressure p. As a result, the plastic loading function f related to porous solids depends on both σ_{ij} and p so that the domain $\mathcal{D}$ of admissible stresses and pore pressures is defined by

$$(\sigma_{ij}, p) \in \mathcal{D}: \quad f(\sigma_{ij}, p) \leq 0. \tag{10.64}$$

If the porous solid obeys the principle of maximum plastic work extended to the contribution of the pore pressure, the normality of the flow rule expressed in Equation (10.52) extends to $d\varphi^p$ in the form

$$d\varepsilon_{ij}^p = d\Lambda \frac{\partial f}{\partial \sigma_{ij}}; \quad d\varphi^p = d\Lambda \frac{\partial f}{\partial p}. \tag{10.65}$$

For a nonassociated flow rule, f has to be replaced by the nonassociated potential g in Equation (10.65).

The equations above can be made more specific in the case where the solid matrix undergoes negligible plastic volumetric changes. For instance, for granular porous solids, the plastic changes are caused by the irreversible relative sliding of the solid grains forming the matrix, so that the plastic volume change of the matrix comes out to be negligible in the absence of any occluded porosity. For a porous solid the observable macroscopic volumetric plastic dilation ϵ^p undergone by the porous solid is due both to the plastic porosity change φ^p and to the irreversible volumetric dilation ϵ_S^p undergone by the solid matrix. Applying Equation (3.53) at the end of a complete unloading process, where the stresses σ_{ij} and the pore pressure

p recover zero values, we derive the volume balance

$$\epsilon^p = (1 - \phi_0)\epsilon_S^p + \varphi^p. \tag{10.66}$$

If the irreversible volumetric dilation ϵ_S^p undergone by the solid matrix is negligible, taking $\epsilon_S^p \simeq 0$ in the volume balance Equation (10.66) we obtain

$$\epsilon^p = \varphi^p. \tag{10.67}$$

Recalling that $\epsilon^p = \varepsilon_{kk}^p$, substitution of Equation (10.67) in Equation (10.63) allows us to rewrite the plastic work δW^p in the form

$$\delta W^p = (\sigma_{ij} + p\delta_{ij})\, d\varepsilon_{ij}^p = \sigma'_{ij} d\varepsilon_{ij}^p. \tag{10.68}$$

According to Equation (10.68) the effective stress σ'_{ij}, which we have already encountered in previous chapters such as, in particular, in Section 3.4.5, is then found to be the unique force driving the plastic strain. This requires the plastic loading function f to depend only on the effective stress σ'_{ij} and allows us to define the domain of admissible stresses and pore pressures in the more specific form

$$(\sigma_{ij}, p) \in \mathcal{D}: \quad f(\sigma_{ij} + p\delta_{ij}) \leq 0. \tag{10.69}$$

In turn, the validity of Equation (10.69) can be checked experimentally as is shown in Figure 10.8. Path ab corresponds to a plastic shear loading by increasing the axial pressure $-\sigma_\mathrm{I}$ (from 28 MPa to 32 MPa), while holding constant both the confining pressure $-\sigma_\mathrm{III} = -\sigma_\mathrm{II}$ (=

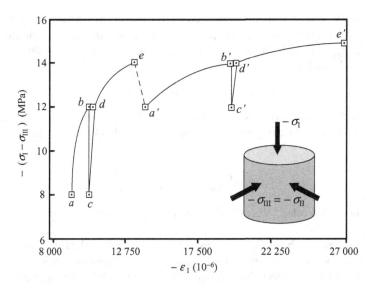

Figure 10.8 Checking the validity of a loading function of the form $f(\sigma'_{ij} = \sigma_{ij} + p\delta_{ij})$; the experimental data correspond to the same limestone as that in Figure 10.7, whose solid matrix was reported to be plastically incompressible, that is, satisfying $\epsilon^p = \varphi^p$ (after Kherbouche, R., Shao, J.-F., Skoczylas, F. and Henry, J.-P. (1995) On the poroplastic behavior of porous rocks. *European Journal of Mechanics, A/Solids*, **14**, 577–587)

20 MPa) and the pore pressure p (= 10 MPa). Path bc corresponds to the exact opposite elastic shear unloading. Path cd corresponds to a new elastic shear loading, but now holding constant the axial pressure $-\sigma_{\mathrm{I}}$ (= 28 MPa) and decreasing both the confining pressure $-\sigma_{\mathrm{III}} = -\sigma_{\mathrm{II}}$ (from 20 MPa to 16 MPa) and the pore pressure p (from 10 MPa to 6 MPa), in order to restore the same effective volumetric stress $-\sigma' = -\frac{1}{3}\sigma'_{kk}$ (that is, 14 MPa), as in the previous loading point b, but now achieved for distinct stresses and pore pressure. The subsequent loading path de corresponds to a new plastic loading, so that it can be concluded that the loading function f must be expressed as a function of the effective stress σ'_{ij} only. This can be confirmed by repeating the procedure (new loading path $a'b'c'd'e'$).

Assuming the plastic incompressibility of the solid matrix which results in Equation (10.68), the usual models accounting for the plastic behavior of solids can then be extended to porous solids simply by replacing the stress tensor $\underline{\underline{\sigma}}$ by the effective stress tensor $\underline{\underline{\sigma}}' = \underline{\underline{\sigma}} + p\underline{\underline{1}}$. For instance, if the plastic loading function of a porous solid subjected to a zero pore pressure is observed experimentally to be the usual cohesive-frictional model given by Equation (10.38), the plastic loading function accounting for a nonzero pore pressure will simply be provided by

$$(\sigma_{ij}, p) \in \mathcal{D}: \quad f(\sigma, \tau) = |\tau| + (\sigma + p)\tan\Phi - C \leq 0, \qquad (10.70)$$

where τ is still defined by Equation (10.54). Again, the relevance of such an extension can be checked through a deviatoric experiment as is shown in Figure 10.9. A deviatoric experiment consists of subjecting the sample to a shear experiment by progressively increasing the axial effective pressure $-\sigma'_{\mathrm{I}}$ beyond a constant confining effective pressure $-\sigma'_{\mathrm{II}} = -\sigma'_{\mathrm{III}}$, so that $\sqrt{3}\tau = |\sigma'_{\mathrm{I}} - \sigma'_{\mathrm{III}}|$. The limestone is the one whose Biot's coefficient was found equal to 0.63

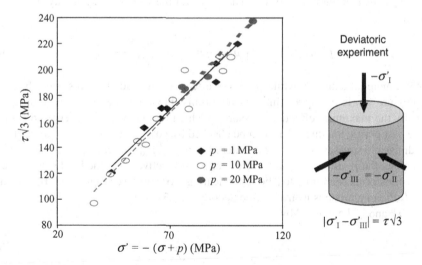

Figure 10.9 Investigation, through a deviatoric experiment, of the relevance of extension (Equation (10.70), of the loading function in order to account for a nonzero pore pressure (data from Vincké, O., Boutéca, M., Piau, J-M. and Fourmaintraux D. (1998). Study of the effective stress at failure *Poromechanics, A tribute to M.A. Biot, Proceedings of the First Biot conference*, eds. Thymus et al., Balkema)

in Figure 4.1. It is then instructive to note that the effective stress $\sigma'_{ij} = \sigma_{ij} + p\delta_{ij}$ governing the plastic strain does not coincide with the stress $\sigma_{ij} + bp\delta_{ij}$ governing the elastic part of the strain in Equation (10.62)

10.2.2 Critical state and the Cam-clay model

The usual model described in the previous sections encompasses a large class of cohesive-frictional porous materials, but experimental observations show its irrelevance to the plastic behavior of soils such as clays. Clays cannot sustain tensile stress, or infinite confining pressure, whereas the observed dilatancy can be either positive (dilation) or negative (contraction), depending on the ratio between the shear τ and the confining effective pressure $-\sigma' = -(\sigma + p)$. The mechanical behavior of soils and rocks can be explored through the deviatoric experiment shown in Figure 10.9. In soil and rock mechanics it is usual to note

$$p' = -\sigma'; \quad q = \tau\sqrt{3}, \quad (10.71)$$

so that only positive values are considered, that is $p' \geq 0$, and to investigate the plastic loading function in the form $f(p', q)$.

Investigating the behavior of clays, in particular the clay of the Cam River, the Cambridge School[2] designed the so-called 'Cam-clay' model. This model was originally carried out to account for the plastic behavior of saturated clays but, nowadays, it is successfully used for a much broader class of materials, and has been explored to account for unsaturated situations.[3] This model is particularly attractive since it covers the whole range of possible behaviors of plastic materials: ideal, hardening and softening plasticity. Adopting the notations of Equation (10.71), the plastic loading function which accounts for the yield surface of clays is expressed in the form

$$f(p', q, p_{co}) = \frac{1}{2}\left(p' - \frac{1}{2}p_{co}\right)^2 + \frac{q^2}{2M^2} - \frac{1}{8}p_{co}^2, \quad (10.72)$$

where M is a material constant, while p_{co} is the effective consolidation pressure. The consolidation pressure p_{co} is the upper limit for the admissible current effective pressure p', and is eventually the maximum effective pressure to which the material has ever been subjected during the past plastic loadings. This can be checked by noting that the surface delimiting the current admissible stresses is an ellipse centered at $p' = \frac{1}{2}p_{co}$ and having $\frac{1}{2}p_{co}$ and $\frac{1}{2}Mp_{co}$ as half-axes along the p' direction and the q direction, respectively. As the hardening force p_{co}, the ellipse transforms according to a homothetic transformation (see Figure 10.10), so that the hardening/softening mode is isotropic (see Section 10.1.3).

Then let ε^p and η^p be defined by

$$\varepsilon^p = -\epsilon^p; \quad \eta^p = \frac{1}{\sqrt{3}}\gamma^p, \quad (10.73)$$

---

[2] A founding paper is Roscoe, K. H., Schofield A. N. and Wroth C. P. (1958) On the yielding of soils, *Géotechnique*, **8**, 22–53.

[3] See Alonso, E. E., Gens A. and Josa A. (1990) A constitutive model for partially saturated soils. *Géotechnique*, **40**, 405–430.

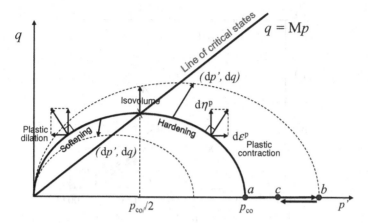

Figure 10.10 The poroplastic Cam-clay model

so that ε^p stands for the plastic volumetric contraction, and η^p is equal to $\frac{2}{3}|\varepsilon_I - \varepsilon_{III}|$ in the deviatoric experiment. The particles forming the solid part of a clay do not undergo any volumetric strain. The volumetric plastic strain ϵ_S^p undergone by the solid matrix is then zero and Equation (10.67) applies. Using the notations of Equations (10.71) and (10.73), the expression of Equation (10.68) for the plastic work δW^p can then be rewritten in the form

$$\delta W^p = p' d\varepsilon^p + q d\eta^p. \tag{10.74}$$

Experimental observations on clays confirm the validity of the principle of maximum plastic work, so that the potential governing the flow rule merges with the plastic loading function f. The flow rule for the Cam-clay model is then expressed through the relations

$$d\varepsilon^p = d\Lambda \frac{\partial f}{\partial p'} = d\Lambda \left(p' - \frac{1}{2} p_{co} \right); \quad d\eta^p = d\Lambda \frac{\partial f}{\partial q} = d\Lambda \frac{q}{M^2}; \quad d\Lambda > 0. \tag{10.75}$$

Similarly to Equation (10.33), based on the consistency condition $df = 0$, the plastic multiplier $d\Lambda$ is expressed in the form

$$d\Lambda = \frac{1}{H} \left(\frac{\partial f}{\partial p'} dp' + \frac{\partial f}{\partial q} dq \right) = \frac{1}{H} \left[\left(p' - \frac{1}{2} p_{co} \right) dp' + \frac{q \, dq}{M^2} \right]. \tag{10.76}$$

According to Equation (10.75), owing to the positivity of the plastic multiplier $d\Lambda$, the material undergoes a plastic dilation ($d\varepsilon^p < 0$) for effective pressures lower than half the consolidation pressure ($p' < \frac{1}{2} p_{co}$), whereas it undergoes a plastic contraction ($d\varepsilon^p > 0$) in the opposite case ($p' > \frac{1}{2} p_{co}$). For $p' = \frac{1}{2} p_{co}$, the material undergoes no volumetric plastic strain ($d\varepsilon^p = 0$) (see Figure 10.10).

It remains to determine how the hardening force p_{co} which accounts for the current strength of the material varies with plastic changes. The strength of clays is determined by the tangle and the interaction of the platelets which constitute the solid matrix. As a result, the irreversible increase in the platelet density governs the hardening state and therefore the changes in the hardening force p_{co}. Since the platelets do not undergo any volume change, the current irreversible increase of their density is appropriately accounted for by the current plastic

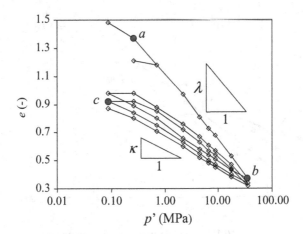

Figure 10.11 Poroplastic behavior of an artificial compacted clay in an experiment where $q = 0$, and where the total void ratio e is plotted against the effective pressure p'

volumetric contraction ε^p. The determination of the hardening modulus H then requires the knowledge of the state equation linking ε^p to the hardening force p_{co}. The latter can be obtained from a standard experiment performed at $q = 0$, where the total void ratio e defined by Equation (3.55) is plotted against the effective pressure p'. Usual experimental observations, such as those reported in Figure 10.11, are well accounted for according to the exponential law

$$p_{co} = p_{co}^0 \exp\left(-\frac{e^p}{\lambda - \kappa}\right); \quad \lambda > \kappa. \tag{10.77}$$

Indeed, the clay behaves elastically up to some effective pressure threshold representing the maximum effective pressure to which the material has been subjected in all the past plastic loadings (point a in Figures 10.10 and 10.11). Along a poroplastic loading path $de = -\lambda \, dp'/p'$ (path ab in Figures 10.10 and 10.11), while $de - de^p = -\kappa \, dp'/p'$ along an elastic unloading path (path bc in Figures 10.10 and 10.11). It results in $de^p = -(\lambda - \kappa)dp_{co}/p_{co}$ since $p' = p_{co}$ along a plastic loading path performed at $q = 0$. The integration of this relation gives Equation (10.77).

Besides, use of Equation (3.56) provides us with the relation linking the plastic volumetric contraction ε^p to the plastic void ratio e^p:

$$e^p = -(1 + e_0)\varepsilon^p. \tag{10.78}$$

Then let $\alpha \equiv p_{co}$ and $\beta \equiv \varepsilon^p$, so that the function $h(\varepsilon^p)$ in Equation (10.27) can be determined by substituting Equation (10.78) in Equation (10.77). Applying Equation (10.34), the hardening modulus H can be expressed in the explicit form

$$H = -\frac{\partial f}{\partial p_{co}} \frac{dp_{co}}{d\varepsilon^p} \frac{d\varepsilon^p}{d\Lambda} = \frac{1}{2}(\lambda - \kappa) \frac{1 + e_0}{\lambda - \kappa} p_{co} p' \left(p' - \frac{1}{2} p_{co}\right). \tag{10.79}$$

When $p' > \frac{1}{2} p_{co}$, the plastic contraction ($d\varepsilon^p > 0$) causes the porous material to harden ($H > 0$). Conversely, when $p' < \frac{1}{2} p_{co}$, the plastic dilation ($d\varepsilon^p < 0$) causes the porous material to soften ($H < 0$). When the loading is held constant at $p' = \frac{1}{2}p_{co}$ and, therefore, $q = \frac{1}{2}Mp_{co}$, an isovolumetric ($d\varepsilon^p = 0$) plastic flow occurs indefinitely: the porous material obeys ideal plasticity ($H = 0$). For this reason the line $q = Mp'$ is commonly termed the line of critical states, while $\frac{1}{2}p_{co}$ is the critical effective consolidation pressure (see Figure 10.10).

If, according to Equation (10.28), $\alpha \equiv p_{co}$ and $\beta \equiv \varepsilon^p$ are energy conjugate hardening variables, they are linked to the locked energy a_H according to

$$p_{co} = \frac{da_H}{d\varepsilon^p}, \qquad (10.80)$$

while the condition of Equation (10.30) stating the positivity of the dissipated energy becomes

$$f = 0: \quad p'd\varepsilon^p + qd\eta^p - p_{co}d\varepsilon^p \geq 0. \qquad (10.81)$$

Use of Equations (10.72) and (10.75) allows us to rewrite Equation (10.81) in the form

$$f = 0: \quad p'd\varepsilon^p + qd\eta^p - p_{co}d\varepsilon^p = d\Lambda \times \frac{1}{2}p_{co}(p_{co} - p') \geq 0, \qquad (10.82)$$

which is satisfied for the whole range of admissible effective pressures $p' \leq p_{co}$. As a result, the assumption that $\alpha \equiv p_{co}$ and $\beta \equiv \varepsilon^p$ are energy conjugate hardening variables ensures the thermodynamic consistency of the Cam-clay model. Nevertheless, only calorimetric experiments could definitively support this assumption, by showing independently that the dissipated energy $p'd\varepsilon^p + qd\eta^p - p_{co}d\varepsilon^p$ is actually that spontaneously transformed into heat through plastic loading.

10.2.3 Capillary hardening and capillary collapse

Up to now we have only considered saturated poroplastic solids. When the porous space is saturated by a nonwetting gas (subscript G) and a wetting liquid (subscript L) the expression of Equation (10.63) for the plastic work can be extended in the form

$$\delta W^p = \sigma_{ij} d\varepsilon_{ij}^p + p_G d\varphi_G^p + p_L d\varphi_L^p, \qquad (10.83)$$

so that the domain $\mathcal{D}$ of admissible stresses σ_{ij} and pore pressures p_G and p_L will be defined as

$$(\sigma_{ij}, p_G, p_L) \in \mathcal{D}: \quad f(\sigma_{ij}, p_G, p_L) \leq 0. \qquad (10.84)$$

If, in addition, the solid matrix undergoes negligible plastic volumetric changes, similar to Equation (10.67), we can write

$$\varepsilon^p = \varphi_G^p + \varphi_L^p, \qquad (10.85)$$

whose substitution in Equation (10.83) allows us to rewrite the plastic work δW^p in the form

$$\delta W^p = (\sigma_{ij} + p_G \delta_{ij}) d\varepsilon_{ij}^p + (p_L - p_G) d\varphi_L^p. \qquad (10.86)$$

According to Equation (10.86), the 'net' stress $\sigma_{ij} + p_G \delta_{ij}$ and the capillary pressure $p_{cap} = p_G - p_L$ are found to be the forces driving the plastic strains. This requires the plastic loading function f to depend only on the 'net' stress $\sigma_{ij} + p_G \delta_{ij}$ and the capillary pressure p_{cap}.

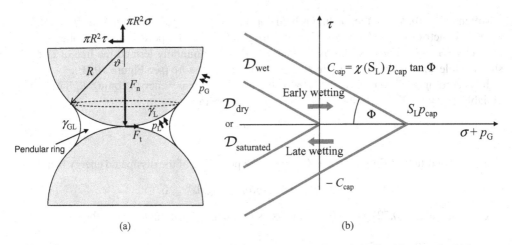

Figure 10.12 Capillary hardening induced by the depressurization of the liquid trapped within two solid grains during early wetting

This allows us to define the domain of admissible stresses and pore pressures in the more specific form

$$(\sigma_{ij}, p_G, p_L) \in \mathcal{D}: \quad f\left(\sigma_{ij} + p_G \delta_{ij}, p_{cap}\right) \leq 0. \tag{10.87}$$

As stated by Equation (10.87), the capillary pressure p_{cap} appears now as a hardening force with regard to the domain of admissible net stresses $\sigma_{ij} + p_G \delta_{ij}$. This capillary hardening force accounts for the extra cohesion which a wet sand gains when its solid grains become stuck together by the depressurization of the little water trapped within the slits between them. In order to address this issue further let us consider the simplified symmetric situation of Figure 10.12, where two spherical grains are held together by means of a pendular ring formed by the trapped liquid. Let F_σ and F_τ denote the normal and tangent components of the contact force F between the two grains, respectively. Assuming no 'dry' cohesion and adopting a plastic loading function of the Coulomb type (Equation (10.38)), the domain $\mathcal{D}$ of admissible contact forces is defined by

$$F \in \mathcal{D}: \quad f(F_\sigma, F_\tau) = |F_\tau| + F_\sigma \tan \Phi \leq 0. \tag{10.88}$$

Because of the existence of the interface stress, according to the Laplace Equation (6.52) where the curvature is changed in $-1/R$, the pressure finally transmitted to the solid grain by the fluid J is not p_J but $p_J + 2\sigma_{SJ}^A/R$. Nevertheless, according to the results of Section 6.1.5, if the effect of the surface deformation ϵ_A is only to spread out the surface energy on the current surface regarding the solid, from Equation (6.40) of σ_{SJ}^A we obtain $\sigma_{SG}^A = \gamma_G = 0$ and $\sigma_{SL}^A = \gamma_L$. Assuming a zero contact angle, so that there is no point force at the triple line exerted on the solid grain, the vertical equilibrium of the spherical solid grains then requires

$$F_\sigma = \pi R^2 \sigma + \left(\pi R^2 - \pi R^2 \sin^2 \vartheta\right) p_G + \pi R^2 \sin^2 \vartheta \left(p_L + \frac{2\gamma_L}{R}\right), \tag{10.89}$$

where σ is the vertical stress exerted on half the solid grain, while angle ϑ represents half the opening angle of the cone delimiting the extension of the pendular ring as illustrated in Figure 10.12(a). Since the current value of ϑ is determined by the current value S_L of the liquid saturation, a rearrangement of Equation (10.89) allows us to write

$$F_\sigma / \pi R^2 = \sigma + p_G - \chi(S_L)\left(p_G - p_L + \frac{2\gamma_L}{R}\right), \qquad (10.90)$$

where the function $\chi(S_L)$ has to satisfy $\chi(0) = 0$ and $\chi(1) = 1$. The horizontal equilibrium of the two solid grains of Figure 10.12 requires the horizontal component F_τ to be equal to $\pi R^2 \tau$, where τ is the shear stress exerting on half the solid grain. Substituting Equation (10.89) and $F_\tau = \pi R^2 \tau$ in Equation (10.88), once given the capillary pressure $p_{cap} = p_G - p_L$, we finally obtain that the domain of admissible stresses $\sigma_{ij} + p_G \delta_{ij}$ is defined by

$$\sigma_{ij} + p_G \delta_{ij} \in \mathcal{D}: \ f(\sigma + p_G, \tau, p_{cap}) = |\tau| + (\sigma + p_G) \tan \Phi - \chi(S_L) p_{cap} \tan \Phi \leq 0, \qquad (10.91)$$

where the effect of the liquid surface energy γ_L has been neglected, assuming sufficiently large grains. In Chapter 6 we saw that p_{cap} can be expressed as a state function of the liquid saturation S_L. A comparison of Equations (10.38) and (10.91) shows that the 'wet' cohesion C_{cap} induced by capillary hardening can be expressed as

$$C_{cap} = \chi(S_L) p_{cap} \tan \Phi, \qquad (10.92)$$

where p_{cap} depends on the current liquid saturation S_L.

It is worthwhile noting that C_{cap} becomes zero both for a dry porous material ($S_L = 0$, $\chi = 0$) and for a wetted material ($S_L = 1$, $p_{cap} = 0$). As a result, the cohesion is a nonmonotonic function of the liquid saturation S_L. For early wetting, where S_L is still close to zero, the porous material becomes wet; capillary menisci form between the solid grains, the cohesion increases, and the porous material strengthens; for late wetting, where the liquid saturation becomes close to $S_L = 1$, the porous material becomes completely wet; the menisci previously formed disappear, the cohesion decreases and the porous material weakens. The capillary hardening corresponding to Equation (10.91) is illustrated in Figure 10.12(b).

The capillary cohesion C_{cap}, whose expression is given by Equation (10.92), is relative to a given liquid saturation S_L. As the liquid saturation S_L increases, the capillary pressure p_{cap} decreases. Conversely to the drying shrinkage which was analyzed in Chapter 9, the constitutive Equation (9.76) predicts that a poroelastic solid such as a sponge, when subjected to rewetting and to a constant confining pressure σ, will undergo a swelling due to the increase of the liquid saturation S_L. However, if the porous solid exhibits a poroplastic behavior, instead of the expected swelling a contraction can be observed for most plastic clays. This phenomenon can be termed capillary collapse. The capillary collapse can qualitatively be explained as follows. For small values of the liquid saturation S_L, as shown in Figure 10.12, the liquid phase is still disconnected in the form of pendular rings trapped in the slits between the particles forming the solid matrix.[4] The granular porous solid can then be viewed as a poroplastic solid saturated by air, whose current strength is governed by capillarity. A significant increase of the liquid saturation will induce a lowering of the capillary pressure and, as an effect, a decrease of the

---

[4] This is the pendular regime as opposed to the funicular regime where the liquid phase is connected as illustrated in Figure 7.4.

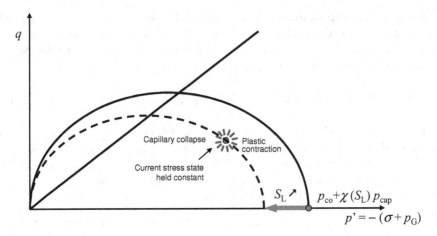

Figure 10.13 Illustration of capillary collapse for the Cam-clay model where, according to Figure 10.10, a plastic contraction accompanies the plastic collapse

current strength. At constant confining 'net' pressure $p' = -(\sigma + p_G)$, this capillary softening can ultimately make the latter meet the updated plastic threshold where, depending on the flow rule, either a plastic dilation or a plastic contraction will occur. Plastic dilation corresponds to the case of sands and, for instance, will be obtained at stress p' held constant during late wetting for the Coulomb model represented in Figure 10.12. The plastic contraction occurring during the capillary collapse of clays is illustrated in Figure 10.13, when adopting the plastic loading function of Figure 10.10 and replacing p_{co} by $p_{co} + \chi(S_L) p_{cap}$.

10.2.4 Stability of wet sandpiles

To give a quantitative illustration of capillary hardening, let us revisit the stability of a sandpile, whose constitutive sand is still in a virgin state, that is, with no cohesion due to compaction. However, the sand exhibits an apparent cohesion due to capillary hardening. Then let H and L be the height and the length of the base, respectively, of the sandpile represented in Figure 10.6. Noting ℓ the length AB, the expression of the weight Ω of the block ABC takes the form

$$\Omega = \rho g \frac{L}{2}(H - \ell \sin \psi), \qquad (10.93)$$

where ρg stands for the specific weight of the wet sand forming the sandpile. Geometrical relations

$$\frac{\ell}{\sin \Psi} = \frac{L}{\sin(\psi + \Psi)}; \quad \frac{H}{L} = \frac{1}{2}\tan \Psi, \qquad (10.94)$$

allow us to rewrite the weight Ω in the form

$$\Omega = \frac{\rho g L \ell}{4 \cos \Psi} \sin(\Psi - \psi). \qquad (10.95)$$

With respect to the dry case, according to Equation (10.86) for the plastic work δW^p the role of the stress is now played by the 'net' stress $\sigma_{ij} + p_G\delta_{ij}$. For the collapse mechanism displayed in Figure 10.6 we have $|d\gamma^p|/dt = V\cos\Phi$. Using this value, together with Equation (10.92) for the cohesion $C = C_{cap}$ induced by capillary hardening, Equation (10.59) can be applied to the collapse mechanism of Figure 10.6 in the form[5]

$$\sup_{f(\sigma_{ij}+p_G\delta_{ij})\leq 0, \Delta=\Phi} \int_\Omega (\sigma_{ij}+p_G\delta_{ij}) \frac{d\varepsilon_{ij}}{dt} d\Omega = \ell V \chi(S_L) p_{cap} \sin\Phi. \qquad (10.96)$$

Combining Equations (10.58), (10.60), (10.95) and (10.96) leads to the statement

$$\sin(\Psi - \psi)\sin(\psi - \Phi) > \frac{4\chi(S_L) p_{cap}}{\rho g L} \sin\Phi \cos\Psi \implies \text{collapse of the wet sandpile.} \qquad (10.97)$$

The minimum of the left-hand side of Equation (10.97) is obtained for

$$\psi = \frac{1}{2}(\Psi + \Phi). \qquad (10.98)$$

Substituting Equation (10.98) in Equation (10.97), we can conclude that the collapse of a wet sandpile will occur as soon as

$$\cos\Psi < \frac{1+8\chi(S_L) p_{cap} \tan\Phi/\rho g L - \left[(1+8\chi(S_L) p_{cap} \tan\Phi/\rho g L)^2 - 1\right]^{1/2} \sin\Phi}{(1+8\chi(S_L) p_{cap} \tan\Phi/\rho g L)^2 \cos^2\Phi + \sin^2\Phi} \times \cos\Phi. \qquad (10.99)$$

Taking $\chi(S_L) p_{cap} = 0$ in Equation (10.99) we find, as expected, that the collapse of a sandpile, either dry ($S_L = 0$) or saturated ($S_L = 1$, $p_{cap} = 0$), will occur as soon as $\Psi > \Phi$. Equation (10.99) shows, in a quantitative way, how capillary hardening induced by wetting increases the possible angle of the sandpile slope. This is the reason why wetting the sand (but not too much to avoid the capillary collapse) increases the strength of sandcastles and makes possible the beautiful reproduction of Gaudi's Sagrada Familia displayed in Figure 1.3.

10.3 From Material to Structure

The sandpile illustrated in Figure 10.6 is a structure made up of an assembly of contiguous elementary volumes, each of them being made of the same porous material and possibly exchanging fluid mass with adjacent volumes. Section 10.2.4 aimed at directly determining the limit slope angle of the sandpile, without analyzing the complex deformation of the structure prior to its breakdown. In contrast, this final section of Chapter 10 exemplifies how poroplasticity can be useful in the prediction and the evolution of the stress and deformation of a structure, by briefly revisiting the mechanical behavior of a weight dam prior to its final failure.

---

[5] In the pendular regime the liquid phase remains trapped in the slits between the particles forming the solid matrix. As a result, the plastic porosity φ_L^p in Equation (10.86) can be taken equal to zero. Therefore there is no contribution of the liquid phase to the plastic work δW^p.

On April 25 1895, the catastrophic failure of the Bouzey Dam, located in the east of France, caused the immediate death of 88 people, and finally that of a total of 200 people, because of the subsequent shortage of drinking water in the vicinity. This failure had been preceded by the appearance of cracks and significant deformation. Could the dam failure have been anticipated? As early as August 5 1895, the French engineer Maurice Levy (1838–1910) published in the *Comptes Rendus à l'Académie des Sciences* a memoir explaining the destabilizing role of the uplift pressure of water, which might apply on the base of a weight dam. Water can seep at the interface between the base of the dam and the foundation, or at any level of the upstream side because of an imperfection or a tensile crack caused by the water pressure. The pressure of the upstream water infiltrating the crack can then efficiently work against the dam weight, and finally can induce a tensile stress causing the crack to propagate from the upstream face to the downstream face.

However, at the time of the failure of the Bouzey Dam, there was no way to determine the location of the first crack on the upstream face under the action of the uplift pressure. Nowadays, this can be done by combining continuum mechanics of porous solids with the finite element method, in order to provide a numerical determination of stresses and strains prevailing in the structure, similarly to that we were able to perform analytically in the consolidation problem examined in Section 5.3. Assimilating the material constituting the structure to a poroplastic solid governed by the loading function of Equation (10.38), while adopting a structure geometry close to that of the Bouzey Dam, Figure 10.14(a) shows

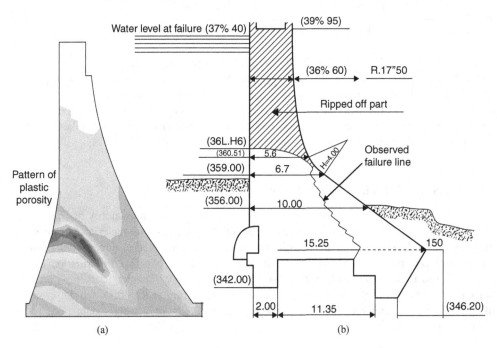

Figure 10.14 (a) Predicted failure line of a typical weight dam by combining continuum mechanics of porous solids and the finite element method (courtesy of Patrick Dangla); (b) observed failure line of the Bouzey Dam in 1895

the calculated pattern of plastic porosity forming progressively, as the height of the upstream water increases. This pattern, that accounts fairly well for the failure line observed in the tragedy of the dam of Bouzey and shown in Figure 10.14(b), is caused by a feedback process: (i) water penetrates at the weak mechanical point of the upstream face; (ii) the buildup of the pore pressure thus generated induces a significant plastic porosity there; (iii) this extra porosity increases the permeability; (iv) because of the increase in permeability the water penetrates further, and so on. It is finally interesting to note that the front of the failure line propagates parallel to the downstream face: this direction of propagation is preferred because of the absence of compressive stress on the downstream face.

Basically, although not as precise as a modern numerical approach, the early calculations of Maurice Levy can still be used efficiently in a simplified assessment of the stability of a weight dam. Under the combined action of the upstream water pressure resultant and of its own weight, a weight dam is a statically determined structure: the overall reaction exerted by the foundation and the moment of the latter are known irrespective of the mechanical behavior of the constitutive material. Assuming in addition that the normal stress under the foundation is linear, its maximum tensile value σ_N at the upstream base can therefore also be determined. Denoting then by σ_{max} the tensile strength of the dam-foundation interface, and letting p be the maximum water pressure prevailing at the upstream base, the stability of the dam is ensured as long as the effective stress $\sigma_N + p$ does not exceed σ_{max}. In contrast to a weight dam, arch dams are statically undetermined, and no simplified approach to the dam stability exists. Numerical methods based on the continuum mechanics of porous solids can then do the job, either in the design of a dam project, or in a back-analysis of the stability of an existing dam which exhibits an unexpected cracking.[6]

Further Reading

Chen, W. F. (1975) *Limit Analysis and Soil Plasticity*, Elsevier Science & Technology.
Coussy, O. (1995) *Mechanics of Porous Continua*, John Wiley & Sons, Ltd.
Coussy, O. (2004) *Poromechanics*, John Wiley & Sons Ltd.
Hill, R. (1998) *The Mathematical Theory of Plasticity*, Oxford University Press.
Lewis, R. W. and Schrefler, B. A. (1998) *The Finite Element Method in the Static and Dynamic Deformation and Consolidation of Porous Media*, 2nd edn, John Wiley & Sons, Ltd.
Salençon, J. (1977) *Application of the Theory of Plasticity in Soil Mechanics*, John Wiley & Sons, Ltd.
Schofield, A. N. and Wroth, C. P. (1968) *Critical State Soil Mechanics*, McGraw-Hill.
Ulm, F.-J. and Coussy, O. (2002) *Mechanics and Durability of Solids, I.: Mechanics of Solids*, Prentice Hall.

---

[6] For a detailed analysis see Fauchet, B., Coussy, O., Carrère, A. and Tardieu, B. (1991) Poroplastic analysis of concrete dams and their foundations. *Dam Engineering*, **2**, 165–192.

11

By Way of Conclusion

This is the end of the journey offered by this book in the strange world of porous solids. However, for the reader who is further interested by porous solids, this end can only be a pause. In this book the energy balances, the constitutive equations and the transport laws relative to a porous solid, have mainly been formulated at the scale of an ideal so-called representative elementary volume (REV). This scale, familiar in continuum mechanics, is the scale where average macroscopic properties can be defined and experimentally determined. The values of these properties are not supposed to be affected whenever they are determined on a sample having a larger size than the REV. For instance, this is the scale at which the concept of the pore size distribution makes sense. This scale is the smallest scale at which the analysis of the mechanical behavior of a large structure is relevant. The structure is then viewed as an assembly of REVs exchanging fluid mass and energy, as in the consolidation of a layer investigated in Section 5.3, or in the safety analysis of a dam considered in Section 10.3.

The analysis of the failure of a structure does not require a detailed knowledge of the porous network of the constitutive material. As input data, the failure analysis only needs an assessment of the macroscopic properties of the material such as, for instance, its Biot coefficient, its cohesion or its friction angle. Instead of structural design considerations, the issue may be the optimization of these material properties. Alternatively, no information may be available about the value of the macroscopic properties. Intensively developed during the last decades, upscaling methods, whose main goal is to provide an assessment of the macroscopic properties at the REV scale, from the microsopic properties at the pore scale, can become the relevant tool. Concerning the determination of macroscopic poroelastic properties from microscopic considerations, we gave a first flavor of these upscaling methods in Sections 3.4.2, 4.1.2 and 7.3. In Section 6.2.4 we also gave a first insight into the analysis of capillary hysteresis, from the knowledge of the morphology and topology of the porous network. However, in these basic examples the geometry used was simplistic, in order to obtain analytical results. Some engineering applications, such as storage optimization or advanced nanotechnologies, require such an accuracy that accounting for the actual geometry may become a major issue. Nowadays, the investigation tools of tomography, as for instance X-ray microtomography, can provide a closer representation of the network geometry. Applying sophisticated numerical tools can prove to be quite useful, in order to obtain precise assessments of macroscopic properties based upon the actual geometry previously determined by imaging. For instance, the surface energy U, as a function of the saturation S_W of the wetting fluid, cannot be determined by macroscopic experiments in the case of capillary hysteresis. Using Equation (6.71) it can be determined from the actual geometry by performing a numerical search of the minimum of the surface energy $U(S_W)$, under the constraint of a fixed volume of wetting fluid made explicit in Equation (6.69). No doubt these types of strategies will be carried out routinely in the near future.

In Chapter 6, we analyzed how surface energy originated from intermolecular forces. However, the pore scale was always assumed to be much greater than the molecular scale, allowing us to perform a continuous description of matter, where the molecules have no extent compared with the geometry of the problems under consideration. For instance, in Section 9.1.3, when dealing with the effects of the disjoining pressure upon confined phase transitions, the pore diameter was assumed to be much greater than the characteristic length ξ, capturing the range of intermolecular forces able to affect the interface energy value. In addition, since this characteristic length was associated with a continuous phenomenological model, this precluded any theoretical assessment of it, its possible determination being only experimental as in Section 8.5.2 and Figure 8.4.

Nanotechnologies involve pores having a radius of a few nanometers. The accuracy of models in which matter is considered as a continuum, such as the ones developed in this book, then becomes questionable. More accurate models require us to resort to statistical physics and molecular numerical simulations. Concepts of statistical physics involve a large number of molecules and molecular numerical simulations are needed to make these concepts quantitatively operational. There is no possibility here of presenting these methods, even superficially. In the following paragraphs, after a few words about statistical physics we will restrict ourselves to providing an illustration connected with confined phase transition.

In statistical physics, the value of a quantity A is known only through its expectation $\langle A \rangle$. If A_i is the value taken by A in the microstate i having the probability p_i, we may then write

$$\langle A \rangle = \sum_i p_i A_i. \qquad (11.1)$$

The question then is what is a microstate i, what is the value of p_i? For this purpose let us consider the so-called grand canonical ensemble, consisting of the assembly of the system under concern, and of an infinitely large reservoir of energy and particles. The system and the reservoir can exchange energy and particles. The temperature and the volume of the reservoir are fixed, as well as the particle chemical potential μ because of the large number of particles the reservoir contains. The grand canonical ensemble is an isolated system, so that both its total energy E_T and its total number of particles N_T are fixed. As a result, when the energy and the number of particles of a specific microstate i of the system are respectively E_i and N_i, the energy of the reservoir is $E = E_T - E_i$, while its number of particles is $N = N_T - N_i$. Since the reservoir is large, we have $E_i \ll E_T$ and $N_i \ll N$. It can then be shown (see the Further Reading Section) that the probability of a microstate i is given by the Boltzmann–Gibbs distribution

$$p_i = \frac{1}{\Xi} \exp\left(-\frac{E_i - \mu N_i}{kT}\right), \qquad (11.2)$$

where the normalizing function Ξ, such as $\sum p_i = 1$, is the grand canonical partition function defined by

$$\Xi = \sum_i \exp\left(-\frac{E_i - \mu N_i}{kT}\right). \qquad (11.3)$$

Figure 11.1 Typical configurations of adsorbed argon at 77 K as predicted by molecular simulations in a silica cylindrical nanopore with diameter $D = 4.8$ nm. (a) $p_V/p_{0V} = 0.62$; (b) $p_V/p_{0V} = 0.65$, where p_{0V} is the bulk saturating vapor pressure for argon (see Equation (8.34)). Gray spheres are argon atoms, while black spheres are hydrogen atoms at the pore surface, which saturate the oxygen dangling bonds of the substrate in order to ensure the electroneutrality of the simulation box (courtesy of Benoît Coasne)

The goal of any molecular simulation is then to sample the space of possible microstates. For an actual system, as for instance a fluid in a pore, the fluid–fluid and the fluid–wall potentials then have to be specified, in order to determine the energy E_i and the number of particles N_i related to the microstate i at a given temperature. However, relevant simulations have to consider such a large number of particles that specific methods, such as the celebrated Grand Canonical Monte Carlo simulation method, have to be carried out to optimize the sampling of the most relevant configurations.

With their background approximately reported, let us now give a flavor of the results which can be expected from molecular simulations, when the system considered is a porous network whose particles interact with the surrounding solid matrix. For this purpose let us briefly revisit the in-pore phase transitions which we explored in Chapter 9. We consider the adsorption of a wetting fluid in a cylindrical silica nanopore of finite length. The pore is opened at both ends towards a bulk reservoir to mimic the conditions of adsorption of real porous systems. The wetting fluid is argon. Both the fluid–fluid and adsorbate–substrate interactions are taken into account.[1] Typical results that can then be obtained from Grand Canonical Monte Carlo simulations are illustrated in Figure 11.1. When the intensity $|\Delta\mu|$ of the supersaturation

---

[1] For the detailed analysis see Coasne, B., Galarneau, A., Di Renzo, F., and Pellenq, R. J .M. (2008) Molecular simulation of adsorption and intrusion in nanopores. *Adsorption*, **14**, 215–221, from which the example is extracted. Let us just say here that the argon/argon interaction is calculated using a Lennard–Jones potential. The interaction between the adsorbate and the atoms of the silica substrate is written as the sum of a dispersion term (see Section

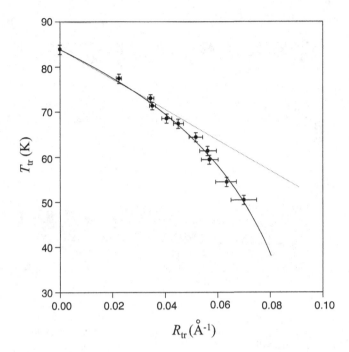

Figure 11.2 Transition temperature T_{tr} plotted against the inverse of the radius R_{tr} of the solid argon core, as predicted by molecular simulations. The curved line accounting well for the molecular simulations corresponds to the modified Gibbs–Thomson Equation (9.28) when we let $\xi = 4.15$ Å, while the straight line corresponds to the standard Gibbs–Thomson equation, when we let $\xi = 0$ in Equation (9.28) (courtesy of Roland Pellenq)

increases or, equivalently, when the ratio p_V/p_{0V} of the vapor pressure p_V to the saturating vapor bulk pressure p_{0V} defined by Equation (8.39) increases, an annular precondensed film of increasing thickness can be seen forming at the pore surface. At some critical value of p_V/p_{0V}, here lying between 0.62 and 0.65 for a pore with diameter $D = 4.8$ nm, the full condensation of the pore occurs. These results were already provided by the phenomenological model developed in Section 9.1.3 but molecular simulations can now allow us to check their validity more quantitatively, as in particular the validity of the modified Gibbs–Thomson Equation (9.28). This is illustrated in Figure 11.2 for the freezing of argon in spherical nanopores of various radii.[2] The transition temperature T_{tr}, at which in-pore solidification suddenly occurs, is plotted against the inverse of the radius R_{tr} of the core of argon actually transformed into solid form, whereas a premelted thin film remains unfrozen at the interface between the solid matrix and solid argon as analyzed in Section 9.1.3 and illustrated in Figure 9.3. Interestingly, molecular simulations accurately account for the modified Gibbs–Thomson Equation (9.28),

---

6.1.2), a repulsive short-range contribution and an induction term due to the interaction of the adsorbed atom with the local field created by the partial charges of the atom in the substrate.

[2] For the complete analysis see Celestini, F., Pellenq, R. J. M, Bordarier, P. and Rousseau, B. (1996) Melting of Lennard Jones clusters in confined geometries, *Zeitshrift für Physik*, D. **37**, 49–53, from which the data reported in Figure 11.2 and determined from molecular simulations is extracted.

when adopting here for the characteristic length scaling the argon–wall molecular interactions the value $\xi = 4.15\,\mathring{A}$. As an illustration of their efficiency, molecular numerical simulations thus provide an attractive alternative theoretical way to the experimental determination of ξ shown in Figure 8.4. They give convincing support to the modified Gibbs-Thomson Equation (9.28), which can in turn be used to investigate the mechanical behavior of a porous solid subjected to freezing, whose large number of pores make the systematic use of molecular simulations impossible at the scale of the REV.

It is now time for the final word, both for these concluding remarks and this book. It should be hoped that the reader, through the theoretical developments, the experimental evidence and the examples provided in this book, is now convinced that physical chemistry and the mechanics of solids can simply and harmoniously combine to provide us with some sound answers to questions involving complex couplings. This concluding chapter has stressed again that no scale is an island. When dealing with the physical mechanics of porous solids, a question formulated at a given scale generally involves an answer at a lower scale. As a result, we definitively need a global strategy, where well-trained specialists of a given scale will be able to pass the baton on to specialists of the adjacent scales in order for the latter to pick up the baton appropriately. May this modest book contribute to this interdisciplinary dialog.

Further Reading

Allen, M. P. and Tildesley, D. J. (1987) *Computer Simulation of Liquids*, Clarendon Press, Oxford.

Frenkel, D. and Smit, B. (2002) *Understanding Molecular Simulation: From Algorithms to Applications*, 2nd edn, Academic Press, London.

Leach, A. (2001) *Molecular Modelling: Principles and Applications*, 2nd edn, Prentice Hall.

McQuarrie, D. A. (2000) *Statistical Mechanics*, 2nd edn, University Science Books.

Nicholson, D. and Parsonage, N. (1982) *Statistical Mechanics and Computer Simulation of Adsorption*, Academic Press, London.

Index

activity
   chemical, 29
adsorption, 125, 153
   isotherm, 127–8, 193
angle
   contact, 118
   dilatancy, 252
   friction, 251

Biot, 41, 61, 86
   coefficient, 61–2
   secant coefficient, 69–70
Boltzmann
   –Gibbs distribution, 272
   entropy equation, 20
   factor, 128, 188
bubble
   formation, 22, 185
   pressure, 72
Buckley–Leverett equation, 143

capillary
   cohesion, 265
   collapse, 263, 265
   hardening, 263–4
   hardening force, 250
   hysteresis, 249
   length, 120, 134
   pressure, 132, 250
   rise, 121, 133

Carnot, 167
Clausius–Clapeyron equation, 175
Clausius–Duhem equality, 170
Clausius–Duhem equation, 15, 33
clay
   as a plastic material, 239
   Cam-clay model, 260
   floculation, 26
   swelling, 25
cohesion, 113–14
consistency condition, 241, 248, 261
consolidation
   and apparent creep, 104
   and effective stress, 59
   equation, 102
   of a layer, 58
contact
   angle, 118
   force, 45
Correns equation, 184, 235
Coulomb, 239, 251
critical state, 240, 263
cryosuction, 223–4
crystallization, 223
   pressure, 184

Dalton's law, 18
Darcy, 85
   law, 85–6, 90, 98–9, 140
      and dimensional analysis, 90

Debye
    forces, 111
    length, 30
diffusion
    and drying, 211
    coefficient, 99, 214
    equation, 98–9, 142, 218
    molecular, 211, 220
        characteristic time of, 216
dilatancy
    angle, 252
    coefficient, 252–3
    phenomenon, 251
disjoining pressure, 129, 131, 190–1
distortion, 45, 253
Donnan equation, 29
drying, 211
    capillary, 211, 218
    kinetics, 215
    molecular diffusion, 211
    shrinkage, 221
    stiffening, 222
Duhem, 11
Dupré formula, 177

energy
    and stability analysis, 40
    balance, 12–13
        for the fluid–solid mixture, 55
        for the porous solid, 56
    balance of free, 12
    balance of mechanical, 51
    blocked, 246, 250
    conjugate variables, 13
    dissipated
        through diffusion, 105
    elastic, 82, 105, 158
    Euler surface, 115
    free, 11
        Gibbs, 12, 37
        Helmholtz, 15, 36
        solid, 61
    interaction, 35, 37
    interface, 113, 123, 133
    interface and hysteresis, 138
    internal, 12
    Lagrange surface, 115
    of a mixture, 13
    of mixing, 36–8
    of substitution, 37
    potential, 120
    supply and chemical potential, 12
    surface, 108–9, 113
ensemble
    grand canonical, 272
entropy
    and daughter phase, 171
    criterion, 144
    melting, 180
    mother phase, 171
    of mixing, 19, 36, 179
    of nuclei, 187
    partial molar, 17
    production, 87, 239, 243, 252
    weak solution, 144
equilibrium
    liquid–solid, 179
    liquid–vapor, 175, 212
        and solute, 178
    mechanical, 97, 101, 110, 202
        disjoining pressure, 205
        local, 50
    of a curved interface, 121
    of meniscus, 107
    of sap, 201
    of triple line, 107
    metastable, 207
    of mixture composition, 13
    phase, 168, 171, 202–4
        and supersaturation, 171
    puddle, 120
    thermodynamic, 21, 24, 204, 210–11
Euler, 41
evapotranspiration of trees, 200

Faraday, 167
Fick, 85
filtration vector, 90, 99
force
    body, 45
    contact, 45–6
    electrostatic, 22, 27

Index

gravity, 58, 100
intermolecular, 16, 85
local, 45
surface, 45–6
van der Waals, 26, 37
viscous, 99
friction
   angle, 251
   element, 240
funicular regime, 265

Gibbs, 11, 167
   –Duhem equation, 11, 13–14
   –Thomson equation, 199, 224
   adsorption isotherm, 127, 193
   free energy of dissolution, 183
   free energy of formation, 34

Hamaker constant, 112
hardening, 245–6
   force, 246, 250
   isotropic, 247
   kinematic, 246
   modulus, 248
   state variable, 247
head of fluid particle, 92
Henry's law, 21
Hill, 239
Hooke, 61
hydrogen bond, 113
hysteresis
   and phase transition, 208
   capillary, 136, 138, 249
      and the ink bottle effect, 139

injection, 141

Jurin height, 121

Keesom forces, 111
Kelvin, 167
   –Laplace equation, 199
   equation, 175–7
      in presence of a solute, 179
Klinkenberg formula, 96
Knudsen number, 96

Kozeny–Carman formula, 91
Kuhn–Tucker conditions, 242

Lagrange, 41
   Green– strain tensor, 43
   multiplier, 125
   saturation, 154
Laplace, 107
   equation, 121
Levy, 268
logarithmic strain, 115
London, 107

Maxwell's rule, 174, 209
mean field theory, 32, 68, 82
mean free path, 95
metastability
   meta-, 40, 171, 175, 188
Mises, 239
mixture
   ideal, 17
   reactive, 32
modulus
   bulk, 62, 67
      matrix, 64, 66–7
   secant, 69
   shear, 62
      matrix, 67
   undrained bulk, 75, 99
molar property, 16
Monte Carlo method, 273

Navier, 85
   –Stokes equations, 86
   equation, 97
nucleation
   and precondensed film, 192
   heterogeneous, 188
   homogeneous, 186

partition function, 272
pendular regime, 265, 267
permeability
   hydraulic, 92
   intrinsic, 91, 95
   relative, 140

permeability (*continued*)
  to the fluid, 90
plastic
  and stress state uniqueness, 244
  consistency, 241
  distortion, 253
  flow rule, 239
  loading function, 240, 252–3, 264
  multiplier, 241, 248
  strain, 239–41
  threshold, 240
  void ratio, 262
  volumetric strain, 253
  work, 10, 239, 240, 242, 251–3
Poisson, 61
  equation, 29
pore
  double porosity, 162
  entry radius, 135, 199
  isodeformation, 160
  size distribution, 1
  volume fraction, 135, 199
potential
  associated, 252
  chemical, 11–12
    of a compressible liquid, 180
    of an elastic solid, 179
  electric, 30
  nonassociated, 252, 254, 257
pressure
  bubble, 72, 75
  capillary, 132
  crystallization, 184
  disjoining, 129, 131, 190–1
  equivalent thermodynamic, 28, 30, 130–1, 192, 204
  mean, 63
  mixture, 18–19
  osmotic, 23–5
    excess of, 25–6, 29, 131
  partial, 18, 22, 24
  pore, 42, 52, 54, 57
pressuremeter, 28, 130

Rankine–Hugoniot jump condition, 144
Raoult's law, 21

representative elementary volume, 271, 275
Roscoe, 239

salt crystallization, 182
Schofield, 239
Shuttleworth equation, 115
Skempton coefficient, 76
softening, 249
spreading coefficient, 118, 129
stability
  condition, 203
  dam, 269
  meta-, 40, 175, 188
  of dry sandpiles, 254, 256
  of ideal mixtures, 34
  of in-pore crystallization, 201
  of regular solutions, 37, 40
  of transition layer, 144
  of wetted sandpiles, 266
  phase in-, 172
strain, 42
  and displacement, 44
  deviatoric, 45, 62
  Green–Lagrange tensor, 43
  logarithmic, 115
  plastic, 240–1
    rate, 239
  principal, 44
  surface, 115, 150, 152
  volumetric, 45
    matrix, 67
  work, 51, 115–17, 128
stress
  Cauchy, 47–8
  deviatoric, 48, 254
  effective, 57, 59, 71, 164
  interface, 114, 117, 150–1, 153, 264
  mean, 48
  net, 263–4
  partition, 54, 55
  pre-, 118, 150–1
  shear, 41
  surface, 114–15
  symmetry, 50
  tensor, 46, 48, 54
  vector, 48

supercooled, 171, 194, 199, 204–5, 207, 210, 224
superheated, 171, 174, 191, 199, 207
supersaturated, 171, 175, 184–7, 189–90, 192
supersaturation, 171, 198
   and liquid–solid transition, 182
   and liquid–vapor transition, 178
   and salt crystallization, 184
surface
   energy, 107–8, 111, 113, 149
     and nucleation, 186
     and precondensed or premelted film, 197
     and saturated poroelasticity, 150
   force, 45–8, 58
   strain, 115, 150, 152
   stress, 114–16
   tension, 107
swelling, 2, 25
   cryogenic, 223

Terzaghi, 57, 71, 86, 164
Thomson, 167
   equation, 181
tree
   embolism, 201
   evapotranspiration, 200
Tresca, 239
triple line, 120

undersaturated, 190–2

van 't Hoff's law, 24
van der Waals, 107, 167
void ratio, 53, 262

wettability, 118
Wroth, 239

Young, 61, 107
   –Dupré equation, 119